Coordination Compounds and Complex Ions

Coordination Compounds and Complex Ions

Dr. Kamlesh Bilgaiyan

ARCLER PRESS

www.arclerpress.com

Coordination Compounds and Complex Ions

Dr. Kamlesh Bilgaiyan

Arcler Press
4164 Lakeshore Road
Burlington, ON L7L 1A4
Canada
www.arclerpress.com
Email: orders@arclereducation.com

ISBN: 978-1-77956-956-1 (Hardcover)

ABOUT THE AUTHOR

Dr. Kamlesh Bilgaiyan (Ph D & NET) a well known widely acclaimed assistant professor of Uttar Pradesh top rated Bundelkhand University, Jhansi (NAAC A+ Accredited University). Besides 10 U K design patents and two published patent, several publications. He has contributed pool of research by guiding two PhDs and more than 27 M Phil dissertations. He has published more than 20 papers in international and national reputed journals. Dr Bilgaiyan is having vast administrative experience of more than 20 years in various administrative positions. Presently he has working as Coordinator of B Sc, institute of Basic Science. He is also subject experts for Board of studies of various universities. Presently working on kinetics, material science and physicochemical properties of various plant extracts. Having more than 23 Years research experience.

TABLE OF CONTENTS

LIST OF FIGURES

PREFACE

Coordination compounds and complex ions represent an important and intriguing subdiscipline of chemistry. They not only have a central role to play in inorganic chemistry but also have pivotal functions in biology, medicine, catalysis, materials chemistry, and environmental chemistry. From gemstone hues to car catalytic converters, from hemoglobin in our blood which preserves our lives to complex chemotherapeutic drugs like cisplatin, coordination chemistry provides a context for the understanding of much of the science that directly affects our lives. Their research integrates the fields of bonding, structure, reactivity, and application, relating theoretical pure chemistry to applied use to the real world.

Their structural diversity, bonding interactions, and diversity of reactivity are the sources of the complexity of coordination compounds. They are often transition metals linked to proximate ligands that yield new magnetic, electronic, and optical properties. Their actions are explained by advanced theories such as Crystal Field Theory and Molecular Orbital Theory, and modern techniques allow scientists to design complexes for specific uses from industrial catalysts to targeted therapy. Their principles are what allow scientists, students, and professionals to address both simple scientific problems and practical technological needs.

The book *Coordination Compounds and Complex Ions* has been organized to provide a balanced introduction and coherent study of the subject. It follows the important principles, theoretical constructs, experimental techniques, and applications of coordination chemistry in an easy-to-follow manner by students and useful for scientists. Each chapter is designed to follow sequentially from the previous one with an increase in level of fundamental concepts to advanced topics and useful applications. The material includes both traditional understanding as well as recent developments, making readers an all-around aware of the topic.

Chapter 1 is an introduction to coordination chemistry. It provides the context in terms of history, introduces coordination compounds and complex ions, and discusses their significance in science and industry. Vocabulary like ligands, coordination number, and geometries is introduced in the chapter, providing the basic vocabulary necessary to understand the following content.

Chapter 2 focuses on coordination compound bonding. There is debate over various models of bonding, including valence bond theory, electrostatic interactions, and sigma-donation and pi-backbonding principles. The chapter points out how the nature of metal–ligand bonding gives rise to structure, stability, and reactivity, setting the stage for more advanced bonding theories.

Chapter 3 is all about Crystal Field Theory (CFT). The chapter explains how the fields of ligands influence the energy of transition metals' d-orbitals, leading to effects such as color, magnetism, and stability trends. Octahedral, tetrahedral, and square planar fields are explained using diagrams and examples in the chapter, and crystal field stabilization energy (CFSE) among other topics is also discussed.

Chapter 4 goes on to discuss Ligand Field Theory and Molecular Orbital Theory. These theories combine crystal field theory with molecular orbital theory to arrive at a deeper understanding of bonding and electronic behavior. The chapter outlines an integrated model describing spectroscopy, magnetism, and reactivity of coordination compounds.

Chapter 5 deals with isomerism in coordination compounds. Structural, geometrical, optical, and linkage isomers are explained with examples, demonstrating how variations in arrangement affect physical and chemical properties. The chapter demonstrates the importance of stereochemistry in coordination chemistry and its applications in catalysis, biological activity, and materials design.

Chapter 6 covers stability of complexes. It introduces concepts of thermodynamic and kinetic stability, stepwise formation constants, and stability factors such as type of ligand, charge, and chelate effect. The chapter also outlines experimental procedures for the determination of stability constants, including theory and practical determination.

Chapter 7 covers the electronic spectra and magnetic properties of coordination compounds. Electronic transitions, spectrochemical series, and the interpretation of UV-Visible spectra are described. Magnetic properties in terms of unpaired electrons and spin states are covered, including providing experimental tools to analyze experimental data and compare it to theoretical predictions.

Chapter 8 discusses the reactions of coordination compounds. Substitution reactions, electron-transfer reactions, and isomerization reactions are covered along with their mechanisms. The chapter addresses how the reactions are affected by factors such as ligand nature, metal oxidation number, and reaction medium.

Chapter 9 focuses on bioinorganic chemistry, the examination of the role of metal complexes within biological systems. Chapter 9 describes the metalloproteins, the enzyme cofactors, and the metal ion transport and their uses in medicine such as metal-based drugs. The chapter shows how coordination chemistry has direct applications in life processes and health care remedies.

Chapter 10 concludes with applications of coordination compounds. Their industrial applications in catalysis, water purification, electroplating, and photography are discussed, along with advanced applications in nanotechnology, materials science, and medicine. The chapter emphasizes the ways in which coordination chemistry concepts are translated into actual advantages in these fields.

This book has been specifically tailored for undergraduate and postgraduate level chemistry students will find it an excellent book to study the concepts and principles of coordination chemistry in a systematic way. Researchers will appreciate its overall coverage and references towards recent applications. Teachers will find it useful as a teaching manual, and professionals in allied disciplines—materials science, pharmacology, and environmental science—will have useful information available to them relating coordination chemistry to their professional pursuits. It is accessible as well to science-literate general readers who wish to acquaint themselves with one of the liveliest and most impactful fields of modern chemistry.

CHAPTER 1

INTRODUCTION TO COORDINATION CHEMISTRY

Contents

INTRODUCTION

Coordination chemistry is a subdiscipline of chemistry that involves the science of coordination compounds, which comprise a central metal atom coordinated to surrounding molecules or ions referred to as ligands. This research closes the gap between inorganic chemistry and organic chemistry because it entails the reaction of metallic ions with several types of ligands like anions, neutral molecules, as well as organic complexing reagents. One of the distinguishing features of coordination compounds is the ability to accommodate diverse geometric architectures and electron configurations, which are crucial in catalysis, materials science, and biochemistry. Familiarity with coordination chemistry fundamentals allows controlling these compounds in order to produce new solutions for complex chemical problems and thereby extend the scope of existing chemistry. The uses of this branch span as broad as medicinal chemistry and environmental science, indicating its broader scope beyond traditional limits. The scope of coordination chemistry is broad, with the synthesis, characterization, and reactivity of coordination compounds encompassed within. It involves the study of how these compounds are active under various conditions, from biological to industrial. For instance, the role of metal ions in biology, such as hemoglobins oxygen transport, is an example of the coordination compound's essential contribution to life maintenance. Again, coordination complex synthesis and design for catalysis reflect improvements in synthetic methods to shape reactions toward more efficiency and selectivity. The capacity of ligands to modify coordination complexes' properties further expands the field's applications, and thus it is imperative to have ongoing research in material development and environmental clean-up methods. Coordination chemistry thus becomes the arena for interdiscipline approaches solving contemporary scientific challenges. Research in coordination chemistry has also stimulated progress in other fields, and as such, there is a need for further study of complex systems. Uniqueness of properties of coordination compounds, such as their ability to stabilize structures and carry out electron transfer processes, has strong technology and science significance. Such is the case with the synthesis of covalent organic frameworks (COFs) that incorporates material science with the concepts of coordination chemistry (Diercks et al., 2017). These complexes possess outstanding rigidity due to coordination and covalent bond synergy that leads to new uses in gas separation and storage. Furthermore, the outcomes of coordination chemistry enable interdisciplinary interactions, and researchers can tackle pressing issues such as sustainable development and climate change. Accordingly, the scope of coordination chemistry continues to expand, solidifying its role as a foundation of scientific innovation and discovery.

Explanation of Coordination Compounds

Coordination compounds, characterized by central metal atoms coordinated with ligating environments, are focal points of synthetic and biological processes. The coordination number is focal to coordination chemistry, describing the number of ligand atoms that are attached to the metal center. Transition metals and lanthanides are typical metals present in coordination compounds due to their ability to form numerous complexes of many geometries. The ligand character—molecules or ions that bond to the metal—is also highly significant as their electronic character may be a huge factor in the stability and reactivity of the compound. By manipulation of these interactions, desired materials can be engineered, a task growing more important in fields such as catalysis and material science, where functionality has to be tailored (Amaral et al., 2025).

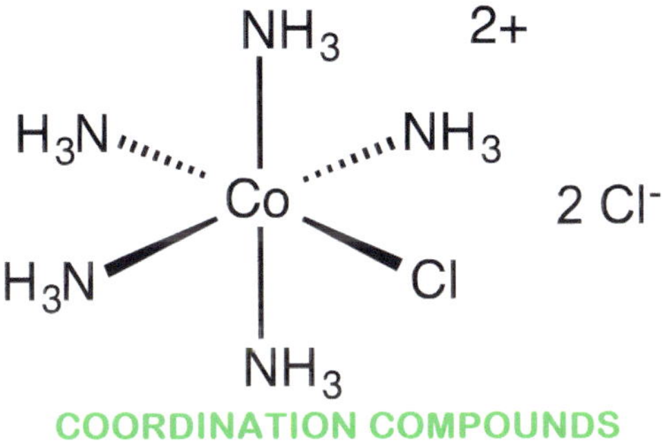

Figure 1.1. *The coordination compounds.*

Source: https://blogger.googleusercontent.com/img/b/R29vZ2xl/AV-vXsEin_Fzjs6FE6lSrUB0t2F68GT11W5JbZ384RVUgQ1EFbk75s-gI6D1ve6zyFhK9o0aIfOgmo9I5ry63D5jqQfT6ExmU2f1RIIl_hK85V6losyhJjQ4-F65GpvXYcwFZCnCy_D76FTrEFlA8O/s1600/COORDINATION+COMPOUNDS+for+Class+12+-+12th+class+chemistry+notes.png.

Coordination compounds are of interest that carries over to a myriad of applications, particularly when considering functionally designing materials. For instance, lanthanide complex formation has been vital in the invention of novel photonic devices owing to the attractive photoluminescent characteristics. As proven in current studies, the lanthanide ions can be accurately adjusted to respond to specific ligands, hence controlling properties such as photoluminescence. Coordination complexes, as exemplified by species formulated with picolinic and glutaric acids, illustrate this by revealing isostructural behavior that makes photophysical properties adjustable (Amaral et al., 2025). Besides, finding hybrid materials, which incorporate organic and inorganic constituents with the help of coordination chemistry, has given new avenues for research, enhancing the prospects of new applications in semiconductor technology and catalysis. Moreover, complexity of coordination chemistry is not merely founded on the primary coordination sphere—the metal-ligand direct interactions—but also on the secondary coordination sphere, covering non-covalent interactions potentially directing the overall properties of the compound. Latest developments encourage the central role of such interactions in achieving bimetallic systems offering tunable selectivity and reactivity. Researchers use molecular engineering methods to alter the properties of rare earth metal complexes and allow them to participate in redox processes and asymmetric catalysis. This effort highlights the dynamical character of coordination compounds, where the trade-off between interactions both within and outside the primary coordination sphere must be well understood in order to create new functionalities and applications for contemporary chemistry (Figure 1.1).

Importance in Various Fields

Coordination chemistry plays a vital role in most areas such as catalysis, pharmaceutical chemistry, and materials science, bearing witness to its vast applicability and utility. In catalysis, transition metal complexes are essential in accelerating reactions and facilitating the production of significant chemicals. Recent breakthroughs in synthesis of such complexes, particularly with varying reaction conditions and metal classes, have reported significant improvements in catalytic performance (Patel et al., 2024). Catalyst engineering using coordination complexes allows for enhanced selectivity and activity, which is essential in industrial applications. Thus, the function of coordination chemistry extends beyond the realm of theory, with direct implications on chemical production efficiency and sustainability. In pharmaceutical chemistry, coordination compounds are now the focal points of drug design and therapy. The selectivity with which metal-complexes can interact with biologically active molecules has made possible the development of effective drugs against many diseases, including cancer and bacterial infections. For instance, transition metal-containing drugs take advantage of the unique properties of coordination geometry and ligand structure to optimize bioavailability and target specificity. Notably, sustained research in metal-organic coordination polymers has opened doors to new therapeutic strategies (Wang et al., 2023). These advances not only provide solutions for health problems at hand but also shed light on the mechanisms underlying metal interactions in organisms. Moreover, materials science benefits immensely from the generality of coordination chemistry, particularly in creating advanced functional materials. Future-oriented research has proven it is feasible to design metal-organic coordination polymers for specific applications, for example, creating luminescent materials that can emit color-controllable light. The ability to engineer materials in this model has far-reaching implications for optoelectronics and sensing technologies, where optical properties need to be highly controlled (Wang et al., 2023). As researchers continue to explore the relationship between structure and function in coordination complexes, new avenues for new applications emerge, promising a great future for interdisciplinarity between chemistry, technology, and industry.

Historical Development of the Discipline

Historical development of coordination chemistry testifies to the formation of scientific thinking and experimentation since the previous century. Rising in the early 19th century, the subject of research gained pace through seminal discoveries that underscored the metal ions' role in the formation of sophisticated molecules. The research on coordination compounds began with attempts at explaining the characteristics of minerals and dyes before quickly developing into a systematic study of the connection between ligands and transition metals. This change paved the way for a greater discussion of metal-ligand interaction, leading to the creation of basic principles of coordination chemistry that continue to be researched today. Of special interest, the 20th century was the breakpoint with the development of better equipment, which allowed for more precise examination of these interactions and opened the door to a yet greater understanding of noncovalent interactions in systems of complexity (Hans-Jörg Schneider, 2022). The development of coordination chemistry also coincides with unparalleled advances in neighboring fields, including organometallic and supramolecular chemistry. The emergence of new theoretical tools and experimental techniques supported scientists' abilities to investigate such fertile interactions. In particular, the invention of organic crown ethers in mid-20th century is commonly cited as a milestone; it was the institutionalization of host-guest chemistry in coordination chemistry. This period witnessed the emergence of synthetic techniques for building up complex molecules and enlarged the research scope in scholarly as well as industrial contexts. Interestingly, whereas significant interest was paid to organic crown ethers, coordination chemistry based on siloxanes began receiving less attention initially. However, there has been renewed interest in siloxanes in recent years due to the unique properties that differ significantly from traditional organic coordination compounds (Dankert et al., 2021). Through the course of history, the development of coordination chemistry also reflects the growing awareness of the importance of noncovalent interactions within biological systems. Researchers have discovered that the basic fundamentals of coordination chemistry are not exclusive to synthesis but extend to biological processes where metal ions play vital roles in enzymes and metabolism. The integration of biological applications into coordination chemistry has established an interdisciplined setting, obliterating gaps between chemistry, biology, and materials science. Therefore, as instrumentation and methodology continue to advance, the field is similarly poised for increasing exploration of new materials and applications. With the inclusion of both interdisciplinary and supramolecular work within the field, continued advancement in coordination chemistry serves as an ongoing demonstration of how precedents shape current and future scientific inquiry and, ultimately, our understanding of complex chemical systems (Hans-Jörg Schneider, 2022).

1.1. FUNDAMENTAL CONCEPTS

The fundamental concepts in coordination chemistry revolve around the binding interactions of metal ions and other ligands. Protagonists of coordination compounds are metals, which possess distinctive electronic properties that can be significantly altered through the coordination of ligands. Ligands, neutral or charged molecules, provide electron pairs to metal ions, thus influencing stability and reactivity of the new formed structure. Identification of these underlying concepts allows chemists to manipulate various parameters like coordination number and geometry in the development of novel materials and applications. One such example is the building of metal–organic frameworks (MOFs), where these concepts are utilized to the extent that the choice of building blocks is critical to the accessibility of desired physical properties. The complex chemistry of these materials is a result of the interaction between ligands and metal ions, displaying the predominant role of fundamental coordination chemistry in modern scientific breakthroughs (Kirlikovali et al., 2022). In addition to the straightforward interaction of metals with ligands, the coordination structure dimension plays a predominant role in determining their functional properties. By controlling ligand sterics and electronics through modulation, scientists can prepare one-dimensional, two-dimensional, or three-dimensional frameworks of complexes. With such capability, scientists are enabled to attain new properties, such as magnetism and luminescence, depending on the geometric fashion and spatial organization of the metal centers. Lanthanide dimer synthesis, for example, demonstrates the capability in coordination chemistry to utilize the established protocols to synthesize molecular materials with purpose-built functionalities. The coordination sphere of lanthanide ions not only enhances their optical and magnetic properties but also helps overcome challenges related to molecular stability in materials engineering (Kévin Bernot et al., 2021). The fact that it is now possible to design these structural motifs is a reflection of the robustness of underlying principles of coordination chemistry. It is essential to learn these basic principles in coordination chemistry if future applications in materials science, pharmaceuticals, and nanotechnology are going to be enhanced. Ultimately, the ability to design coordination complexes with specific properties is largely dependent on fundamental principles, which dictate how such complexes are able to transition from molecular species to bulk and ultimately to devices. For instance, the creation of luminescent devices based on lanthanide-based dimers shows how intricately coordinated networks can be tailored for technological advancements. The researchers focus on creating stable films with the capability to hold their properties even when subjected to varying environmental conditions, demonstrating the indispensable role of ligand selection and coordination techniques to achieve such ends. This overlap between coordination chemistry and emerging materials has strong potential in driving future research and development, revealing how building blocks drive interdisciplinary breakthroughs (Kévin Bernot et al., 2021; Kirlikovali et al., 2022) (Figure 1.2).

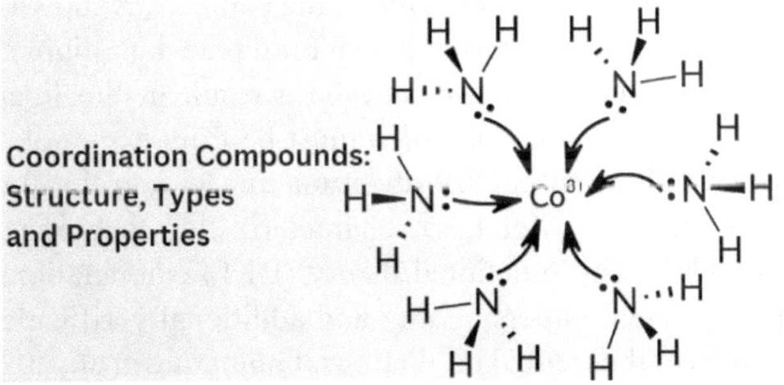

Figure 1.2. *The fundamental concepts in coordination chemistry.*

Source: https://article.study24x7.com/live/published/7ed66490-66b8-11ef-97dc-479fba2331ab.png.

1.1.1. Coordination Number and Geometry

In coordination chemistry, an understanding of coordination number and geometry is crucial because they determine the reactivity and physical properties of metal complexes. Coordination number refers to the count of ligand atoms that are directly bonded to a metal atom or ion at the center. This number can be fairly wide, normally from two and six, based on the metal under consideration and the nature of the ligands. An example is that coordination number four may lead to various geometries like tetrahedral and square planar geometries, while six leads to octahedral geometries. It is crucial to note the differences as the geometry influences considerably the electronic properties of the complex along with its application in the field of catalysis, medicine, and material science. Ligand geometry is in relation to steric as well as electronic characteristics pertaining to the metal center and also the coordinating ligands. For instance, metal ions with d-orbitals, i.e., transition metals, exhibit highly complex coordination geometries due to their capability of undergoing multiple bonding interactions. The nature of ligands—whether monodentate, bidentate, or polydentate—also affects the geometric outcome of the coordination sphere. As explained by the study on Ag(I) complexes, the occurrence of tetrahedral and binuclear complexes points out how different ligands can lead to complex geometries. The duality of the coordination mode of Ag(I), which may be trigonal or distorted trigonal-pyramidal when unconstrained, reveals the finesse of the interdependence between ligand location and metal coordination (Komlyagina et al., 2025). Coordination number and geometry are more than academic interest because they control the stability and activity of complexes. Study of the metal-ligand contacts provides insight into geometry, as evidenced in numerous reported structures. For example, the geometries shown for copper complex coordination were primarily pentacoordinate but highly distorted. Geometrical distortions such as these may be a result of the ligand's steric demands and electronic interactions, which must be fully explored. Having an understanding of how these affect coordination can lead to the formation of new materials and catalysts. Moreover, characterization techniques, such as X-ray diffraction and density functional theory (DFT) computations, enhance coordination environments' understanding and additional verification through experimental findings of theoretical predictions (Guimaraes et al., 2024) (Figure 1.3).

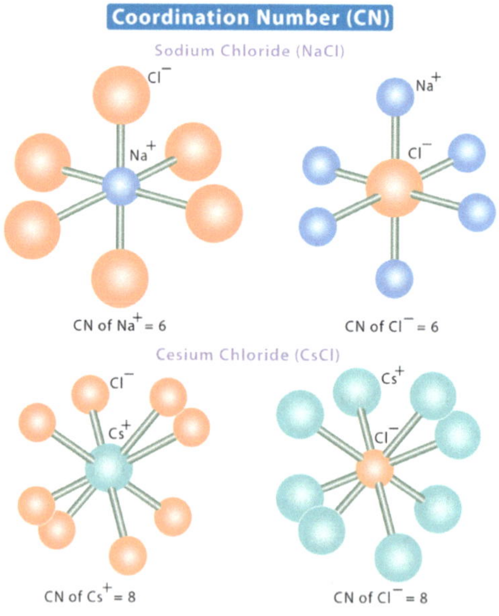

Figure 1.3. *Coordination number and geometry.*

Source: https://www.chemistrylearner.com/wp-content/uploads/2023/05/Coordination-Number.jpg.

1.1.2. Ligands and Their Classifications

One of the most important aspects of coordination chemistry, ligands play a crucial role in coordinating metal ions and forming coordination complexes. Ligands can be generally classified into several classes based on the number of donor atoms they have and how they coordinate. The most basic classifications include monodentate ligands with one donor atom and bidentate ligands which coordinate with two donor atoms. Bidentate ligands have a tendency to form chelate complexes, in which the metal ion is enclosed within a ring formed by the ligand. The chelation stabilizes the complex, increasing its solubility and reactivity, providing evidence of the extent to which metal coordination relies on ligand architecture.

Understanding these donor atom classifications is at the heart of the design of ligands that can selectively coordinate metal ions and thereby modify properties for application ranging from catalysis to drug discovery. Beyond the general donor atom classifications, ligands can be grouped on the basis of their charge and sterics. Neutral ligands like ammonia and water tend to act as Lewis bases, and anionic ligands are negatively charged, e.g., chloride and acetate ions. Each ligand class interacts with metal ions differently, determining both the overall stability and geometry of the resulting complex. Larger ligands can influence the electronic environment of the metal center, introducing selective catalytic activity and reactivity patterns. Moreover, the electronic properties of the ligand, i.e., π-accepting or σ-donating behavior, are accountable for establishing the electronic pattern in the metal-ligand bond. This subtle classification system enables chemists to predict and dictate the behavior of coordination compounds with great accuracy, directing theoretical models and practical application in virtually every chemical process. Recent advances in ligand design have substantially enhanced the promise of coordination complexes, particularly in lanthanide-based material synthesis. New ligands such as hexamethylhexacyclen and other aminopolycarboxylic acids have been synthesized to study their coordination chemistry towards divalent lanthanide ions. Study of these complexes reveals how, with the choice of ligand, the properties of the resultant coordination complex can be tuned to a significant extent, for instance, stability and photophysics. Hexamethylhexacyclen ligands, for instance, allow stabilization of lanthanide metals through sterically congested coordination environments. This is significant to many applications, e.g., luminescent materials and catalysts. Therefore, the understanding of how ligands and their types connect with the properties they impart to metal complexes is essential to advance the field of coordination chemistry towards advances in material science and molecular engineering (Bokouende et al., 2024).

1.1.3. Metal Centers and Their Properties

The properties of metal centers are the fundamental foundation for the study of coordination chemistry. Such metal centers, from transition metals to lanthanides, possess typical electronic structures which result in different oxidation states and coordination geometries. Transition metals, for example, possess incompletely filled d-orbitals, which are able to hold different ligands around them. This ability of holding different ligands is the cause of stable coordination complexes, which makes them useful in biological systems as well as synthetic pathways. Furthermore, their characteristic properties design redox properties, magnetic properties, and catalytic properties, stimulating material science advancements and enzymatic action. These properties bring into focus the significance of metalligand interactions and their crucial functions in the achievement of molecular rearrangements in coordination compounds, which ensures avenues for future expansion in this dynamic field (Soroceanu et al., 2022). Metal centers are to be interpreted with precise attention to their geometric arrangement and electron configuration, which control their stability and reactivity. Coordination number plays an essential role in defining the geometries such that four-coordinate complexes tend to adopt tetrahedral geometries and six-coordinate complexes tend to adopt octahedral geometries. These geometric frameworks not only influence the space position of ligands but also manage electronic surroundings at the metal center, altering its reactivity with various substrates and intermediates. For example, stabilizing low oxidation states in certain coordination complexes might lead to enhanced catalytic activity, particularly in electrochemistry. Changes in geometrical geometries and ligand types have been shown to alter the electronic nature of metal centers, ultimately affecting what types of chemical reactions they can catalyze. This knowledge is valuable for researchers desiring to design more useful catalysts and materials (Cheng et al., 2021). Furthermore, the biological relevance of metal centers highlights their key role in various metabolic processes. Metal ions, such as iron, cobalt, and zinc, are also essential co-factors in enzymes that facilitate significant biochemical reactions through their unique electronic characteristics. Hemoglobin, for instance, contains an iron metal center that is oxidizable and reducible reversibly, which facilitates efficient oxygen transport in biological systems. Their role in metal ion functionality is also exhibited in their application in medicinal chemistry, where metal complexes are used in drug design and therapeutic treatment, typically leveraging their characteristic reactivity and ligand interaction. Recent advancements of Schiff-base ligands have widened the horizon of metal-focused applications, where they have been found to be highly versatile and promising as bioactive cores in medicinal chemistry (Soroceanu et al., 2022). With the discovery of metal center properties and behavior, scientists can make new research fields accessible to applied as well as fundamental coordination chemistry (Figure 1.4).

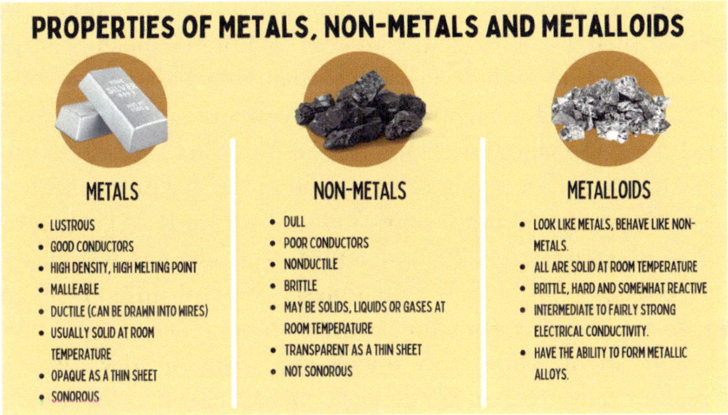

Figure 1.4. *Metal centers and their properties.*

Source: https://i0.wp.com/www.theengineeringchoice.com/wp-content/uploads/2024/06/Properties-of-Metals-Non-metals-And-Metalloids.webp.

1.2. TYPES OF COORDINATION COMPOUNDS

Coordination compounds, or simply coordination complexes, are formed due to interaction between ligands surrounding a central metal ion or atom and them, which could be neutral molecules or an anion. Other coordination compounds differ based on numerous factors such as the type of metal center applied, the nature of the ligand, as well as the geometric structures. For instance, transition metals possess numerous oxidation states and coordination numbers, leading to numerous complex types such as octahedral and tetrahedral geometries. The unusual properties of such complexes are particularly influential in their reactivity and stability, with the wide-ranging consequences of ligand-metal interactions for coordination chemistry serving to underscore the richness of coordination compounds with uses in catalysis and materials science. One of the most frequent classifications of coordination compounds is by the number and type of ligands surrounding the central metal ion. Monodentate ligands with a single donor atom will merely bind to the metal, whereas polydentate ligands, or chelating agents, have multiple potential binding sites to result in more stable complexes. Stability is derived primarily from the chelate effect, where chelating ligands produce superior thermodynamic conditions compared to non-chelating monodentate ligands. An appreciation of such differences is significant since they influence not only the coordinating chemistry itself but also the physical and chemical features of the resulting compounds. Sterics and electronic factors become significant considerations, highlighting the significance of ligand categorization in coordinating chemistry. Furthermore, coordination compound synthesis is also regulated by external conditions such as solvent, pressure, and temperature, which subsequently affect stability and properties of the complexes. For example, variations in temperature can change the stereochemical nature of coordination complexes with consequent phase transitions as documented in different studies. The study of such compounds, e.g., $BiFeO_3$, illustrates coordination and crystal chemistry's interaction in complex systems, e.g., variable coordination spheres with different magnetic and electrical responses at different temperatures (Atanasov et al., 2011). Structural variations, e.g., in alkaline earth nitridogermanate nitrides, also illustrate diversity within coordination compounds where peculiar ion arrangements and coordination numbers exist. These varied characteristics highlight the comprehensive approach necessary to completely appreciate the significance of coordination compounds in present-day chemistry (Johrendt et al., 2008).

1.2.1. Homoleptic vs. Heteroleptic Complexes

Metal complexes are classified as homoleptic or heteroleptic by coordination chemistry, and such classification is extremely significant for understanding their electronic and steric characteristics. Homoleptic complexes are those that possess multiple equivalent ligands coordinated to a central metal ion. This homogeneity may lead to a strong symmetry in the complex, often simplifying its properties to analyze. Such complexes are typically observed to possess well-defined geometrical shapes and manageable electronic properties, rendering them very useful in such applications as catalysis and materials science. The stability inherent in homoleptic complexes is often facilitated through the effective overlap of the metal d-orbitals and the ligand orbitals, allowing for strong metal-ligand interactions. Heteroleptic complexes, however, consist of different types of ligands with the central metal atom, which allows them to possess greater diversity in both structure and function. The difference in donor types and numbers can determine the coordination geometry surrounding the complex and the electronic distribution within the complex. The variability allows heteroleptic complexes to adapt to diverse coordination environments and undergo intricate reaction mechanisms. For instance, coordination chemistry research involving new ligands, such as those described in René Frank et al. (2024), reveals that heteroleptic complexes can stabilize low-valent metal species and show unusual reactivity, underscoring their importance in preparing complex materials and electronic devices. The applications of these two classes are relevant far beyond laboratory reports. Heteroleptic complexes, for instance, are more soluble and reactive in various solvent systems, as revealed in the case of lanthanide complexes in Chowdhury et al. (2023). The unique characteristics from mixed ligand environments enable the development of luminescent materials and catalysts with unique features. In contrast, while homoleptic complexes offer lesser versatility, their simplicity in structure can be of benefit in the understanding of fundamental coordination bonding principles. Ultimately, both homoleptic and heteroleptic complexes are crucial for the development of the coordination chemistry field, each in its own way contributing to the development of new chemical systems (Figure 1.5).

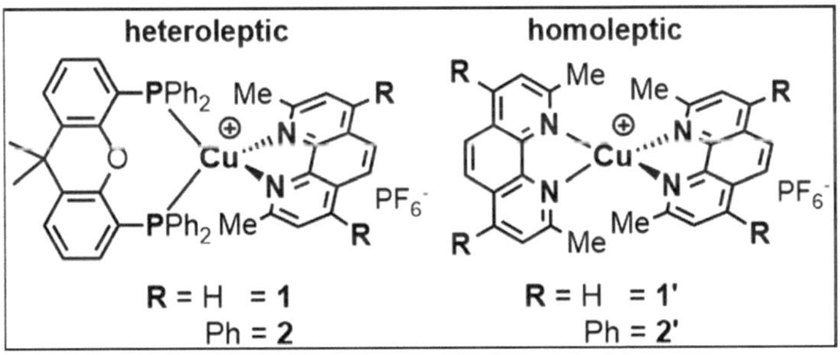

Figure 1.5. *Homoleptic vs. heteroleptic complexes.*

Source: https://www.researchgate.net/profile/Michael-Karnahl/publication/309474424/figure/fig1/AS:948657557880841@1603188995997/General-structure-of-the-heteroleptic-CuI-complexes-1-and-2-and-their-homoleptic.ppm.

1.2.2. Neutral, Cationic, and Anionic Complexes

The classification of coordination complexes as neutral, cationic, and anionic species is a key issue in understanding and utilizing coordination chemistry. Neutral complexes, which are devoid of net charge, are stable and generally useful, particularly as catalysts for many reactions. Neutral transition metal complexes, for instance, are likely to exhibit multiple coordination geometries, and they form a key component of homogeneous catalysis where molecular interactions can be precisely controlled. Studies on such species aid in the description of complex reaction mechanisms that underlie most synthesis routes, thus advancing knowledge in organic and inorganic synthesis (Cheng et al., 2021). Cationic complexes, which are charged positively, are formed through the coordination of transition metals with ligands capable of supporting the net positive charge. These complexes are likely to be more reactive, particularly for electrophilic substitution reactions, where the positive charge is used to attract the nucleophiles. Their capacity to react with various anionic reactants makes them significant in organic transformations and catalytic reactions. Particularly, cationic ligands find applications in addition to their traditional synthesis; they also play a significant role in electrochemical systems to enable electron transfer, as observed with recent homogeneous catalysis developments (Cheng et al., 2021). The importance of the study of cationic complexes in coordination chemistry and their applicability in the design of novel approaches to synthesis is highlighted by these developments. Anionic complexes, however, carry a net negative charge and are often stabilized by strong electrostatic interactions with counterions. Anionic complexes add to the versatility of the chemistry of transition metals, enabling them to display anomalous patterns of reactivity frequently unobserved in their neutral and cationic counterparts. The catalytic activity displayed by anionic complexes is important in that anionic complexes can serve as nucleophiles or stabilize reactive intermediates in most chemical reactions. Current advancements in metal-organic frameworks (MOFs) underscore the potential of anionic complexes for use in drug delivery systems, where their ability to be tailored enhances biocompatibility and enables controlled release mechanisms. Deciphering such interactions is important for unlocking the potential of anionic species in synthetic and medicinal chemistry toward new applications and technological advancement.

1.2.3. Examples of Common Coordination Compounds

Coordination compounds are widespread throughout all fields of chemistry, with a diverse range of structural and functional properties. One such common occurrence is heme, the iron coordination compound found in hemoglobin. Heme is a critical oxygen transport molecule for vertebrates, enabling efficient delivery of oxygen from lungs to tissue and removal of carbon dioxide. The characteristic heme structure, a porphyrin ring coordinated to an iron atom, allows for the reversible binding of oxygen. Such a function is essential to enable aerobic life, showing coordination compounds' capability to affect biological activity significantly. Interestingly enough, heme derivatives, or metalloporphyrins, are found in other living organisms as well and serve various roles, such as electron transfer and catalysis, thus again proving the diversity of coordination chemistry in the biological world. One other significant class of coordination compounds is transition metal complexes, which find extensive use in catalysis. An example is the complex $[Co(NH_3)6]Cl_3$ with cobalt coordinated to six ammonia ligands, illustrating the prevalence of octahedral geometry in transition metal complexes. The compounds are also important in industrial processes, especially chemical synthesis and petroleum refining. Transition metal complexes have the ability to catalyze various chemical reactions by providing lower energy reaction routes having less activation energy and consequently inducing greater reaction rates. Moreover, the majority of transition metal complexes possess colorful properties due to d-d electronic transitions, making them applicable in dye technology and photochemical applications. Colorful properties provide significant applications other than catalysis, and the significance of these coordination compounds in real-world applications is apparent. Coordination compounds also account for vast function in materials science, particularly in coordination polymer formation. Such polymers, typically formed through metal-ligand coordination behavior, have tremendous stability and tunable attributes. A huge instance includes coordination networks constructed of organic ligands and metallic ions, which can provide frameworks of preferred porosities. Such structures are utilized in crucial ways in gas storage, catalysis, and separations. Advances in computational aids, such as the ToposPro computer program, have enhanced understanding and design of these compounds by making possible the study of their crystal structures (Blatov et al., 2014). With further development of coordination chemistry, particularly with respect to the synthesis of new compounds, the potential for additional uses is vast and fascinating, and motivating ongoing research into this constantly evolving area of science (Figure 1.6).

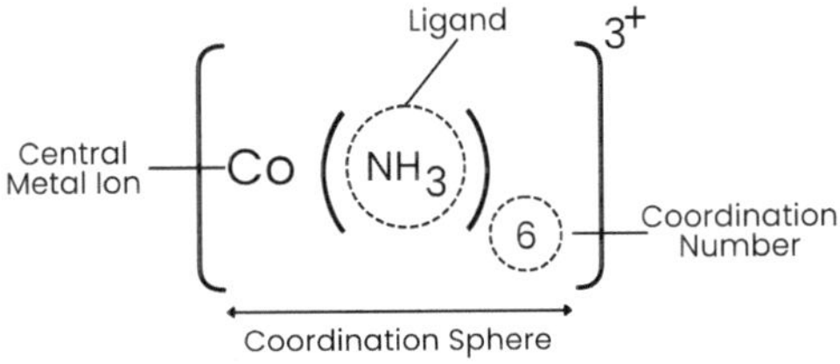

Figure 1.6. *Examples of common coordination compounds.*

Source: https://www.vedantu.com/seo/content-images/d0bd6087-6074-48d2-9f5a-31ce9548a292_coordination_sphere.png.

1.3. BONDING IN COORDINATION COMPLEXES

The intricacy of bonding in coordination complexes in coordination chemistry lies at the heart of defining the properties and applications of these substances. At the center of these complexes are metal ions or atoms at their core, typically transition metals, which bond to the ligands that cover them. These ligands can be neutral molecules or anions and can donate electron pairs to the metal center with the formation of coordinate covalent bonds. This interaction stabilizes the complex and determines its geometric shape and reactivity. Ligands can have different coordinating capacities, which are typically bidentate or polydentate agents, which allows them to create chelate complexes. The chelation effectively enhances the stability of the metal-ligand bond to position it at the heart of use ranging from catalysis to biology. The nature of ligands in coordination compounds is beyond mere donation of electron pairs; they have a decisive effect on the physical and chemical characteristics of the entire complex.

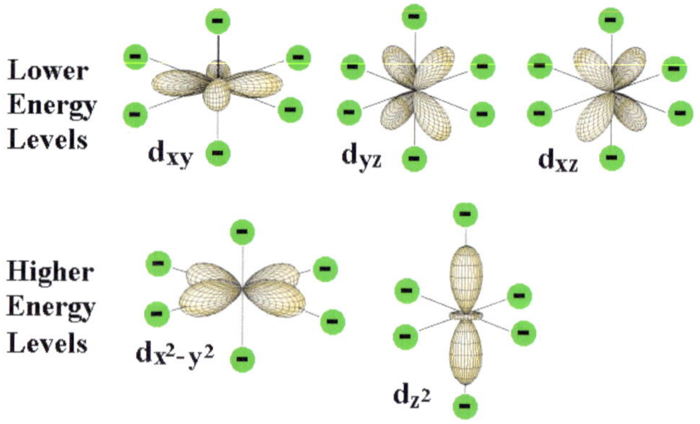

Figure 1.7. *Bonding in coordination complexes.*

Source: https://files.mtstatic.com/site_4334/65805/0?Expires=1756723461&Signature=l6DWwGIXAWZOHCFoKStG7eYtmDjQCBHv5hZw8apr1HVLNsXIoud82YbMWfr25J17sAaNWOARDYTR-5dXvL3LkdVv3FI9BxsA8ERdPMVy-GOGGfqNQ-AmL80xo6spYV65JIgkVdjRdFY-Iea66rA~dA3LsOvfrHjTxkfO6ic9vG3Q_&Key-Pair-Id=APKAJ5Y6AV4GI7A555NA.

For example, ligands are able to convey distinctive electronic and steric effects on reaction mechanisms and the electronic environment of the central metal. A good case in point is in 2-(2′-Pyridyl)imidazole, where there are two nitrogen atoms that are able to coordinate to metal centers. The bidentate ligand topology allows for the creation of diverse coordination complexes from discrete units to one-dimensional polymers. Furthermore, structural characteristics of such complexes tend to be complex because they can have several topologies stabilized by intermolecular interactions such as hydrogen bonding and π-stacking, thereby allowing for more insight into supramolecular assembly and stability processes within coordination chemistry (Singha et al., 2023). Beyond usual bonding mechanisms, bismuth species interaction with transition metals is one of the examples highlighting the complexity of bonding in coordination complexes. The diversity of bonding, transfer and covalent, is highlighted in the advanced nature of such metal interactions. Bismuth ions are either donor or acceptor and have strong ligand-metal interactions to enforce unique reactivity profiles in catalytic and stoichiometric reactions. This variation is not only relevant from a chemical point of view but also in terms of practical applications such as material science and catalysis, where coordination complex properties contribute significantly toward performance. Applying this understanding of bonding can lead to the synthesis of novel coordination compounds with tailored functions, which ultimately means progressing areas such as medicinal chemistry and environmental science (Ramler et al., 2021) (Figure 1.7).

1.3.1. Theories of Bonding: Valence Bond Theory (VBT) and Crystal Field Theory (CFT)

The VBT and CFT theories are the fundamental basis of coordination compound bonding, both accounts describing details of electronic structure and complex behavior in a different manner. VBT emphasizes orbital overlap and hybridization as important features of bond formation, proposing that bonding occurs because of the pairing of electrons within the atomic orbitals. This method is quite effective in the description of coordination complex geometries since it accounts for metal orbital hybridization that accommodates ligand interactions. For instance, ligands have the ability to induce d-orbital hybridization at metal centers with resulting geometric structures such as octahedral or tetrahedral geometries. Hence, VBT provides a fundamental framework for predicting the spatial disposition of ligands around metal ions, which is significant in explaining the reactivity and properties of coordination compounds. CFT, on the other hand, investigates the effect of ligands in the surrounding environment on d-orbital energy levels of transition metal ions, with a focus on electrostatic interaction between the metal ion and the ligands. CFT postulates that ligands create an electrostatic field which results in the degeneration of d-orbitals splitting into various energy levels. This splitting of d-orbitals is critical to color, magnetism, and stability of coordination complexes. Splitting of d-orbitals varies with the metal ion character, type of the ligand, and their geometry. All ligands, as classified under the spectrochemical series, produce greater splitting, leading to variations in electronic transitions that are accountable for the colors of coordination compounds. CFT is, therefore, essential to account for the spectroscopic behavior and properties of such complexes. Besides VBT and CFT, coordination chemistry also involves the synthesis and utilization of coordination compounds, particularly analytical chemistry and in medicine. For example, Schiff bases, obtained from condensation reactions among aldehydes and amines, are exhibited to possess tremendous potential in coordinating with transition metals. These compounds can exhibit enhanced biological activity and tailoring for specific applications, such as drug delivery and antibacterial compounds, showing the relevance of bonding theories in practical applications. Advancement in the research of coordination chemistry highlights the significance of understanding VBT and CFT because they provide the ability to design new materials through coordination bond manipulation. In all, use of these theories enhances not just our fundamental knowledge concerning coordination compounds but also their applications in various scientific disciplines, showing how significant they are.

1.3.2. Role of d-orbitals in Bonding

d-orbitals' contribution to bonding is central to coordination chemistry and the character of transition metals. The d-orbital electrons participate in bond formation with ligands, in influencing the geometry and reactivity of coordination complexes. These orbitals are capable of accommodating varying amounts of electrons, and therefore, transition metals are able to display varying oxidation states. Coordinate bonds are created through d-orbitals by donation of lone electron pairs by the ligand into vacant d-orbitals of the metal center. This interaction is significant in the determination of coordination compound stability and structure and therefore plays a significant role in their chemical nature. Notably, the structure of such complexes tends to be dictated by the unique d-orbital configurations and energies present, hence generating the typical coordination geometries such as octahedral, tetrahedral, and square planar arrangements. The electronic structure of the transition metals involves partially occupied d-orbitals that are vital in enabling straightforward electronic interactions during bonding. When ligands approach a transition metal, they can influence the energy levels of the d-orbitals significantly through the process of ligand field stabilization. This process splits the d-orbitals into different levels, such as in octahedral or tetrahedral coordination complexes. The extent of such splitting is extremely crucial in determining the coloration of the complex, its magnetic property, and its overall stability. For instance, changes between split d-orbitals can result in intense colorations within complexes due to electronic transitions combined with d-d excitation, as further elucidated through electronic spectra analysis (Song et al., 2020). Such association points to the importance of understanding contributions from d-orbitals to bond stability and complex formation in coordination chemistry complexes. In coordination chemistry, the role of d-orbitals is not confined to participation in bonding alone but also in influencing magnetic properties as well as reacting substrates. The spin state of d-electrons will determine whether the complex is diamagnetic or paramagnetic, which can significantly alter its interaction with other substrates and ligands. Moreover, d-orbital involvement in π-backbonding, particularly in strong field ligand complexes, enhances the bond order and decreases the metal-ligand bond lengths. In cases of systems containing dative bonds, e.g., in the process of crown ether complex formation, the role of d-orbital involvement is crucial in stabilizing unique oxidation states of metals like plutonium (Wang et al., 2020). This combination between d-orbitals and bonding interactions provides understanding of the behavior and applications of coordination complexes in numerous fields, including catalysis and materials science (Figure 1.8).

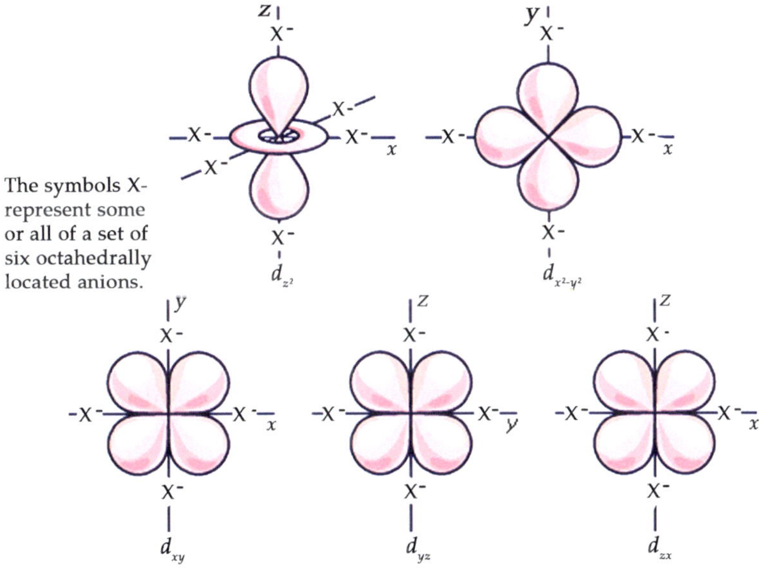

Figure 1.8. *Role of d-orbitals in bonding.*

Source: https://cdn.britannica.com/47/7447-004-142D2365/shapes-d-orbitals-designations-symbols-some-anions.jpg.

1.3.3. Influence of Ligand Field Strength

The ligand field strength concept is of utmost importance in coordination chemistry since it directly influences the electronic structure as well as metal complex energetics. Ligands surrounding a central metal ion induce an electrostatic environment that will alter the energy of the d-orbitals. Such an interaction allows for variation in d-orbital splitting, the term used to describe the variation between strong field and weak field ligands. Strong field ligands such as CN^- or CO induce greater splitting of the d-orbitals, which will form low-spin configurations and significant stabilization of specific electronic states. Poor field ligands like I^- bring about minimal splitting, which generally results in high-spin configurations. Ligand field strength plays a crucial role in determining the magnetic properties, color, and reactivity of the metal complexes, which are critical aspects of coordination chemistry. Ligand field strength manifests itself beyond theoretical concepts to actual applications in coordination complex synthesis and functionality. For instance, ligand field theory can account for the transition metal acetylacetonates complexes behavior. It has been found that the presence of strong ligands has the potential to actively affect transition metal-ligand coordination bonds by changing their geometries and chemical properties. Interestingly, unusual effects such as ligand displacement are feasible through competing interactions, for example, halogen bonding, hence underlining the subtle interplay between ligand field strength and molecular design (Torubaev et al., 2025). This interplay underscores the necessity for wise care in ligand identity and field strength when designing new metal complexes for applications that range from catalysis to materials science. In addition, computational methodologies have enlarged our knowledge of ligand field effects, providing insights that supplement those gained from experiment. Sophisticated methodologies such as ab initio ligand field theory (aiLFT) allow d-orbital energies and electronic configurations in complex settings to be calculated and interpreted. These computational models have shown organometallic systems, when viewed in the context of ligand field theory, to display behavior similar to that of classic coordination complexes. For instance, planar $[Cu(CR_3)_4]^-$ complexes can exhibit typical low-spin d8 structures, which describe how ligand field parameters influence their stability and reactivity (Deeth, 2024). Such computational models enhance our understanding of how various types of ligands regulate bonding interactions differently, which allows for more accurate predictions and leads to the development of novel coordination compounds in chemistry.

1.4. SYNTHESIS OF COORDINATION COMPOUNDS

Preparation of coordination compound is the focal process in coordination chemistry, connecting the coordination of organic ligands and metal ions to form complex structures. A number of methodologies have been developed to synthesize it so that scientists can produce tailored compounds for innovative applications. The process typically begins with the selection of an appropriate metal precursor, followed by the introduction of an appropriate ligand with appropriate donor atoms. For instance, there were the new developments that proved the successful synthesis of the coordination compound with the BSNP ligand by using spectroscopic means for structural characterization and stability measurements. The isolated complexes, for instance, those involving Fe(III) and Cu(II), exhibited well-defined electrochemical behavior and coordination geometries, hence proving the variety of the interactions that are possible based on purposeful ligand selection and synthesis methodology (Omar et al., 2025). Further, coordination compounds are studied beyond traditional applications to areas such as catalysis and bioactivity. Catalytic reactivity and selectivity have been enhanced by interaction of substrates with synthesized metal complexes. Of special interest are single-atom catalysts, which are characterized through their minimized synthesis processes and maximized catalytic activity. The coordination environment around the metal centers and their interactions with one another are crucial because they influence the availability of active sites and enable the reorganization of reaction intermediates. Thus, not only does the synthesis of coordination compounds offer materials with intricate geometries and properties but also provides scope for novel catalytic solutions that expedite organic synthetic protocols (Ye et al., 2024). Integration of computational investigations has also become a significant trend in the analysis and synthesis of coordination compounds. The calculations using Density Functional Theory, for example, give a clearer understanding of the electronic interactions in these complexes. Computational investigations aid in the identification of active sites, enhance the prediction of reactivity, and offer geometric preference information of synthesized compounds. As exemplified by the study of BSNP coordination compounds, DFT yielded critical information for explaining the structure-activity relationship pertaining to their antimicrobial and anticancer activity. Such incorporation of computational analysis into synthesis process legitimates the dynamic aspect of coordination chemistry to predict and optimize compound behavior across applications—representing a notable advance in the discipline (Omar et al., 2025) (Figure 1.9).

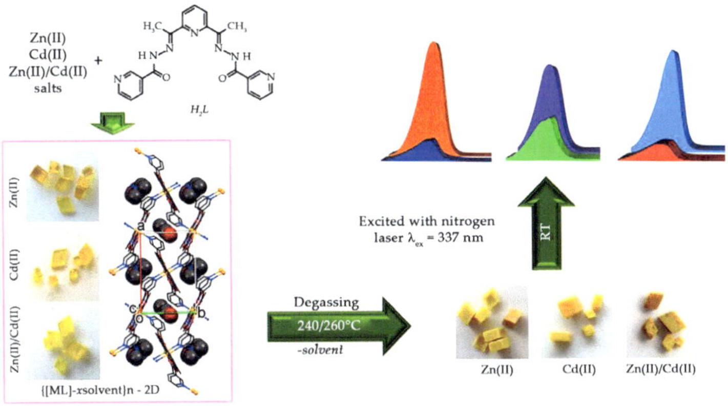

Figure 1.9. *The synthesis of coordination compounds.*

Source: https://www.researchgate.net/profile/Lilia-Croitor/publication/355652414/figure/fig1/AS:1083409292103680@1635316312805/Schematic-representation-of-the-synthesis-of-coordination-compounds-1-3-photos-of.jpg.

1.4.1. Methods of Synthesis: Direct and Indirect

Direct and indirect synthesis are the two broad categories into which the synthesis of coordination compounds can be grouped. Direct synthesis is the interaction of metal ions and ligands in a single-step process, often in a suitable solvent under controlled conditions. This approach is simple and typically yields a high percentage of the target complex, which is why many chemists prefer it.

One of the most common examples of direct synthesis is when metal ion intermediates are directly reacted with chelating ligands, resulting in stable coordination complexes. However, there are difficulties, such as the formation of by-products or the presence of parallel reactions, that may reduce the efficiency of this method. Information regarding the dynamics and conditions that allow favorable direct synthesis is crucial in the synthesis of certain coordination compounds with specific characteristics, particularly in catalysis and material science applications. On the other hand, indirect synthesis is one that begins with the formation of more straightforward coordination complexes and can be converted into more complex structures. Direct synthesis is convenient when it is not possible due to thermodynamic or kinetic limitations.

One of the most typical methods in indirect synthesis is the use of solvent-mediated methodologies, where metal ions are reacted with ligands within a solvent that is favorable for sequential assembly of the target complex. The stepwise approach gives better control over the conditions of the reaction, which allows the synthesis of complex coordination structures that might prove difficult to achieve in direct synthesis alone. Furthermore, indirect synthesis is responsible in cases where the stabilization of reactive intermediates or separation of special coordination geometries is necessary, hence expanding the scope of coordination compounds accessible.

Both direct and indirect syntheses underscore the richness and diversity of coordination chemistry. This diversity is particularly crucial in the formation of new materials with specified properties and applications in fields such as optoelectronics and catalysis. The study of such synthetic approaches may also shed light on structure-function relationships in coordination compounds. For instance, studies of lead halide perovskites demonstrate how refining synthetic pathways—both direct and indirect ones—can lead to extraordinary advances in their application to solar cells and lighting-emitting devices. Investigating and mastering these methods not just extends theoretical knowledge but also powers real-world advances within the field, allowing chemists to develop materials to react to specific technological demands and thus confirming the critical importance of synthesis methods in coordination chemistry.

1.4.2. Factors Affecting Synthesis

Preparation of coordination compounds is impacted by many factors, extending to both physical as well as chemical conditions. Among the most important of these is the choice of ligands because their electronic and steric features directly affect the stability and shape of the resulting complexes. Biguanides are an example of bidentate ligands that exhibit a tendency towards strong chelation of metal ions that facilitates the establishment of stable metal-biguanide complexes. Sterics-electronics interaction can result in a range of coordination modes, which are accessible to study by spectroscopic and computational means. These analyses not only disclose the structural richness of such complexes but also offer insight into their reactivity and catalytic ability, as is discovered in studies of biguanide-based transition metal complexes (Rani, 2025). Synthesis reaction temperature is another important parameter that controls the outcome of coordination compound formation. Reactions performed at varied temperatures have the tendency to favor either kinetically or thermodynamically controlled pathways to produce varied products.

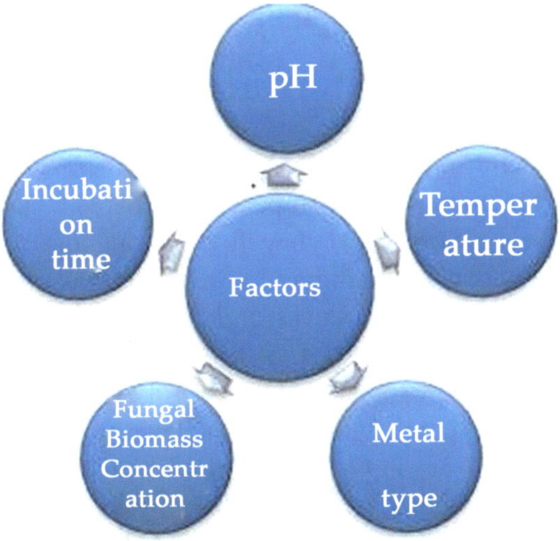

Figure 1.10. *Factors affecting synthesis.*

Source: https://www.researchgate.net/profile/Khushbu-Gupta/publication/327597252/figure/fig2/AS:886817708589056@1588445226024/Fig-4-Top-five-factors-which-affect-the-synthesis-of-nanoparticles.jpg.

As an example, the synthesis of polyoxotungstate-based transition metal complexes demonstrates how temperature control facilitates the construction of intricate supramolecular complexes to provide a means for anion encapsulation within metal-ligand architecture. Specific compounds such as [Cu_5(2,2-bpy)5(μ2-$Cl)_2(PO_4)_2(H_2O)_2$][$HPW_{12}O_{40}$]·$2H_2O$ and [Zn(phen)$_3$]$_2$[PWVWVI$_{11}O_{40}$]·$5H_2O$ show temperature-dependent metal-ligand interactions. Further, electrochemical characteristics and capacitance behavior shown in these compounds also underscore how synthesis conditions can regulate ensuing physicochemical properties (caihong peng et al., 2024). Lastly, the concentration of the reactants plays a crucial role in regulating the pathways of synthesis as well as final coordination product properties. The metal ion to ligand ratios influence the complex formation directly since low or above required concentrations can lead to incomplete reactions or the presence of unwanted side products. The presence of oxidizing or reducing agents during the synthesis also significantly contributes towards influencing the metal ion oxidation states and, in turn, the coordination sphere of the ligands. The study of variations in concentrations of reactants is significant, particularly in research on biguanide-based complexes and polyoxotungstate architecture, since these factors largely determine the materials' functional properties and applications in medicine and catalysis, emphasizing their relevance within the broader context of coordination chemistry (caihong peng et al., 2024; Rani, 2025) (Figure 1.10).

1.4.3. Purification Techniques

In coordination chemistry, purification techniques are crucial to preserve the integrity and functionality of compounds prepared. Such techniques are required to remove impurities that may form throughout synthesis, such as unreacted starting materials, by-products, or solvents. Recrystallization, which takes advantage of solubility differences, and chromatography, which separates compounds due to their interactions with a stationary phase, are common techniques used for purification. Techniques like high-performance liquid chromatography (HPLC) are particularly valuable due to their ability to produce high resolution and efficiency in separating complex mixtures. The effectiveness of such purification methods has a direct implication on the characterization of coordination compounds because impurities could alter their chemical and physical properties, thus leading to incorrect inference as to their stability and reactivity (Omar et al., 2025). Coordination compounds are not only needed to be synthesized but also need to be purified to achieve the desired level of purity for their biological and industrial applications. Following purification, techniques such as spectroscopy (IR, NMR) and mass spectrometry are used to determine the identity as well as purity of synthesized metal complexes. These analytical techniques enable chemists to define the electronic and structural properties of the compounds and disclose their potential uses, e.g., in the design of drugs or catalysis. The effectiveness of the purification process also dictates the biological testing of these compounds, as contaminants could prevent them from exerting their antimicrobial, antifungal, or anticancer activities (Marinova et al., 2025). Therefore, proper purification processes require detailed attention for further progressing the utilization of coordination compounds in numerous scientific and medical applications. Recent advancements in purification processes, including the utilization of green chemistry techniques, reflect a growing emphasis on sustainability in coordination chemistry. For instance, microwave-assisted recrystallization is a rapid and energy-saving process of purification compared to traditional methods, reducing the amount of solvent used and waste created. Also, techniques like flash column chromatography have been enhanced using new stationary phases, achieving the highest level of separation efficiency. As the subject keeps developing and expanding, there is greater and greater appreciation for the contribution such purification methods make in not just enhancing yield but also ensuring safety and efficacy of coordination compounds when used in biomedical applications. These developments bring into focus the interconnected character of purification processes and advancements in coordination chemistry, driving innovation and studies in line with prevailing scientific goals (Omar et al., 2025) (Figure 1.11).

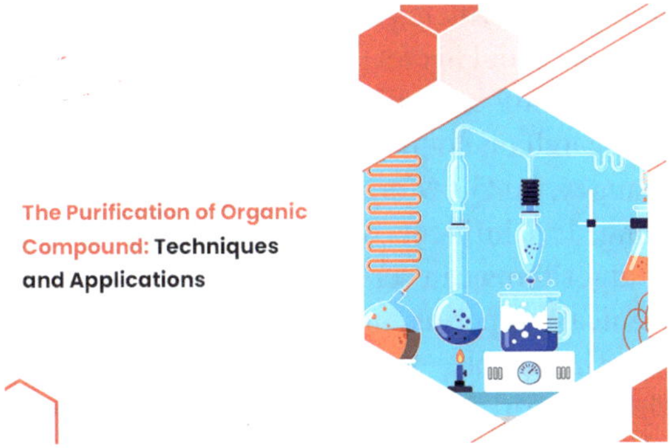

Figure 1.11. *Purification techniques.*

Source: https://reachemchemicalsblog.b-cdn.net/wp-content/uploads/2024/08/the_purification_of_organic_compound.jpg.

1.5. APPLICATIONS OF COORDINATION CHEMISTRY

Applications of coordination chemistry cut across most fields of science, underscoring its relevance and versatility. Among the main fields where coordination complexes have made huge contributions is catalysis. Transition metal complexes are common catalysts and mediate a wide range of chemical reactions, from the formation of organic compounds to polymerization. Transition metal catalysts are more effective and selective due to the new coordination environments provided by ligands that are capable of lowering the activation energy required for a reaction substantially. By modulating these ligands, catalytic efficiency for some transformation is made possible, and coordination chemistry becomes imperative in the development of sustainable and efficient chemical processes. Furthermore, new catalytic system design has been enhanced through the integration of computational models and machine learning algorithms, which enhance our understanding of reactivity in coordination chemistry (Rani, 2025). Besides catalysis, coordination chemistry plays a crucial role in materials science with coordination polymers as a major interest. Coordination polymers are materials with metal centers coordinated to organic ligands with the formation of one-, two-, or three-dimensional networks. This structural variance allows for fine tunability of porosity, stability, and reactivity, which is vital in uses like energy storage, sensing, and environmental clean-up. Advanced improvements in synthetic techniques, like mechanochemical and microwave techniques, have also broadened the array and efficacy of coordination polymers. With scientists striving to uncover more on these materials, their potential in creating sustainable solutions for various fields becomes clearer. Scalability and efficiency for commercial application are still a challenge, with research on coordination polymers needing to keep exploring and innovating (Pandey, 2025). Moreover, coupling with artificial intelligence, as machine learning, has influenced coordination chemistry in recent years. Machine learning algorithms can scrutinize data sets with detailed information regarding the chemical properties and interactions not obvious to traditional analytical methods. This enables greater prediction of the behavior of chemicals as well as acceleration of discovery of new coordination compounds and catalysts. By applying machine learning in coordination chemistry, researchers are able to traverse the vast landscape of chemicals more easily, accelerating the process of designing novel compounds with desired functionalities. Breaking the conventional constraints of empirical study, machine learning is a new instrument that ushers new frontiers for coordination chemistry to explore new and unexpected ranges of innovation and application (Rani, 2025). The possible implications for pharmaceutical discovery and development and materials science are profound, opening avenues of revolutionary potential for both.

1.5.1. Role in Catalysis

The role of coordination complexes in catalysis is paramount, particularly in homogeneous catalysis, in which transition metal complexes significantly enhance reaction efficiencies. Transition metals, due to their variable oxidation states and coordination geometries, can facilitate a range of chemical processes through stabilization of transition complexes and activation energy reduction. In fact, recent advances in electrochemical homogeneous catalysis demonstrate the transformatory potential of such complexes. In particular, the development of noble-transition-metal catalysts has provided fresh avenues for organic synthesis, catalyzing C–H functionalization and carbon-carbon bond formation under mild conditions (Cheng et al., 2021).

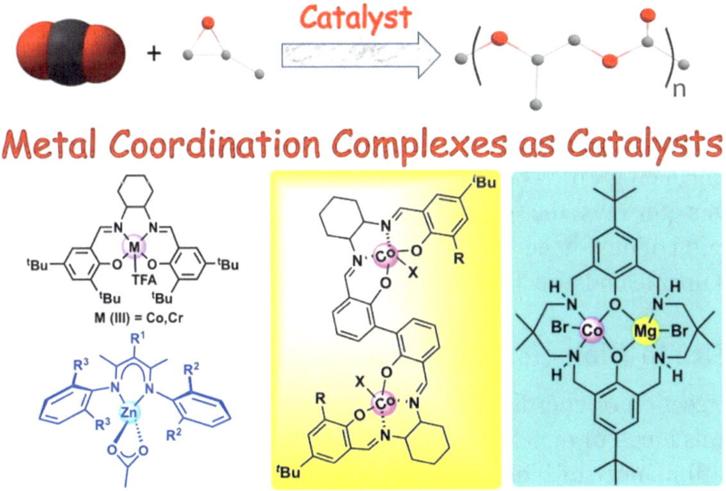

Figure 1.12. *The role of coordination complexes in catalysis.*

Source: https://ars.els-cdn.com/content/image/1-s2.0-S0010854523002667-ga1_lrg.jpg.

Such dynamic capability has the tendency to outperform traditional catalytic techniques, showing the way that coordination chemistry enables leading-edge strategies in synthetic application and competes with challenges in efficiency and selectivity. Incorporation of photochemistry into catalysis has, in turn, further enhanced the operational efficiency of coordination complexes, particularly in photoelectrochemistry applications. Incorporation of photocatalysts with electrochemical reactions enables it to achieve the reaction selectively at lower potentials, thus circumventing the constraint caused by the high-energy inputs. For instance, recent work highlights the use of photoelectrocatalysis for the acceleration of challenging redox reactions using photosensitizers that absorb light and facilitate electron transfer. This not only minimizes the likelihood of undesirable side reactions but also facilitates better stabilization of crucial reactive intermediates in catalytic processes (Freund et al., 2021). Coordination complexes are thereby intermediates and catalysts of prime importance in these advanced methodologies, attesting to the richness of coordination chemistry as it meets the boundaries of modern catalytic applications. The advances in materials science through coordination chemistry have also been remarkable with the emergence of metal-organic frameworks (MOFs) and covalent organic frameworks (COFs). These reticular materials with their adjustable surface chemistry and porosity present exciting possibilities for catalytic processes in a broad array of applications, from environmental remediation to energy conversion. The ability to fine-tune structure at the molecular level enables reaction environments to be customized, with enhanced substrate accessibility and selectivity (Freund et al., 2021). Moreover, such matrices can stabilize catalytic species beyond those that can be addressed by traditional catalysts, further enhancing efficiency and reusability. The ongoing discovery of MOFs and COFs portrays the pivotal position of coordination chemistry in the construction of next-generation catalytic materials that are not only efficient but also sustainable and effective for a range of industrial processes (Figure 1.12).

1.5.2. Use in Biological Systems

The intersection of coordination chemistry and biological systems is revealed and reveals much of metallobiology, wherein metal complexes play significant roles in all manner of biochemical processes. Coordination compounds make possible a vast array of biological processes, including but not limited to oxygen transport, electron transfer, and catalysis. By being complexed with organic ligands, metal ions form complexes that are able to influence the structural and functional behavior of biomolecules. For example, hemoglobins' oxygen-binding ability is an immediate result of its heme groups having iron as the stabilizing and coordinating entity in accordance with coordination chemistry principles. Such a coordination relationship reveals the nature of life processes as it reinforces the position of coordination chemistry as the foundation of biological macromolecules' functionalism and metabolic processes. Coordination chemistry keeps offering new drug design strategies in drug design. Metal drugs are gaining popularity as their efficacy in treating various diseases, including cancer and inflammation, has been well established. One such unique group of drugs, oxicam derivatives, exhibit the ability to coordinate with multiple metal ions monodentately and bidentately, enhancing their potential for drug action. Configurational adaptability in coordination geometries, as seen with Mn(II), Fe(II), and Cu(II) complexes, facilitates stabilization for the benefit of drug action in biological systems. Specifically, the metal complex DNA interactions of oxicam-based complexes, rationalized from studies on their antibacterial and anticancer activity, are observed to represent the profound impact of coordination chemistry on therapeutic applications. The preparation of these complexes showcases how metallic components may be controlled to yield compounds with higher bioreactivity than their non-coordinated analogs, as a demonstration of the strategic use of coordination chemistry in drug therapy. Moreover, the use of coordination compounds extends far beyond classical pharmaceutical chemistry to cutting-edge fields such as biomimetics and environmental chemistry. New avenues to the synthesis of metalloligands that bind to biological receptors with high efficiency are opened up by the study of triscatecholate binding motifs. Their preparation and use illustrate one example of the potential for designing biomimetic catalysts, which could be able to carry out biological processes as in nature without the shortcomings of existing methodologies. These advances are directly relevant in metabolic engineering and biocatalysis, where some metal-ligand relationships can fine-tune catalytic processes mimicking enzymatic processes. As research gets better, such compounds create new opportunities for new interventions in biological systems, highlighting the unique contribution of coordination chemistry to fashioning innovations closely resonating with our concept of life and processes, as evidenced by research in the selective synthesis of novel metalloligands and their functional relevance to applications in biology (Medley et al., 2025; Naeem et al., 2024).

1.5.3. Applications in Materials Science

Materials science has been richly supplemented by the advancement in coordination chemistry, particularly metal nanoparticles (NPs) synthesis and functionalization. These NPs are particularly given remarkable properties based on their high surface area-to-volume ratio and quantum size effect. A working assumption underlying their value is the coordination environment control, and thus they find application in catalysis, electronics, and nanomedicine. For instance, stabilized metal nanoparticles are utilized in catalytic converters to enhance rates of reaction and reduce emissions. Experimental data for such NPs can prove tricky to analyze. Adsorbate and substrate interactions provide dynamic variables that complicate the picture. Observations today indicate that conventional methods to describe bulk materials may not be appropriate for NPs; *in situ* methods and artificial neural networks are emerging as new tools to obtain valid insights into the properties of such materials (Ahmadi et al., 2019). Incorporating coordination compounds into structural materials has also unveiled new horizons in the development of advanced technologies, especially for energy-related purposes. Research on reduced activation ferritic/martensitic steels indicates that they are highly promising to be utilized in fusion reactors because of their high mechanical properties at temperatures greater than 550°C.

Research on atomic-scale evolution of such materials under extreme conditions is crucial for making them reliable and long-lasting in high-temperature applications. Techniques such as transmission electron microscopy and atom probe tomography are significant in the characterization of the material, demonstrating nanoparticle morphology, grain boundary chemistry, and void distribution.

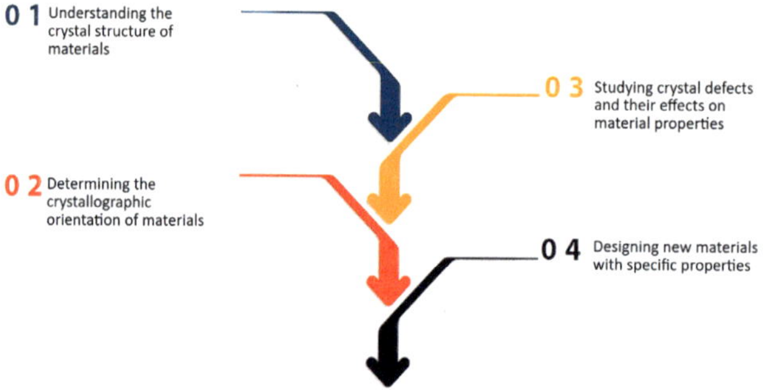

Figure 1.13. *Applications in materials science.*

Source: https://fastercapital.co/i/Crystallography--Decoding-the-Language-of-Crystallization--Applications-of-Crystallography-in-Materials-Science.webp.

The advanced analysis provided by such innovative techniques highlights the pivotal position of coordination chemistry in the fine tuning of the material properties to facilitate future energy solutions, representing a turning point of intersection of chemistry and engineering disciplines (Auger et al., 2011). Moreover, coordination chemistry enables the synthesis of functional materials with tailored properties for particular applications. For example, control over ligand interactions enables one to design hybrid materials that can exhibit enhanced magnetic, optical, or catalytic properties. New approaches, such as the incorporation of coordination compounds into polymer matrices, have been explored to generate responsive materials that can adapt to changing conditions. This adaptability also has consequences that extend beyond materials science, such as fields as biomedical engineering, where drug delivery or tissue growth can be engineered into smart materials. As research continues to develop, coordination chemistry's synergy with materials science continues to enable new to become available, demonstrating the value of interdisciplinary thinking to develop next-generation materials engineered for specific applications. This intersection makes possible a vision of new applications that reconcile scientific discovery with useful technological progress (Figure 1.13).

1.6. ANALYTICAL TECHNIQUES IN COORDINATION CHEMISTRY

Coordination chemistry, the central branch of chemistry, employs several analytical methods that are key to understanding the behavior and properties of coordination compounds. One of the most significant techniques is spectroscopic analysis, among which methods include UV-Vis, IR, and NMR spectroscopy. These methods help to elucidate structural information, electronic transitions, and ligand environments in coordination compounds. For instance, UV-Vis spectroscopy allows electronic transitions to be calculated that can be used to indicate coordination geometry in metal complexes. Through the study of the absorption spectra, chemists can make deductions about the ligand field strengths and the nature of the metal-ligand interactions, which are crucial in predicting the stability and reactivity. Basically, spectroscopic techniques are a cornerstone in coordination chemistry analysis that provides valuable data used in the characterization and utilization of the complexes. In addition to spectroscopy, electrochemical methods are increasingly employed to investigate the coordination compounds' properties.

Cyclic voltammetry and differential pulse voltammetry are two of the techniques employed to find powerful information concerning the redox conduct of metal sites in coordination complexes. Electrochemical procedures can allow us to find reduction and oxidation potentials, which can inform us concerning the stability of the oxidation state and electron transfer kinetics. Besides, the synergism of electrochemical methods and theoretical modeling has enhanced our understanding of complex interfacial processes in such systems, which is important for catalysis and energy storage applications. The use of these methods, not only enhances the characterization of coordination compounds, but also aids in directing the design of new materials with controlled properties for specific applications.

Along with that, the development of advanced computational techniques has also revolutionized the analytical landscape in coordination chemistry. Computational modeling, through methods such as density functional theory (DFT), enables researchers to predict and model the behavior of coordination compounds at the molecular level.

It includes predictive capability in modeling complex phenomena such as the formation of solid electrolyte interphases (SEI) in battery systems, where coordinated species are crucial. The application of computational methods allows the formation of a deeper understanding of the thermodynamics and kinetics of coordination reactions, therefore complementing standard experiment methodologies. The integration of computational modeling with experimental data underscores an active approach to coordination chemistry, demonstrating how these analytical tools merge to extend further investigation and application of this vital field (Wang et al., 2018; Wu et al., 2017).

1.6.1. Spectroscopic Methods

Spectroscopic methods play a significant role in coordination chemistry, as they provide useful information concerning the structure, bond, and properties of complexes. Among these methods, vibrational spectroscopy such as Fourier-transform infrared (FTIR) and Raman spectroscopy stands out due to its ability to present qualitative and quantitative information about molecular vibrations. These methods facilitate the identification of typical chemical fingerprints and provide indirect measurements of crystal properties, such as coordination and polymorphism. The non-destructive nature of these spectroscopic devices allows for the analysis of small sample sizes so that they are best adapted to the study of complex coordination compounds without compromising their integrity (Makreski et al., 2024). Its unequivocal excellence in rapid analysis and multirelational use makes vibrational spectroscopy an extremely useful tool for explorations in coordination chemistry. Electronic absorption spectroscopy adds to the insight in vibrational detail information on the electronic transitions of coordination complexes. This method is particularly well-suited for illuminating the ligand-metal charge transfer (LMCT) processes and d-d transitions, thus illuminating scientists about the ligand field and metal oxidation numbers in coordination compounds. By the examination of the UV-Vis spectra, scientists are able to discover significant parameters such as molar absorptivity and extent of coordination of the ligand to assist in identifying geometry and electronic structure. The immediate contact of spectroscopic methods with the metal-ligand matrices is an aspect that expands their use beyond structural determination alone; they are of pivotal importance in studies of the reactivity and stability of coordination complexes under various conditions, primarily contributing towards new discoveries in material science and biochemistry (Singha et al., 2023). Secondly, the employment of various spectroscopic methods heightens the quality of analysis that can be derived from coordination chemistry. For instance, the integration of UV-Vis spectroscopy with FTIR and NMR can generate a full picture of a complex's behavior, including its interaction with biomolecules like DNA. By adopting a combination of spectroscopic tools, researchers are able to confirm coordination modes, analyze ligand interactions, and track dynamic changes in the event of reactivity studies. For example, binding studies generally involve the utilization of fluorescence spectroscopy to investigate the metal complex affinities for nucleic acids to explore the mechanisms by which these metal complexes can affect biological systems.

Therefore, the complementary roles of these spectroscopic investigations enable one to take an integrated approach in the exploration of coordination chemistry towards making progress in drug design, materials development, and environmental remediation (Singha et al., 2023) (Figure 1.14).

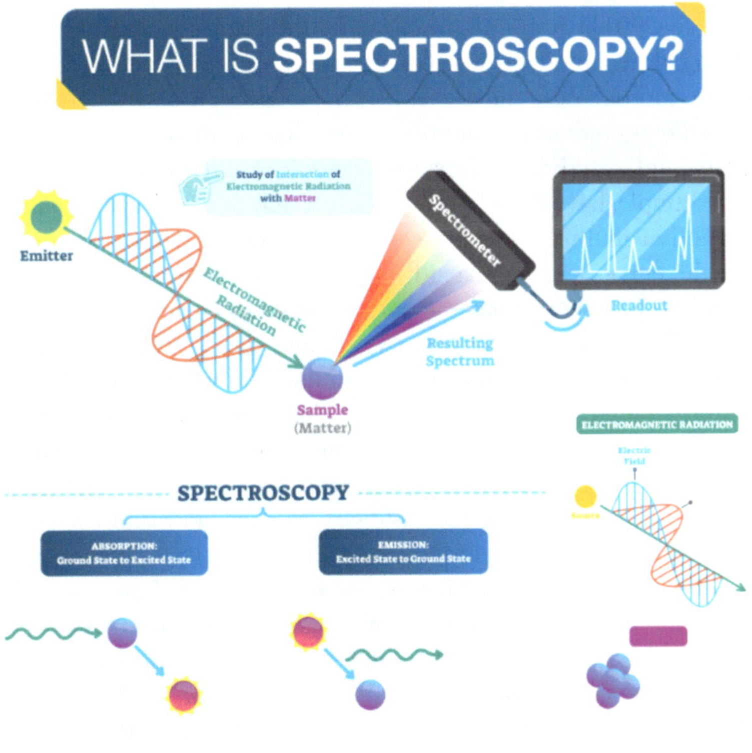

Figure 1.14. *Spectroscopic methods.*

Source: https://www.tilspec.cz/wp-content/uploads/2021/01/sp-what-is-spectros-copy.jpg.

1.6.2. Electrochemical Methods

Electrochemical methods play a crucial role in the study of coordination chemistry, particularly for the characterization of redox characteristics of metal complexes and ligands. These methods provide a means to observe electron transfer processes at the molecular level, providing data about the electronic structure of coordination complexes. For instance, as in the early work on dithiolene complexes, one-electron transfers facilitated an understanding of the roles of ligands in electron behavior. This work showed that oxidation and reduction processes might be occurring largely on the ligands rather than on the metal center, dissolving traditional oxidation-state representations. These observations are also pivotal in coordination chemistry, where metal complex redox reactivity by interaction of coordination environment dictates their applicability and reactivity in fields such as catalysis and materials science (Eisenberg et al., 2011). In addition to the definition of electron transfer processes, electrochemical techniques are also conducive to discovering new ligand structures in coordination complexes. New pincer-type ligand construction is one case, and that produces complex coordination geometries, increasing metal ion stability and reactivity. Bimetallic complexes involving a pentadentate ligand, for example, have been found to possess rich electrochemical phenomena via redox-active multiple sites, according to studies. These complexes not only provide enhanced understanding of the interactions between metals and metals but also show potential for cooperative reactivity, particularly in small-molecule activation reactions. Combining the structural characteristics of the ligands with electrochemistry contributes to the understanding of how coordination geometries can influence the electrochemical behavior of metal sites (Fiedler et al., 2015). These electrochemical investigations have more than fundamental coordination chemistry implications and impact applications in energy storage and conversion technologies. For instance, the regulation and understanding of redox properties by electrochemical means are crucial to designing effective fuel cell and battery catalysts. Detailed analysis of metal-ligand interactions and electron transfer dynamics can inform the design of materials that optimize energy conversion processes. Furthermore, through the application of the spectroscopic and computational techniques that are adjunct to electrochemical analysis, researchers are able to gain an overall picture of those factors that influence the reactivity and stability of coordination complexes. Electrochemical methods are typically a valuable tool in the ongoing quest to promote coordination chemistry through the integration of fundamental studies and application (Eisenberg et al., 2011; Fiedler et al., 2015).

1.6.3. Chromatographic Techniques

The significance of chromatographic techniques to coordination chemistry cannot be overestimated because they are central to the separation, identification, and investigation of metal complexes. Chromatography, in some of its versions—i.e., gas chromatography (GC), liquid chromatography (LC), and thin-layer chromatography (TLC)—permits chemists to effectively separate components of complex mixtures according to specific interactions with stationary and moving phases. This separation is of utmost significance for coordination complexes, which may have different coordination geometries and electronic properties that affect their behavior in the process of chromatography. Besides, the elevated sensitivity and resolution of chromatography techniques render them valuable for investigating products of reaction processes with transition metal compounds to unravel pathways of reactions and guarantee the formation of desired products such as those outlined in complicated investigations of coordination compounds (Hassan et al., 2008). In coordination chemistry, chromatographic techniques assist in probing the stability of metal complexes under different conditions and hence our understanding of their reactivity and stability.

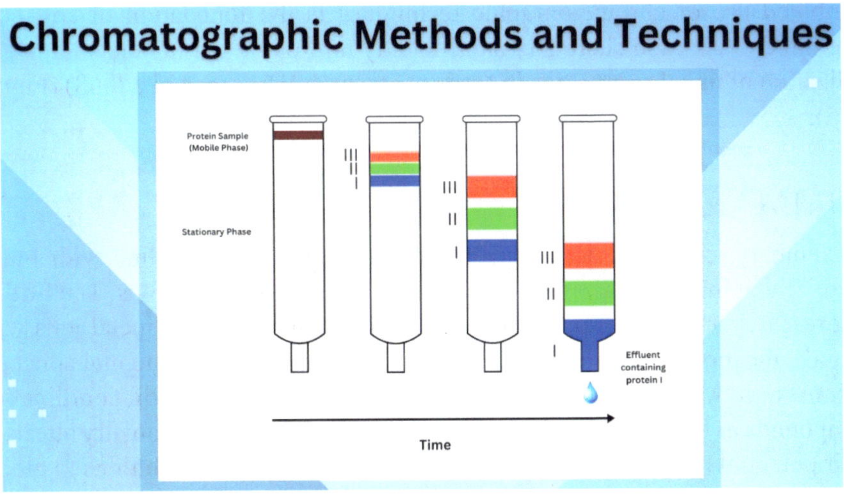

Figure 1.15. *The significance of chromatographic techniques in coordination chemistry.*

Source: https://veeprho.com/Chromatographic-Methods-and-Techniques.jpeg.

For instance, employing high-performance liquid chromatography (HPLC) enables the probing of electronic properties of various metal-cluster complexes, with implications in their coordination chemistry and mechanisms of reaction. It is through these methods that chelating ligands can be proved to exist or through which the stability of certain metal-lic ligand complexes, as in the case of systems characterized by considerable coordination modes such as those involving diphosphines (Bhowmick et al., 2012), is ascertained. The ability to separate and recognize small differences in molecular interactions by chromatographic separation thus aids in the optimization of conditions necessary for the synthesis of desirable coordination complexes and adds to our knowledge on their functioning in various applications. Moreover, chromatography is complemented by various spectroscopic techniques, providing a synergistic approach to the comprehensive study of coordination compounds. Through concurrent applications of chromatography and spectroscopy, scientists can gain detailed information regarding molecular structure, bonding, and coordination. For instance, successful purification of some complexes enables the subsequent structural elucidation through nuclear magnetic resonance (NMR) and infrared (IR) spectroscopy. This coalescence of approaches not only confirms successful synthesis of desired coordination complexes but also makes possible the exploration of their chemical behavior under different conditions. As shown in trailblazing research with difficult clusters and advanced reaction routes, combined use of chromatographic techniques is the foundation of advances listed in coordination chemistry significantly improving our understanding and utilization of metal complexes in areas of research (Hassan et al., 2008) (Figure 1.15).

SUMMARY

In summary, coordination chemistry research offers a discipline with much promise that intersects a variety of scientific fields and applications. The further that researchers delve into the interactions and manners of the metal ions with ligands, the more sophisticated the comprehension of the fundamental concepts becomes as well as their applied effects. The application of coordination compounds in fields like catalysis, materials science, and biochemistry attests to their potential to drive technological innovation and serve as solutions to global challenges. One real-life scenario is the management of the NTDs in Nigeria, which, as cited, calls for a deep understanding of the diverse stakeholders in health systems so that partnerships and resources can be effectively allocated. Strengthening those avenues is similar to the intricate relationships formed in coordination complexes, whereby coordination among elements yields superior capability and performance (Onasanya et al., 2025). Furthermore, the call for collective effort transcends theoretical construction; it aligns with the pragmatic underpinnings of cooperative action among stakeholders across most industries. For instance, where health-related social needs are under examination, geospatial analysis illustrates the extent to which community resources mobilize to fulfill the needed care. In such instances, collaboration between health systems and community-based organizations plays a vital role in optimizing access to needed services such that vulnerable groups can be effectively supported. This reflection highlights the interconnectivity of applied chemistry uses and public health programs, demonstrating how an integrated approach is crucial in achieving global success across scientific and humanitarian endeavors (Fitzpatrick et al., 2024). Summing up, the incorporation of coordination chemistry concepts into different fields is evidence of an integrative comprehension that is relevant for the scientific community and policy-makers alike. Emphasizing cooperative approaches not only accelerates the pace of research but also fosters more healthy societies by virtue of informed decision-making and distribution of resources. Ultimately, the fabric of coordination chemistry weaves through the themes of sustainability, health, and technology to demonstrate that coordinated unity, as with its respective coordination complexes, can yield exemplary outcomes. The future must be a promise to build interdisciplinary connections and shared vision among all actors for the achievement of the general aims of health equity and scientific progress (Fitzpatrick et al., 2024; Onasanya et al., 2025).

Coordination chemistry is the field that plays a pivotal role in establishing the interactions of metal ions with ligands, where the structure of these complexes strongly dictates a variety of applications in the field of materials science and biology. Major fields of discussion here are coordination number and geometry rules, that dictate the manner in which atoms are arranged in the complex. This foundation offers a solid premise for examining the effects these characteristics have upon stability and reactivity within coordination compounds. Furthermore, advances in synthetic techniques of forming an enormous variety of coordination complexes have expanded the boundaries of inquiry, opening up possibilities for catalysis, sensor manufacturing, and electronic materials. By investigating such dynamics, scientists can understand more about the behavior of molecular systems since such coordination is critical to facilitate enzymic processes and facilitate electron transfer reactions, leading us ultimately to a deeper understanding of chemical and biological interactions, as revealed through synthesis progress and applications of complex molecules in Intergovernmental Panel on Climate Change (2023) and Qing-Guo et al. (2021). Another critical aspect of coordination chemistry is the identification of ligand design and its impact on the stability and function of metal complexes. Whether the ligands are monodentate or polydentate controls directly how the coordination interacts with the metal, forming different complex formation. Ligands can incorporate new properties in the coordination complex, such as catalytic activity or selectivity in a chemical process. Moreover, the electronic character of the metal and that of the ligand establishes the overall electronic picture, showing the delicate interplay that ultimately determines the properties of coordination compounds. Current research is focused on the rational design of new ligands with an eye towards the building of better catalysts and multifunctional materials. The research identifies the importance of understanding ligand-metal interactions, which is the main impetus driving progress in the field, such as in the findings reported in Intergovernmental Panel on Climate Change (2023) and Qing-Guo et al. (2021). Lastly, the application of coordination chemistry extends to the creation of biomimetic systems and new materials. The knowledge gained from the study of coordination compounds solves basic questions while designing drug-delivery materials, magnetic resonance imaging, and solar technology, for instance. Coordination chemistry examples are also assigned meaning through the functioning of intricate biological systems like metalloenzymes and metalloproteins, where metal ions have a pivotal role to play in various metabolic processes. Researchers are now interested in how synthetic coordination complexes can mimic such biological systems to devise improved therapeutic agents or novel materials. Such interdisciplinary applications reveal the broad importance of coordination chemistry towards technological advances and medical breakthroughs, as discussed in the worthy contributions reported in Intergovernmental Panel on Climate Change (2023) and Qing-Guo et al. (2021).

REFERENCES

1. Ahmadi, M., Roldan Cuenya, B., & Timoshenko, J. (2019). Is there a negative thermal expansion in supported metal nanoparticles? An *in-situ* X-ray absorption study coupled with neural network analysis.
2. Amaral, V. S., Carlos, L. D., Ferreira, R. A. S., Girginova, K., et al. (2025). Synthesis, characterization and luminescent properties of lanthanide-organic polymers with picolinic and glutaric acids. https://core.ac.uk/download/15565296.pdf (accessed on 13th September 2025).
3. Atanasov, A., Palewicz, A., Sani, A., Udovenko, A. A., Harris, A. B., Gavriliuk, A. G., Gavriliuk, A. G., et al. (2011). Magnetoelectric ordering of *$BiFeO_3$* from the perspective of crystal chemistry. http://arxiv.org/abs/1005.3102 (accessed on 13th September 2025).
4. Auger, M. A., Castro Bernal, M. V. de, Leguey Galán, T., Lozano-Pérez, S., et al. (2011). Analytical characterization of oxide dispersion strengthened steels for fusion reactors. https://core.ac.uk/download/30045019.pdf (accessed on 13th September 2025).
5. Bernot, K., Daiguebonne, C., Calvez, G., Suffren, Y., & Guillou, O. (2021). A journey in lanthanide coordination chemistry: From evaporable dimers to magnetic materials and luminescent devices. *Accounts of Chemical Research*, *54*, 427–440. https://doi.org/10.1021/acs.accounts.0c00684 (accessed on 13th September 2025).
6. Bhatt, K., & Upadhyay, S. V. (2020). Review on Schiff base: A scrupulous binding detector for metal ions. *Organic Chemistry Plus*. https://doi.org/10.37256/ocp.112020380 (accessed on 13th September 2025).
7. Bhowmick, A. (2012). Synthesis and characterization of transition metal cluster containing diphosphine ligands. https://core.ac.uk/download/53188216.pdf (accessed on 13th September 2025).
8. Blatov, V. A., Shevchenko, A. P., & Proserpio, D. M. (2014). Applied topological analysis of crystal structures with the program package ToposPro. *Crystal Growth & Design*, *14*, 3576–3586. https://doi.org/10.1021/cg500498k (accessed on 13th September 2025).
9. Bokouende, S. S., Ward, C. L., & Allen, M. J. (2024). Understanding the coordination chemistry and structural and photophysical properties of EuII- and SmII-containing complexes of hexamethylhexacyclen and noncyclic tetradentate amines. *Inorganic Chemistry*. https://www.semanticscholar.org/paper/9f87b1b8b3e2c69d841acfc72c4225ba01a3c863 (accessed on 13th September 2025).

10. Cheng, X., Lei, A., Mei, T. S., Xu, H. C., Xu, K., & Zeng, C. C. (2021). Recent applications of homogeneous catalysis in electrochemical organic synthesis. *CCS Chemistry*, *4*, 1120–1152. https://doi.org/10.31635/ccschem.021.202101451 (accessed on 13th September 2025).
11. Chowdhury, T., Evans, M., Coles, M., Bailey, A. G., Peveler, W., Wilson, C., & Farnaby, J. H. (2023). Reduction chemistry yields stable and soluble divalent lanthanide tris(pyrazolyl)borate complexes. *Chemical Communications*. https://www.semanticscholar.org/paper/f2ac7b32152766883465334d460e57a4cb3b63fd (accessed on 13th September 2025).
12. Dankert, F., & von Hänisch, C. (2021). Siloxane coordination revisited: Si–O bond character, reactivity and magnificent molecular shapes. *European Journal of Inorganic Chemistry*, *2021*, 2907–2927. https://doi.org/10.1002/ejic.202100275 (accessed on 13th September 2025).
13. Deeth, R. (2024). Ligand field theory, Pauli shields and ultra-covalency in organometallic chemistry. *Physical Chemistry Chemical Physics: PCCP*. https://www.semanticscholar.org/paper/22d873c5135982304cdfb238e11d913be9647f20 (accessed on 13th September 2025).
14. Diercks, C. S., & Yaghi, O. M. (2017). The atom, the molecule, and the covalent organic framework. *Science*, *355*. https://doi.org/10.1126/science.aal1585 (accessed on 13th September 2025).
15. Eisenberg, R., & Gray, H. B. (2011). Noninnocence in metal complexes: A dithiolene dawn. https://core.ac.uk/download/4892476.pdf (accessed on 13th September 2025).
16. Fiedler, A. T., Lindeman, S., & Wang, D. (2015). Bimetallic complexes supported by a redox-active ligand with fused pincer-type coordination sites. https://core.ac.uk/download/213061671.pdf (accessed on 13th September 2025).
17. Fitzpatrick, S. L., Banegas, M. P., Mosen, D. M., Voelkel, J. L., Keast, E. M., Betcher, A., & Potter, C. (2024). Establishing a regional health system and community-based organization social care coordination network: An application of geospatial analysis. *The Permanente Journal*, *28*, 157–162. https://www.semanticscholar.org/paper/23c6d89525f4fa6b67186bf762f512a03ecad87b (accessed on 13th September 2025).
18. Frank, R., Denker, L., & Barthen, M. (2024). Amino imidazolin-2-imine ligands in magnesium complexes: Approaches towards low-valent Mg(I) species. *Zeitschrift für Anorganische und Allgemeine Chemie*. https://www.semanticscholar.org/paper/c3fdb0c21ff2ab395421813749eaeef39e040c37 (accessed on 13th September 2025).

19. Freund, R., Zaremba, O., Arnauts, G., Ameloot, R., Skorupskii, G., Dincă, M., Bavykina, A., et al. (2021). The current status of MOF and COF applications. *Angewandte Chemie International Edition, 60*, 23975–24001. https://doi.org/10.1002/anie.202106259 (accessed on 13th September 2025).
20. Guimaraes, S.P., Dos Santos, L.H.R., & Rodrigues, B.L. (2024). Coordination structure and intermolecular interactions in copper(II) acetate complexes with 1,10-phenanthroline and 2,2'-bipyridine. *Acta Crystallographic. Section C, Structural Chemistry*. https://www.semanticscholar.org/paper/a45f20bf788bcbd262c499992eb225af38820dfc (accessed on 13th September 2025).
21. Guo, Q. H., Jiao, Y., Feng, Y., & Stoddart, J. F. (2021). The rise and promise of molecular nano topology. *CCS Chemistry*, *3*, 1542–1572. https://doi.org/10.31635/ccschem.021.202100975 (accessed on 13th September 2025).
22. Hassan, M. R., Kabir, S. E., Karim, M. M., Nicholson, B. K., et al. (2008). An electron-deficient triosmium cluster containing the thianthrene ligand: Synthesis, structure and reactivity of $[Os_3(CO)_9(\mu 3\text{-}\eta^2\text{-}C_{12}H_7S_2)(\mu\text{-}H)]$. https://core.ac.uk/download/29196439.pdf (accessed on 13th September 2025).
23. Intergovernmental Panel on Climate Change. (2023). *Climate Change 2022–Impacts, Adaptation and Vulnerability*. https://doi.org/10.1017/9781009325844 (accessed on 13th September 2025).
24. Johrendt, D., Junggeburth, S. C., Oeckler, O., Schnick, W., et al. (2008). Nitridogermanate nitrides $Sr_7[GeN_4]N_2$ and $Ca_7[GeN_4]N_2$. https://core.ac.uk/download/12174703.pdf (accessed on 13th September 2025).
25. Kirlikovali, K. O., Hanna, S. L., Son, F. A., & Farha, O. K. (2022). Back to the basics: Developing advanced metal–organic frameworks using fundamental chemistry concepts. *ACS Nanoscience Au*, *3*, 37–45. https://doi.org/10.1021/acsnanoscienceau.2c00046 (accessed on 13th September 2025).
26. Komlyagina, V. I., Bakaev, I. V., Romashev, N. F., & Gushchin, A. (2025). Unusual coordination type of the halogenated BIAN ligands in silver(I) complexes. *Chemistry–An Asian Journal*. https://www.semanticscholar.org/paper/e830ee02ff9edb1344f11b5732cc4c6f6360bed2 (accessed on 13th September 2025).
27. Kovalenko, M. V., Proteşescu, L., & Bodnarchuk, M. I. (2017). Properties and potential optoelectronic applications of lead halide perovskite nanocrystals. *Science*, *358*, 745–750. https://doi.org/10.1126/science.

aam7093 (accessed on 13th September 2025).

28. Makreski, P., Pejov, L., & Jovanovski, G. (2024). Crossroads of vibrational (infrared and Raman) spectroscopy and X-ray powder diffraction in identification and characterization of some minerals–advantages and limitations. A review. *Macedonian Journal of Chemistry and Chemical Engineering*. https://www.semanticscholar.org/paper/c7d6aca9e36bda74b573f1140914ba9afff1635c (accessed on 13th September 2025).
29. Marinova, P., & Tamahkyarova, K. (2025). Synthesis, investigation, biological evaluation, and application of coordination compounds with Schiff base—A review. *Compounds*. https://www.semanticscholar.org/paper/1e31a6921d097590397d993c9285d5d55dd80d2a (accessed on 13th September 2025).
30. Medley, A. W., Miller, P. D., Bayse, C. A., Pike, R. D., & Bender, T. (2025). Amphiphilic tripodal Metallo ligands: Synthesis, coordination chemistry and preliminary catalysis. *European Journal of Inorganic Chemistry*. https://www.semanticscholar.org/paper/70b40586313d470d87f1f986280d28a9d4d3af37 (accessed on 13th September 2025).
31. Naeem, M., Ashraf, A., Imran, M., Siddiqui, W., Muhammad, G., Saleem, A., Younas, U., et al. (2024). Oxicams as bioactive ligand system in coordination complexes and their biological applications. *Comments on Inorganic Chemistry*, *45*, 63–87. https://www.semanticscholar.org/paper/d815ad08f0a4c5bd3564c731eafbe52ffb390333 (accessed on 13th September 2025).
32. Nimir, H., Hamza, A., & Hussein, I. A. (2020). Development of greener D-metal inorganic crosslinkers for polymeric gels used in water control in oil and gas applications. *Energies*, *13*, 4262–4262. https://doi.org/10.3390/en13164262 (accessed on 13th September 2025).
33. Omar, I., Alharas, M., Feizi-Dehnayebi, M., Alharbi, S. K., Abo-Dief, H., Qasem, H. A., El-Khatib, R., et al. (2025). Design, synthesis, physic-chemical characterization, stability determination, and biomedical applications of some novel tetra-dentate imine metal chelate supported by theoretical approaches: Bridging coordination chemistry and life sciences. *Applied Organometallic Chemistry*. https://www.semanticscholar.org/paper/236e7237fefb3a995330784d1f84e84bca3c709c (accessed on 13th September 2025).
34. Onasanya, O., Oladepo, G., Forje, G., Abua, E. E., & Amazigo, U. (2025). Evaluating stakeholder coordination and partnerships for NTD elimination in Taraba state, Nigeria: A multi-level analysis.

BMC Infectious Diseases, 25. https://www.semanticscholar.org/paper/d2f72756db1b38e852b411e6508c390da26391a7 (accessed on 13th September 2025).

35. Pandey, S. (2025). Coordination polymers: Bridging inorganic chemistry and materials science. *Journal of Science Innovations and Nature of Earth.* https://www.semanticscholar.org/paper/6fb06d2cd70be55eabea25e33b11f61e5e0b50ba (accessed on 13th September 2025).
36. Patel, H., & Raval, H. (2024). Advancements in coordination chemistry: A comprehensive review on the synthesis and characterization of transition metal complexes with 4-amino-5-pyridyl-4H-1,2,4-triazole-3-thiol ligands. *SPU Journal of Science, Technology and Management Research.* https://www.semanticscholar.org/paper/9ac07fc2695af166a7f24d5d77404ff604d018ff (accessed on 13th September 2025).
37. Peng, C., Li, G., Li, K. C., & Cui, X. B. (2024). Six polyoxotungstate-based transition metal compounds for electrochemical capacitor application and a comparative analysis of factors affecting capacitances. *Dalton Transactions.* https://www.semanticscholar.org/paper/24d794a5831c5c7617322a543f4065a9bff030c3 (accessed on 13th September 2025).
38. Ramler, J., & Lichtenberg, C. (2021). Bismuth species in the coordination sphere of transition metals: Synthesis, bonding, coordination chemistry, and reactivity of molecular complexes. *Dalton Transactions.* https://www.semanticscholar.org/paper/d9cc0e49a4bbb375f999124655bbe91932ee97b8 (accessed on 13th September 2025).
39. Rani, B. (2025). Coordination chemistry of cobalt biguanide complex: Ligand interaction and metal complex stability. *International Journal for Research in Applied Science and Engineering Technology.* https://www.semanticscholar.org/paper/bcfb4b8de5ab3174c71c8f082afb4ccb19b24337 (accessed on 13th September 2025).
40. Rani, M. (2025). Role of machine learning in inorganic and coordination chemistry: Applications and challenges. *International Journal on Science and Technology.* https://www.semanticscholar.org/paper/6b5dbf915a293075d82ba6eb583b30db6ec63b63 (accessed on 13th September 2025).
41. Robinson, J. R. (2014). *F-Element Binolate Complexes: Control of Secondary Structure for Applications in Redox Chemistry, Complex Design, and Asymmetric Catalysis.* https://core.ac.uk/download/76395967.pdf (accessed on 13th September 2025).

42. Schneider, H. J. (2022). Noncovalent interactions: A brief account of a long history. *Journal of Physical Organic Chemistry, 35*. https://doi.org/10.1002/poc.4340 (accessed on 13th September 2025).
43. Schulz, M., Poddelskii, A., Bohnert, H. A., Eckhardt, F., Hirner, L., Leingang, S., Kaifer, E., et al. (2025). Variation of the redox properties and coordination chemistry of redox-active aromatic diguanidine ligands with the substitution pattern. *European Journal of Inorganic Chemistry*. https://www.semanticscholar.org/paper/5563f195433d4d37628504eab254e223f3582922 (accessed on 13th September 2025).
44. Singha, D., Halder, S. C., Jana, A. D., & Pal, N. (2023). The coordination chemistry and supramolecular interactions of 2-(2′-pyridyl) imidazole ligand: A comprehensive review with theoretical insight. *Reviews in Inorganic Chemistry, 44*, 231–253. https://www.semanticscholar.org/paper/5aa37cfa9db0ef79a02a8c565037a732cb1a33ec (accessed on 13th September 2025).
45. Singha, U. K., Pradhan, S., Mishra, D. K., Gurung, P., Chettri, A., & Sinha, B. (2023). Synthesis, physicochemical characterization and DNA binding study of a novel azo Schiff base Ni(II) complex. *European Journal of Chemistry*. https://www.semanticscholar.org/paper/5825080d8f197eb75d5ac98d0df8bdb0529229de (accessed on 13th September 2025).
46. Slater, M., Kim, D., Lee, E., & Johnson, C. S. (2012). Sodium-ion batteries. *Advanced Functional Materials, 23*, 947–958. https://doi.org/10.1002/adfm.201200691 (accessed on 13th September 2025).
47. Song, J., Wei, C., Huang, Z. F., Liu, C., Zeng, L., Wang, X., & Xu, Z. J. (2020). A review on fundamentals for designing oxygen evolution electrocatalysts. *Chemical Society Reviews, 49*, 2196–2214. https://doi.org/10.1039/c9cs00607a (accessed on 13th September 2025).
48. Soroceanu, A., & Bargan, A. (2022). Advanced and biomedical applications of Schiff-base ligands and their metal complexes: A review. *Crystals, 12*, 1436–1436. https://doi.org/10.3390/cryst12101436 (accessed on 13th September 2025).
49. Torubaev, Y., Shaashua, O., Braunstein, S., & Pappo, D. (2025). Halogen bond-driven ligand displacement: Co-crystal lattice versus coordination bonds. *Chemistry (Weinheim an der Bergstrasse, Germany), 31*. https://www.semanticscholar.org/paper/afa020cb0cf993d36c7e059002b00af9c13b4186 (accessed on 13th September 2025).
50. Wang, A., Kadam, S., Li, H., Shi, S., & Qi, Y. (2018). Review on modeling of the anode solid electrolyte interphase (SEI) for lithium-ion batteries.

npj Computational Materials, *4*. https://doi.org/10.1038/s41524-018-0064-0 (accessed on 13th September 2025).

51. Wang, S. D., Xue, Z., Mu, Y., Li, J. H., & Wang, G. (2023). Multicolor phosphorescence modulated by excitation and temperature in Zn-based coordination polymers. *Inorganic Chemistry*. https://www.semanticscholar.org/paper/c3b885de9823f376aad53f96495ef7d6619b9a5f (accessed on 13th September 2025).
52. Wang, Y., Hu, S. X., Cheng, L., Liang, C., Yin, X., Zhang, H., Li, A., et al. (2020). Stabilization of plutonium(V) within a crown ether inclusion complex. *CCS Chemistry*, *2*, 425–431. https://doi.org/10.31635/ccschem.020.202000152 (accessed on 13th September 2025).
53. Wu, D., Sedgwick, A. C., Gunnlaugsson, T., Akkaya, E. U., Yoon, J., & James, T. D. (2017). Fluorescent chemo sensors: The past, present and future. *Chemical Society Reviews*, *46*, 7105–7123. https://doi.org/10.1039/c7cs00240h (accessed on 13th September 2025).
54. Wu, M. X., & Yang, Y. W. (2017). Metal–organic framework (MOF)-based drug/cargo delivery and cancer therapy. *Advanced Materials*, *29*. https://doi.org/10.1002/adma.201606134 (accessed on 13th September 2025).
55. Ye, B. C., Li, W. H., Zhang, X., Chen, J., Gao, Y., Wang, D., & Pan, H. (2024). Advancing heterogeneous organic synthesis with coordination chemistry-empowered single-atom catalysts. *Advanced Materials*, *36*. https://www.semanticscholar.org/paper/178a11139ca0183549d071afdb31d7e57e99a8a0 (accessed on 13th September 2025).

CHAPTER 2

BONDING IN COORDINATION COMPOUNDS

Contents

INTRODUCTION

Coordination chemistry has been of strong interest to scientists due to the unique bonding characteristics of coordination compounds. Coordination compounds consist of a central metal ion or atom coordinated by the surrounding ligands via a variety of bonding interactions, primarily coordinate covalent bonds. The character of these interactions influences the structural nature of the compounds as well as the functional behavior of the compounds in most applications.

Advances in knowing the coordination chemistry of ligand attachment and geometry implications are significant in explaining how coordination compounds behave in different environments, like living organisms. Thus, the study of coordination compound bonding confirms our fundamental knowledge of chemical principles while pushing applications to practical issues like catalysis and materials science, which results in breakthroughs in drug discovery and pollution remediation. A significant area of interest in coordination chemistry is its role in catalytic processes addressing the major global issues. The recent advances in the synthesis of coordination compounds have led to assessing their promise for the sequestration and utilization of carbon dioxide. For example, new copper phosphonate architectures have presented remarkable catalytic efficiency in the conversion of CO_2 into valuable chemical products at mild conditions. These advancements mirror the prevalently influential position of ligand design, sustaining enhanced hydrogen bonding interactions and supporting low-dimensional structures and hence improving catalytic efficiency. Further notable is how inclusion of particular types of ligands can greatly influence performance of metal complexes in catalyzing reactions, mirroring the fine balance between structures of bonding and functional effects within coordination compounds (Ghanghas et al., 2025). These findings reflect the broader applications of coordination chemistry to the creation of green solutions for environmental problems. Additionally, the study of coordination compounds is not only confined to coordination complexes but also its medical applications through oncology. The therapeutic potential of coordination compounds was initially proved by cisplatin, a very common chemotherapeutic drug. Present research has shown that copper coordination compounds contain optimistic antitumor activity with cytotoxicity that is often superior to traditional agents. Specifically, copper complexes of diimine ligands have shown potential drug candidates due to their ability to cause DNA damage preferentially within the cancer cells. Utilizing the distinctive redox properties of copper, researchers have now begun to uncover the mechanisms underpinning their cytotoxicity, which indicates the potential for targeted therapies (Alvarez et al., 2018). The meeting of coordination chemistry with medicine emphasizes the flexibility of coordination bonding in coordination compounds and evokes the significant role of coordination compounds in solving major problems in health and the environment.

Definition of Coordination Compounds

Coordination compounds are an interesting collection of chemical compounds that involve the coordination of metal center atoms or ions with the surrounding molecules or anions, which are called ligands. The unique feature of coordination complexes is that they have the ability to attain the coordinate covalent bond formation, where both the bonded electrons are contributed by the ligand. This results in a specific geometry around the metal center, which accounts for the properties and the reactivity of the compounds. A number of metals, including transition metals, lanthanides, and actinides, serve as central atoms with a high degree of diversity in the types of coordination compounds that may be synthesized and examined. More importantly, the nature of ligands—in terms of varying from neutral molecules to ions—also changes the electronic setting of the metal center, significantly affecting the overall stability and properties of the resulting complex. The coordination compound bonding is governed mainly by the rules of coordination chemistry. Coordination number, or ligand attachments to a central atom, is a critical parameter determining the geometry of the complex. The common coordination geometries are the octahedral, tetrahedral, and square planar geometries, each of which is linked to specific ligand orientations and bond angles. In addition, the metal ion's electronic structure defines its ability to bind, in which d-orbital participation is typically significant for transition metals. These electronic and geometric factors are essential to utilize to be able to predict reactivity and the functioning of coordination compounds in catalysis and materials science. The intricate relationship between the structure and bonding characters of the compounds sets a foundation platform for exploring their diverse applications in research and industry. Contemporary research uses advanced computational tools to study the properties and coordination bonding of the coordination compounds. Computer programs such as Multiwfn facilitate complex wavefunction analyses that enhance our understanding of the electronic distributions and interactions with these complexes. By the application of features like electron density topology analysis, researchers can obtain meaningful information about electronic structure and macroscopic properties such as magnetic properties and reactivity. Magnetic coordination compounds such as lanthanide single-molecule magnets reveal how structure and function can exist side by side, whereby light is shed into their potential use in data storage and quantum computing (Woodruff et al., 2013). Moreover, the ability to effectively visualize electronic properties using computational tools accentuates the significance of theoretical progress in coordination chemistry, leading the way to new material design and synthesis techniques (Lu et al., 2011).

Importance of Bonding in Coordination Chemistry

Coordination chemistry significantly relies upon the bonding nature to determine the behavior and properties of coordination compounds. At the heart of the topic are the metal ion-ligand atom interactions close by, upon which many structures and functions depend. The interactions are ionic, covalent, and coordinate covalent, the latter of which is typical in itself of coordination chemistry. The characteristic geometries that arise from such patterns of bonding influence anything from reactivity to stability, and consequently influence the manner in which such compounds behave in biological processes, catalysis, and materials science. A comprehension of these bonding processes is paramount for rational syntheses of novel coordination complexes with desired chemical characteristics, because the interaction between metal-ligand bonding affects the electronic profile and total stability of the resultant compound directly, emphasizing the unavoidable involvement of bonding in coordination chemistry.

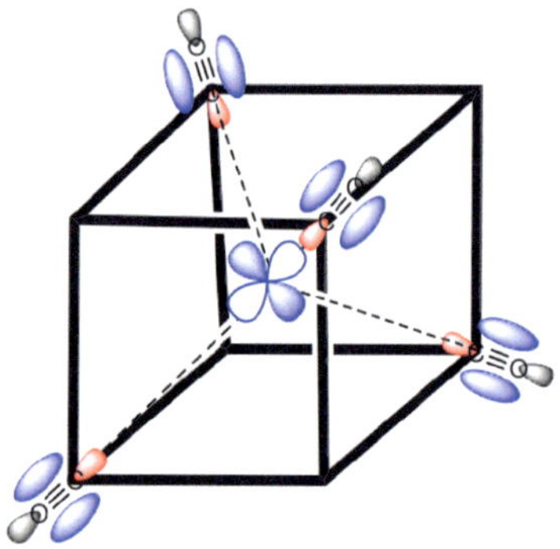

Figure 2.1. *Importance of bonding in coordination chemistry.*

The involvement of bonding in coordination chemistry also has an important function in applications, particularly catalysis and material science. Metal complexes function as catalysts for myriad reactions whose gross activity is determined by electronic and steric properties imparted upon them by the ligands through coordination bonds. Tuning of the type of ligands, for instance, results in improved efficiency in electrocatalytic reactions such as the hydrogen evolution reaction (HER), a critical reaction in clean energy technology. As discovered in modern studies, low-cost but high-performance catalysts are being developed by tuning the coordination environment at metal centers. This influences significant aspects such as active site reactivity and accessibility and leads to major breakthroughs in materials capable of satisfying future energy needs. The ability to design these interactions emphasizes the relationship between bonding strategies and applications in actual coordination chemistry, highlighting its importance. Moreover, coordination chemistry bonding is necessary to better understand more complicated biological systems. The majority of metalloenzymes, which are necessary for biological catalysis, operate based on complicated metal-ligand interactions that define the way they behave.

These bonds enable substrates to be precisely placed and facilitate electronic transfer processes necessary for biochemical transformations. Their geometry and coordination are central to determining the specificity and efficiency of catalysis by these enzymes. Investigations into bonding mechanisms in these biological complexes not only extend our knowledge of cellular processes but also inspire biomimetic catalyst design in synthetic chemistry. This combined approach teaches us that a thorough understanding of bonding is central not only to coordination compounds, but also to larger scientific pursuit, reaffirming the central role that bonding plays in the interconnected spheres of chemistry, biology, and technology (Figure 2.1).

2.1. HISTORICAL BACKGROUND

The coordination theory was initially proposed early in the 19th century, marking a milestone for the history of inorganic chemistry. Initial investigations by scientists like John Dalton and his atomic theory became the foundation of knowledge related to complex metal ion and ligand structures. The coordination number was identified in 1869, which immensely advanced the knowledge about how various metal cations coordinate with the surrounding molecular species. Worthy of mention here is that, in 1913, Alfred Werner received the Nobel Prize in Chemistry and introduced the theory of coordination chemistry, which reasserted that certain metals are capable of coordinating with multiple molecules or ions, referred to as ligands. The foundational work shed light on the nature of bonding in such compounds and opened avenues for further research into the diverse applications of coordination complexes as catalysts and materials. Synthesis of coordination compounds has followed parallel advances in analytical techniques, allowing precise characterization of such species. In the 20th century discovery by spectroscopy, such as ultraviolet-visible (UV-Vis) and infrared (IR) spectroscopy, began to resolve electronic transitions and vibrational modes of coordination complex structures. Further, the use of crystallographic techniques, including X-ray crystallography, assisted in unraveling the geometric shapes of the atoms in the compounds so that the complex architectures could be visualized in three dimensions. This was in effect to emphasize the ability of coordination compounds to exhibit a variety of colors, magnetic properties, and reactivity patterns directly related to their structural geometries and the bonding interaction types. These advancements not only increased scientific knowledge but also led to the application of coordination compounds in several industrial and medicinal applications. With continuous research, it was understood that bonding in coordination compounds relied not only on the metal and ligand nature but also on their surroundings. This insight was associated with the investigation of transition metals' oxidative and reductive character, which played a crucial role in the stability and formation of such complexes. Studies of redox reactions demonstrated that oxidative stress could alter metal centers' electronic states, hence their ligand-binding affinities and coordination modes. This complex interaction signifies the importance of understanding the historical context of coordination compound bonding, as these precursors continue to be a force in current research. Comparison of the efforts of pioneering theorists and current knowledge brings to researchers the awareness of both chemical and environmental forces involved in complex formation and stability, a topic still crucial in current studies of coordination chemistry/cite7/cite8.

2.1.1. Discovery of Coordination Compounds

The discovery of coordination compounds was a giant leap in inorganic chemistry because it enhanced our understanding of metal bonding. Coordination compound is a broad category of materials composed of central metal ions coordinated with neighboring molecules or ions, which are referred to as ligands. This interaction has a tendency to create new geometrical structures and electronic properties, which are crucial in most applications like catalysis, materials science, and drug discovery. A pioneering effort by Alfred Werner in the early 19th century laid a basis for the understanding of such intricate structures through coordination number determination and the geometrical type of such compounds. Werner's pioneering research showed that transition metals could be multiply bonded to ligands, and thus there was the hope for the establishment of systematic research of metal-ligand interactions and the roles they play in chemical reactions. Additional research in coordination compound chemistry has yielded a sequence of varied and complex structures that have diverse functions in most scientific disciplines. This includes the knowledge of how ligands are able to alter metal center electronic states and, therefore, influence stability and reactivity. Nitrogenous ligand-containing coordination complexes such as imidazole-based compounds have been of interest because of their bioactivity. These compounds are usually biomolecule analogs and thus are able to bind certain protein targets, making them potential in drug discovery. For instance, imidazole-containing coordination compounds have displayed an extensive range of biological activities, and are used in anticancer treatment and other biomedical applications. The ability to form stable, reactive coordination environments exemplifies the very close interrelationship between coordination chemistry and biology (Sharma et al., 2021). Synthesis of coordination compounds is facilitated by a range of synthetic procedures that change over time and become more sophisticated. Current advances in design and development of new ligands, particularly boryl radicals, are an expression of the innovative character of this line of inquiry (Peng et al., 2022). The boryl species present new avenues for synthesizing organoboron species through unique radical borylation reactions, evidence that coordination chemistry extends beyond classical ionic or covalent bonding models. Such methodologies rely on electronic and steric metal and ligand manipulation, thereby giving rise to compounds of customized reactivity and selectivity. This active situation brings out the importance of coordination compounds not just in describing simple chemical bonds but also in the design of efficient methods for synthesizing novel materials and drugs, thereby illustrating their paramount position in modern chemical investigation (Figure 2.2).

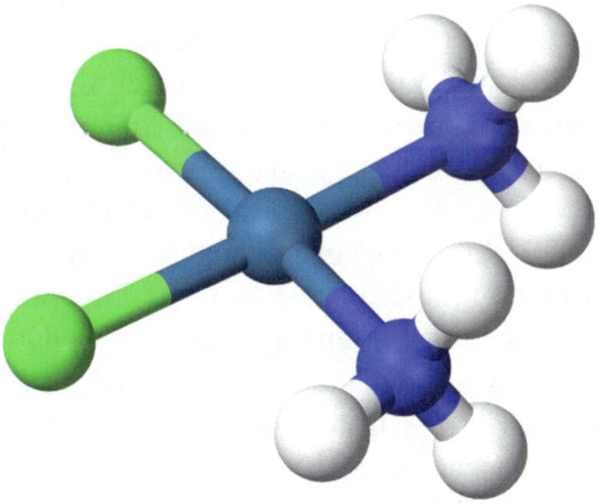

Figure 2.2. *The discovery of coordination compounds.*

Source: https://upload.wikimedia.org/wikipedia/commons/9/96/Cisplatin-3D-balls.png.

2.1.2. Key Figures in Coordination Chemistry

Coordination chemistry has been influenced immensely by the work of pathfinder researchers such as Alfred Werner, whose work provided the basis upon which our understanding of coordination compounds was established. Werner, in 1893, introduced concepts of coordination numbers and established the shapes of these complexes. His innovative empirical formulae, demonstrating the metal complex geometry, provided a scientific way of classifying and describing the coordination compounds. His path-breaking work established the theory of coordination, demonstrating that metals can be coordinated by various ligands in coordinate covalent bonds. Werners becoming the first Nobel Prize laureate in Chemistry in 1913 demonstrates the extent of his contribution to the field, as it set further study of the behavior of ligands and metal coordination dynamics underway, further refining our knowledge of coordination compound bonding. Werners work paved the way for further leading figures, such as Linus Pauling and F. Albert Cotton, to leave their permanent mark on coordination chemistry. Pauling's hypothesis of the hybridization theory and application of molecular orbital theory accounted for coordination compounds' electronic configuration. Through research of the combination of orbitals to form hybrid orbitals, Pauling helped explain the shapes found in transition metal complexes. Furthermore, F. Albert Cotton's early research on cluster compounds extended the field of research, outlining the properties and reactivities of various metal-metal bonded compounds. His description of cluster in chemistry revolutionized the research of metal interactions in coordination compounds, which enhanced their understanding of their complexity and uses in catalysis and material science. These individuals collectively worked on theoretical models that now underpin most aspects of coordination chemistry. In modern times, researchers have gone further from where they were with significant advancement in coordination chemistry. Unfolding breakthroughs in computational chemistry, as highlighted by recent studies (Liu et al., 2022), have facilitated the prediction and design of new metal-ligand interactions, attaining optimized catalysts for applications. Further, breakthroughs in nanotechnology have made it possible to explore coordination compounds at the nanoscale, enhancing their functions across various fields like pharmaceuticals and energy storage. Contemporary scientists also worry about the sustainability and environmental impact of coordination compounds, leading to greener processes in their production and application (Cheng et al., 2021). The influence of illustrious contributors to coordination chemistry therefore continues to feed new studies, attempting to unlock the unique properties of coordination compounds for novel solutions in scientific and technological endeavors.

2.1.3. Evolution of Theories Regarding Bonding

The evolution of bonding theories has undergone substantial transformations, particularly when it comes to coordination compounds. The initial classical theories such as the octet rule constituted the foundation for the description of atomic interaction and chemical bonding. These theories did become increasingly inadequate for the more complex structures, and particularly so for those that involve transition metals. The evolution of ligand field theory offered further understanding by considering the spatial geometry and energy interactions of ligands that surround a metal ion. This model allowed chemists to predict and explain the properties of most coordination compounds, a key landmark in the comprehension of chemical bonding. Coordination between metal ions and ligands was then understood not only to be ionic or covalent but as occurring by an intermixture of many types of interactions, in fact propagating the debate on coordination bonding theories. With more research, the establishment of molecular orbital (MO) theory further clarified the understanding of coordination bonding interactions in coordination compounds. This theory emphasized the importance of overlap between atomic orbitals and thus explained more comprehensively the electronic configurations and resulting stability of coordination complexes.

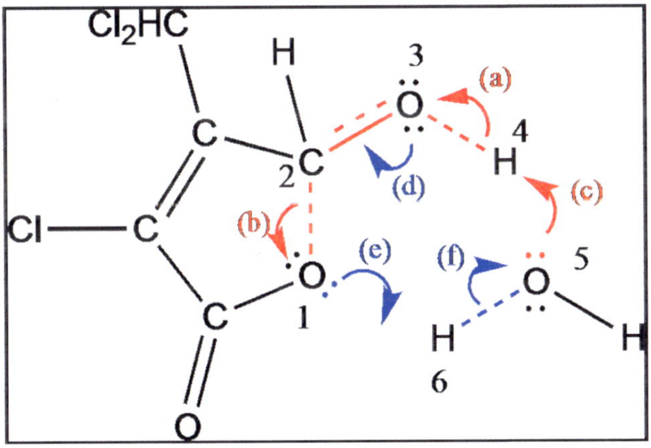

Figure 2.3. *The evolution of theories regarding bonding.*

Source: https://www.researchgate.net/publication/329220133/figure/fig6/AS:960036050309147@1605901839834/The-bonding-evolution-theory-analysis-allows-us-to-easily-reveal-the-sequence-indicated.png.

By considering the molecular orbitals resulting from both metal d-orbitals as well as from ligand orbitals, chemists were further able to refine their models to properly account for the nature of bonding in various complexes. The discovery of concepts such as π-backbonding and σ-donation revealed additional richness in interaction that the previous theories could not adequately account for. For this reason, the molecular orbital theory highlighted the significance of hybridization in determining the geometry and the properties of different coordination compounds in support of the theoretical framework of bonding in this coordination chemistry. In more recent periods, computational resources and experimental design have allowed for the inclusion of emergent models that reflect electron correlation effects and non-adiabatic phenomena. Structural transformation in coordination compounds due to pressure is one such example, reflecting the dynamic modification of bonding under varying conditions, and demonstrating progress in our understanding of chemical interactions.

Experiments on materials like CuP_2Se have shown such complexity, as they undergo alterations that reconstruct atomic interactions in a significant way under pressure and thereby enhance properties like superconductivity and electron conductivity. Such observations indicate the interplay between bonding and environmental factors outside, underlining the importance of context in theories of bonding. As investigation further unfolds the intricacies of coordination chemistry, re-evaluation and refinement of bonding theories remain crucial in the process of designing new catalysts and materials, such as hydrogen fuel generation and other green technologies (Li et al., 2021; Zhu et al., 2019) (Figure 2.3).

2.2. BASIC CONCEPTS OF COORDINATION CHEMISTRY

The fundaments of coordination chemistry involve the understanding of the interactions among metal ions and organic ligands that lead to coordination complexes. The ligands, which can be neutral molecules or anions, possess varying binding capabilities to metal centers depending on their structural features as well as electronic characteristics. For instance, ligands are either monodentate or polydentate, with the latter getting bound multiple times to one metal ion, thereby enhancing the stability of the resulting complex. Investigation of the coordination modes is crucial as they are responsible for determining the geometry and electron distribution in the compound, which in turn determines the physical and chemical properties of the compounds. This interaction of metal coordination and ligand function leads to the creation of diverse structures, making ligand design crucial in coordination chemistry. In coordination compounds, chelation can never be overemphasized. Chelating ligands or chelators have multiple binding sites that can bind metal ions, leading to the creation of stable ring structures. This stabilization is a result of the chelate effect, which is the enhanced coordination of chelating ligands for metal ions compared to their monodentate counterparts. The ability of ligands to be able to form rigid or flexible structures is seen to characterize their role in sequestering metal ions as shown with macrocyclic ligands such as those derived from thiacalix[4]arene. They also permit intriguing interactions, like hydrogen bonding and π-π stacking, that can significantly influence the extraction capabilities and selectivity of metal ions in solid as well as solution phases. The fine rapport between metal coordination and the synthesis of a ligand highlights the necessity of chelation in coordination chemistry, as established by extensive research on cyclic compounds with such efficiencies at extraction. Furthermore, biomolecules such as nucleotides' coordination chemistry reveals the molecular intricacy present in biological processes and their interactions with metal ions. Nucleotides, comprising phosphate groups, sugar, and nucleobases, play pivotal roles in various biochemical processes, facilitated by the coordination of metal ions, which enhances key reactions. The coordination of metal ions to nucleotides is also important in cellular processes, including DNA replication and transcription, in which metal ions stabilize conformations and enhance enzymatic activities. Moreover, structural variations of nucleotides, including acyclic nucleoside phosphonates, emphasize the potential of altering such biomolecules for therapy, including antiviral therapy. The complex interplay of metal coordination and nucleotide function not only bears witness to the underlying principles of coordination chemistry but also testifies to its relevance in understanding biological processes and furthering medicinal chemistry progress, as documented by recent research findings (Sigel et al., 2022).

2.2.1. Definition of Ligands

A fundamental aspect of coordination chemistry involves the understanding of ligands, which are ions or molecules with the capability to donate electron pairs towards a central metal atom or ion to form coordination compounds. As Lewis bases, ligands stabilize metal centers through coordinate covalent bonds. The nature of these ligands, including their ability to donate electrons, geometry, and charge, plays a vital role in determining the nature of the resultant complex. Ligands may be classified as monodentate, bidentate, and polydentate based on the number of donor atoms present. The classification assists in defining the coordination number of the metal ion as well as the geometrical arrangement of the resultant compound. An understanding of these definitions is the basis for the comprehension of bonding and interaction types in coordination complexes, including insight into their reactivity, as well as their applications in catalysis and materials science. The various ligand characteristics, such as their steric and electronic influences, are important in determining stability and reactivity of coordination compounds. Certain ligands create unique environments around the metal center, which can enhance stability of the complex or cause specific reactivity. For example, regioisomeric influences of ligands can cause differences in bonding affinities and structural outcomes, thus affecting physical and chemical behavior. According to recent literature, coordination bonding and corrosion inhibiting performance may be greatly affected by the positioning of functional groups in a ligand (Verma et al., 2024). The same metal ion-ligand specific interactions also regulate a vast array of applications, ranging from catalysis to the construction of novel materials, and consequently highlight the importance of ligand characterization in coordination chemistry. Moreover, the advancements in electrochemical processes have kindled fresh research interest in the development of functional ligands capable of being utilized in electrochemical reactions, particularly in homogeneous catalysis (Cheng et al., 2021). Preparation of novel ligands with specificity to a single transformation allows for greater control over reaction paths and selectivities. Some ligands, for example, can stabilize otherwise unstable metal oxidation states under standard conditions. Designing ligands compatible with aggressive environments or electron transfer facilitators is also significant to enhance catalytic efficiency in industrial processes. The synergy between coordination chemistry and ligand design not only advances the fundamental insight into metal-ligand chemistry but also stimulates innovation in practical applications ranging from energy conversion to material synthesis. The fusion of these concepts is crucial to develop both the theoretical and practical horizons of coordination compounds in contemporary chemistry (Figure 2.4).

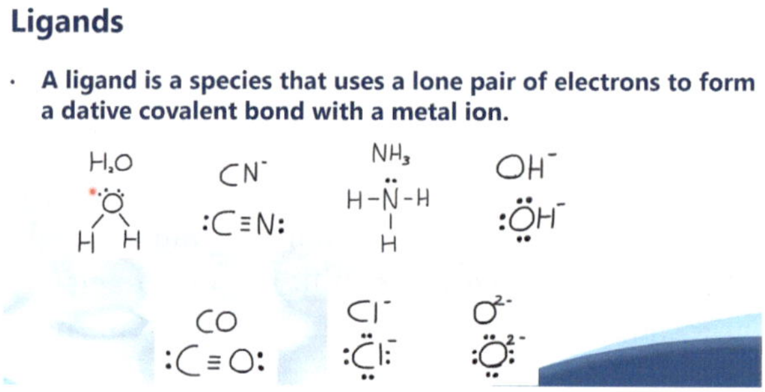

Figure 2.4. *The fundamental aspect of coordination chemistry involves the understanding of ligands.*

Source: https://i.ytimg.com/vi/Bq_q9p10CZc/maxresdefault.jpg.

2.2.2. Coordination Number and Its Significance

Coordination number is the key principle of coordination compound research because it is an important indicator of how many atoms of a ligand are bound to a central metal atom. Coordination number is an indicator of the spatial structure and geometry of the compound and could have an influence on its reactivity and chemical stability. For instance, a coordination number of six is typically associated with an octahedral geometry, one that is unusually good at optimizing ligand-to-metal contact. Coordination number is more than just a classification process; it plays an essential role in the determination of the electronic properties of the metal center. Because coordination number has the capacity to alter orbital hybridization as well as electron distribution, the logical design of coordination compounds with a specific use such as catalysis or medicine delivery is made more convenient with increased knowledge of coordination numbers. Additionally, coordination number discrimination can be used to predict physical properties of coordination compounds such as solubility, color, or magnetism. Increasing coordination number is related to more complex structure, the latter potentially exhibiting anomalous physical properties. Tetrahedral (coordination number four) coordination compounds, for instance, might possess distinct optical properties from those of octahedral analogs and therefore impact their functions in various fields such as materials science and biomedicine. This coordination number vs. physical property relationship illustrates the richness of coordination chemistry and underscores its significance in investigation and application. Awareness of these relationships enhances our capacity to manage these compounds towards desired goals, e.g., the development of novel materials or drugs. Apart from their educational significance, coordination numbers find practical use in nanotechnology and drug delivery systems. Current advances in nonclassical platinum complexes, for example, are exploring coordination geometries and their impact on therapeutic efficacy in cancer treatment (Johnstone et al., 2016). Similarly, new energy storage devices are taking advantage of the role played by coordination compounds and their unique characteristics as defined by their coordination numbers (Jang-Hwang et al., 2017). These examples demonstrate the diversity and value of coordination chemistry where the understanding of underlying principles can lead to new solutions for problems in modern times. The simultaneous occurrence of coordination number, geometry, and application is evidence of the pertinence of this idea within the broad scheme of chemical bonding and materials science.

2.2.3. Types of Coordination Compounds

Coordination compound nomenclature is a complex but unavoidable aspect of understanding their bonding and reactivity. Coordination compounds may be divided on the basis of the number of ligands that are coordinated around a central metal ion or atom. Monodentate ligands coordinate through one atom to yield a simple and usually stereotypical coordination environment. Bidentate and polydentate ligands coordinate through multiple sites to yield more stable complexes due to the chelate effect. This stabilization is an important feature in many applications ranging from catalysis to pharmaceutical uses. Therefore, familiarity with these kinds is not only important for predicting the properties of coordination compounds but also for tailoring them to a particular chemical reaction or pharmaceutical use. The space geometry of coordination compounds is significantly important in deciding their stability and reactivity in terms of metal center and type of ligand. Octahedral, tetrahedral, and square planar are the most common geometries. As an example, the coordination numbers of octahedral complexes are six, yielding a space-symmetric structure that best compromises spatial separation of ligands and minimizes electron repulsion. Tetrahedral and square planar structures are found more frequently with certain transition metals and sets of ligands, impacting electronic properties along with chemical behavior of the compounds. Such geometric variation accounts for the very strong effect of metal character and ligand structure on the properties and functionalities of these coordination complexes. Coordination compounds also have varied interactions and functionalities based on the character of the metal-ligand bond.

The coordination in these complexes is typically coordinate covalent in character, in which ligands donate electron pairs to metal ions. This interaction has the potential to significantly alter the metal's electronic properties, affecting its ensuing chemical reactivity, for instance, its catalytic activity. Advances of more recent times have been attracted to the importance of these coordination compounds to develop more efficient catalytic systems.

For instance, the introduction of mild C–H activation reactions into coordination complex frameworks enhances their applicability in the construction of multifunctional molecules in a sustainable manner, as indicated in the research on the application of chemosensor arrays (Hamedpour et al., 2019) and effective mild conversions (Gensch et al., 2016). Understanding these interactions is crucial for advancing chemical synthesis and functional material design (Figure 2.5).

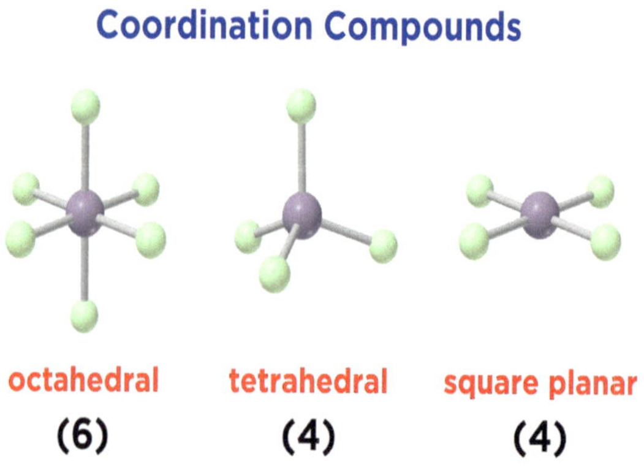

Figure 2.5. *The classification of coordination compounds.*

Source: https://i.ytimg.com/vi/Hs5JBjX51dc/maxresdefault.jpg.

2.3. TYPES OF BONDING IN COORDINATION COMPOUNDS

The bonding nature in coordination compounds is simply characterized by the interactions between metal ions and ligands around them, i.e., various types of bonds. Coordinate covalent bond is typically found in coordination compounds where ligands, being Lewis bases, donate electron pairs to the metal, which results in stable complexes. This process, in addition to stabilizing the metal ions, also influences the electronic properties of the resulting complex. In certain compounds with B ← N coordination, for instance, a special environment is created which optimizes the redox character of the system so that numerous oxidation states can be stabilized by such interaction. Experiments have shown the effect of incorporating B ← N coordination on the molecular shape and electron distribution, showing the complexity and adaptability of bonding for these coordination compounds (Zhu et al., 2018).

Additionally, understanding the various aspects of coordination bonding allows researchers to create compounds with desired properties for exact applications. For example, the insertion of transition metals into coordination complexes can produce intriguing magnetic and catalytic properties, accessible through advanced techniques such as single-crystal X-ray diffraction. Through the observation of structural changes exhibited by these complexes in the course of oxidation and reduction reactions, scientists gain a more in-depth knowledge of the inherent mechanisms that govern their reactivity and stability. Moreover, hydrogen bonding and similar weak interactions also help stabilize the overall structure, thereby their relevance in their study towards the creation of new materials and systems (Stoddart, 2017).

Finally, advancement in synthetic methodology has opened up the possibility of synthesizing complicated architectures that encompass more than a single mode of bonding, thus expanding the nature of coordination chemistry. This encompasses the assembly of mechanically interlocked molecules (MIMs) such as catenanes and rotaxanes, which challenge traditional views of chemical bonding. The identification of the role of mechanical bonds in controlling chemical interactions has opened new avenues in supramolecular chemistry, which places a focus on the importance of weak interactions in the construction of complex architectures. With the progress of research, coordination covalent bonds along with other non-covalent forces will be more significant in the design of the next generation of materials and molecular devices, emphasizing the continued significance and evolution of bonding theories in coordination compounds (Stoddart, 2017).

2.3.1. Ionic Bonding in Coordination Complexes

Study of ionic bonding in coordination complexes illustrates the cation-anion interaction involving the coordination complex. In coordination chemistry, cations, commonly transition metals, are surrounded by a wide variety of ligands that can either be neutral or anions. Such ligands coordinate with the metal using their lone pairs and create stable complexes that have the ability to highly decrease the net energy of the system. The ionic character of these interactions is significant as it directs the electronic and spatial arrangement of the complex. For example, low valent metal complexes such as those of titanium can facilitate revolutionary reactions by triggering small molecules through ionic interaction, a process which ultimately dissolves local bonds and triggers further chemical reactions (Oloyede et al., 2024). Thus, the delicacy of ionic bonding contributes to the stability and reactivity that is so characteristic of such coordination complexes. The coordinative stability of such complexes is largely determined by their ionic bonding properties, which are determined by charge density and ligand field strength. In this case, the metal-ligand interaction is rationalized by the balance of electrostatic attraction and covalent character, frequently by definition of the ligands themselves. Bidentate or polydentate ligands produce more stable complexes through chelation since there are several binding sites around a single metal center that yield stability in terms of enhanced ionic interactions. Such an effect of chelation is particularly noticed in transition metals since their ability to accommodate various coordination numbers enables them to form an enormous number of complex structures. In addition, research into the impact of ionic bonding on reactivity shows its relevance in fields like catalysis and drug design, where the accuracy of coordination determines the function of metal-based drugs in use scenarios (Joshi et al., 2023). Aside from their involvement in stability and reactivity, ionic bonding in coordination complexes is significant in determining their physical properties, such as solubility and complexation affinities. The ionic character of ionic interactions will enhance the solubility of such complexes in polar solvents, thereby influencing their bioavailability in biological processes. This aspect is particularly relevant to the development of anticancer drugs, in which coordination of metal ions with some ligands can facilitate targeted delivery and effectiveness. Moreover, the ionic character can modulate electronic transitions within such complexes, significant to its application in biosensing and bioimaging, thereby further highlighting the significance of ionic bonding in coordination chemistry. Thus, the intricacy of ionic interactions continues to be at the forefront of further developing our understanding of coordination compounds and their application across various scientific fields (Joshi et al., 2023; Oloyede et al., 2024) (Figure 2.6).

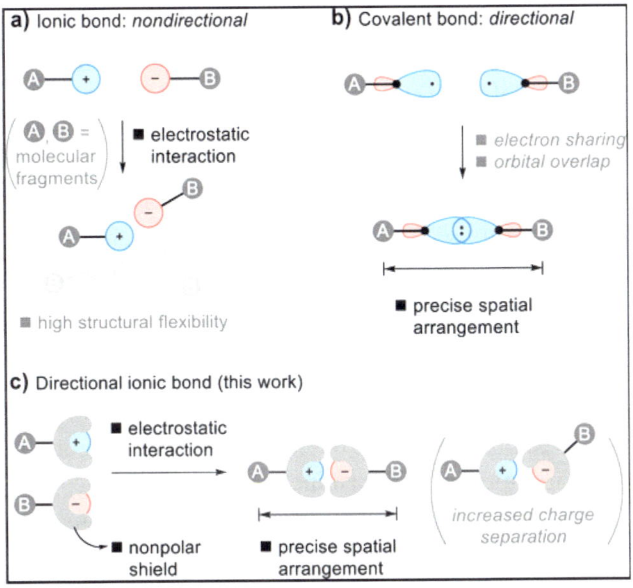

Figure 2.6. *The exploration of ionic bonding within coordination complexes.*

Source: https://pubs.acs.org/cms/10.1021/jacs.3c01030/asset/images/medium/ja3c01030_0001.gif.

2.3.2. Covalent Bonding and Its Role

The structural stability of coordination compounds is the foundation provided by covalent bonding, and it is one of the decisive factors involved in their formation and stability. Metal ions in coordination compounds tend to be Lewis acids, but ligands that are around them are Lewis bases due to the fact that they have lone pair electrons. This gives rise to the development of covalent bonds among metal and ligands, which can either be anions or neutral molecules. Importantly, the stiffness of these covalent bonds controls the geometric shapes and reactivity of the coordination complex. An understanding of such bonds lies at the root of most coordination chemistry concepts like ligand field theory and electronic properties that are associated with the metal center. A consideration of how these covalent bonds control coordination numbers and geometric shapes aids researchers to predict more reliably the properties of coordination complexes they prepare in their labs. The role of covalent bonding is not confined to mere structural support in coordination complexes but plays a prominent role in their chemical reactivity and functional application. For instance, covalent bond formation decides the ability of these complexes to perform redox reactions, as noted in most catalytic processes and bioinorganic systems. Covalent interactions with the ligands can modify the electron density around the central metal ion, thus modifying its oxidation status and reactivity. This is particularly important in fields such as catalysis, where metal-ligand interaction plays essential roles in vital chemical processes. Additionally, the creation of synthesizing ligands with dynamic covalent bonds shows the possibility of the tunability of the interactions to lead to innovations in materials science, particularly through the application of new electrode materials with enhanced performance over the traditional ones (Sun et al., 2023). Besides their chemical reactivity, the covalent nature of bonding in coordination compounds influences their physical properties, such as solubility, thermal stability, and optical properties. The specific interaction between metal ions and covalently bonded ligands is responsible for the overall stability of the compounds against changes in the environment and determines their likely applications in different uses. For example, covalent structures of coordination compounds have been studied in electronic materials due to their enhanced conductivity and structure stability. Furthermore, applying principles of covalent bonding has allowed novel organic-inorganic hybrid materials to be synthesized which can benefit from desirable properties from both components. These advancements, in addition to ongoing work in the field, point to the basic worth of understanding covalent bonding in coordination compounds because it is still an integral source of performance and value in engineering applications (Qing-Guo et al., 2021).

2.3.3. Coordinate Covalent Bonds and Their Characteristics

The investigation of coordinate covalent bonds offers profound understanding into coordination compound bonding, which is characterized by distinctive electronic and structure features. In essence, a coordinate covalent bond is formed when an atom contributes both electrons to share a covalent bond with an atom that lacks an entire electron pair. This unique mode of bonding is a key contributor to coordination complex reactivity and stability, allowing coordination complexes to exhibit diverse types of functional behavior in chemical reactions. Donor-acceptor interactions promoted by these bonds are fundamental to understanding mechanisms such as catalysis and metal-ligand complex formation, which have direct relevance to the physical and chemical properties of the resulting compounds. Consequently, the study of coordinate covalent bonds highlights the dependency of molecular reactivity and structure in coordination chemistry. Properties of coordinate covalent bonds are also highlighted by the operation of metal-ligand cooperation, as noted in a number of transition metal complexes. The complexes are prone to being reliant on cooperative operation between metals and ligands in order to activate strong covalent bonds in substrates. For instance, metal-ligand cooperation is responsible for catalytic reactions that prefer efficient reaction under mild conditions, where half of the substrate interacts with the metal center and coordinated ligand simultaneously. These interactions are fundamental to pincer-type ligand reactions that enhance the overall reactivity of the metal-ligand complex. Besides, the presence of temporary cooperative ligands can add a dynamic element to such systems, which displays the responsiveness and flexibility of coordinate covalent bonding in catalysis (Luo et al., 2024). Coordinate covalent bonds are also relevant beyond structural interest since they manage electronic properties and stability in coordination compounds. One of the interesting aspects of this type of bonding is that it can stabilize distorted coordination geometries, such as the three-center two-electron bond found in ionic carbene complexes. Such complexes exhibit the covalent nature of metal center to bound ligand interactions, revealing how precise electron sharing stabilizes atypical bonding geometries (Cui et al., 2022). By facilitating new electron arrangements, coordinate covalent bonds allow metal complexes to be involved in a variety of reactions, which are critical in catalysis and organic synthesis. As such, studying their nature and behavior makes us more sensitive to the intricate mechanisms that define coordination chemistry (Figure 2.7).

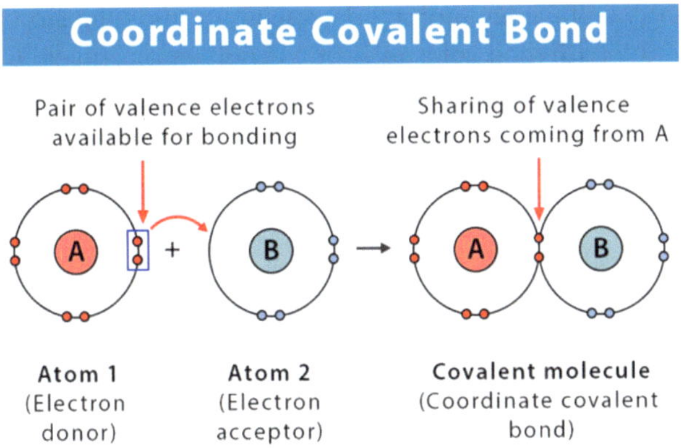

Figure 2.7. *Coordinate covalent bonds and their characteristics.*

Source: https://www.chemistrylearner.com/wp-content/uploads/2020/09/Coordinate-Covalent-Bond.jpg.

2.4. THE ROLE OF METAL IONS

The role of metal ions in coordination complexes is key to the understanding of the bonding natures and reactivity of the complexes. Metal ions, usually transition metals, possess unique electronic structures that allow them to be found in different oxidation states. This is pivotal in the determination of the nature of the metal-ligand interactions. As Lewis acids, the metal centers can coordinate electron-donating ligands to form the complex. For instance, in electrochemical systems, the ions of transition metals can facilitate reactions through the ability to receive or donate electrons in redox reactions efficiently. This multivalence enhances the ease of coordination compounds, making them indispensable in catalytic applications, e.g., in current progress in homogeneous catalysis and electrochemical organic synthesis (Cheng et al., 2021). Furthermore, the spatial arrangement of the metal-ligand bonds and the arrangement geometry influence the coordination compounds' properties. The coordination number, as determined by the capacity of the metals for ligands, tends to give rise to firm structural patterns, e.g., octahedral or tetrahedral. These geometric environments may have an effect on energy levels and ultimately decide the electronic, magnetic, and optical properties of the compounds. For instance, the operation of metal ions within sodium-ion batteries illustrates how their geometric arrangement directly influences material efficiency and electrochemical response. As the demand for renewable energy substitutes continues to grow, understanding the geometric and electronic influences will be key to the development of next-generation energy storage technology. The analogies that the structural features of metal ions draw with their roles in various applications reveal the underlying property of coordination chemistry (Jang-Hwang et al., 2017). The interactions of the metal ions with the ligands also provide basic implications to various disciplines, ranging from catalysis to materials chemistry. The interactions can regulate the reactivity of the ligands and stability of the complex, paving the way for selective catalysis. For example, in recent electrochemical reaction studies, well-defined metal-ligand architectures have enabled the assembly of novel catalytic cycles with the desired selectivity and efficiency. By taking advantage of the unique properties of metal ions, chemists are able to design complex materials with tailored functionalities for drug delivery, for example, to electronic devices. Therefore, metal ions not only signify central species in coordination compounds but also reflect the complicated interaction between bonding and function in these essential chemical entities, pointing to their importance in contemporary chemistry and materials technology (Cheng et al., 2021).

2.4.1. Properties of Transition Metals in Coordination Compounds

The properties of transition metals in coordination compounds are predominantly controlled by their unique electronic structures and the availability of d-orbitals.

Transition metals have an incomplete d-subshell, enabling them to form a variety of bonding interactions with ligands. This incompleteness is responsible for the ability of the metals to form coordinate covalent bonds, in which the ligands act as electron donors. The diversity in oxidation states and coordination geometries controls the chemical behavior of the metals, with a great effect on their reactivity and stability in varying surroundings. Further, the ability to form stable complexes with a range of ligands expands the functional applications of transition metals in such areas as catalysis, materials science, and bioinorganic chemistry, laying the stage for their exploration in coordination chemistry. The coordination geometry of the transition metal complexes, defined by the arrangement of the ligand atoms about the metal center, is central to establishing the chemical properties of the complexes.

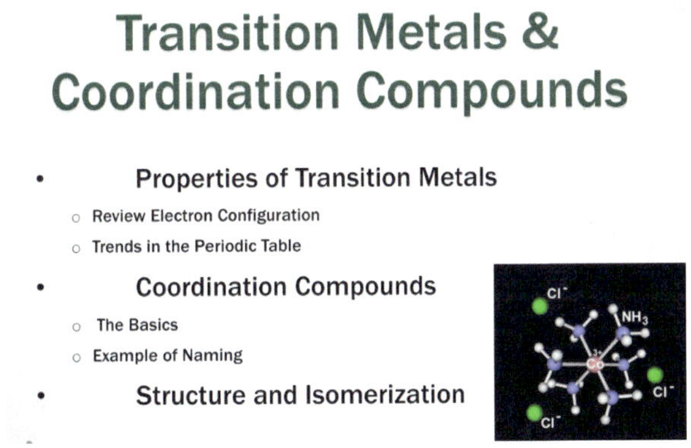

Figure 2.8. *The properties of transition metals in coordination compounds.*

Source: https://cdn1.slideserve.com/2330059/chapter-24-transition-metals-coordination-compounds-l.jpg.

Common geometries include octahedral, tetrahedral, and square planar, each influencing the metal center electronics and sterics. Such diversity in coordination geometries gives rise to different ligand fields that consequently influence the stability and reactivity of the complexes. For example, octahedral complexes typically possess distinctive ligand field splitting patterns that can lead to unique optical and magnetic properties. Smart ligand design as a means of controlling ligand environments is central to coordination compound research, particularly in their applications in molecular magnetism, where the accurate positioning of ligands will significantly impact magnetic interaction and properties (Fu-Guo et al., 2019). In addition to that, spin transition behavior of transition metal complexes is yet another significant aspect of their properties. The reversible spin interconversion between different spin states can greatly tune a compound's magnetic, dielectric, and optical properties. For instance, photoinduced spin transition offers a pathway for the control of multiple properties upon external stimuli, including light or pressure. Such functionality is useful for the synthesis of advanced multifunctional materials for applications in smart devices and nanotechnology. Besides, coordination of photoresponsive units with metal ion complexes allows for new structures with reversible functional modification, which makes contributions to the development of photonics and molecular sensors (Yin-Meng et al., 2019). Hence, it is important to comprehend the coordination between coordination geometry, ligand features, and electron transitions so that one can maximize the potential of transition metals in coordination compounds (Figure 2.8).

2.4.2. Influence of Oxidation States on Bonding

Understanding the oxidation state effect on bonding in coordination compounds is essential in comprehending how they react and chemically behave. Coordination compounds are created through a central metal atom that is bonded to ligands, which are either ionized or neutral molecules. The oxidation state of the metal dictates its electron configuration, and therefore its ability for bonding interactions. For instance, a metal of greater oxidation state will typically have more electrophilicity preference, and it will prefer to bind electron-donating ligands. This is a property that indicates the metal's tendency to stabilize complexes whereby the ligands donate electron pairs to the metal center, stabilizing the overall structure of the coordination compound. Furthermore, the differences in oxidation states will lead to differences in coordination geometries, which have importance with regards to the ligand arrangement in space surrounding the metal and consequently for the spectroscopic properties and reactivity of compounds. The nature of bonding of transition metals in coordination compounds makes importance more pronounced for oxidation states. One significant feature is that transition metals are capable of existing in multiple oxidation states, allowing them to coordinate with many different ligands. Complexes where metals exist in various oxidation states are of particular interest in catalytic reactions where selectivity and efficiency can be controlled by the redox activity of the metals. For example, the transformation from 2^{+}–3^{+} oxidation states in iron complexes can have an important function in their capacity to catalyze oxygen transfer reactions. As an aside, coordination compounds will also have distinctive electronic properties as a consequence of d-orbital involvement, which further complicates the oxidation state-ligand interaction relationship. These dynamics highlight the intricate interplay of oxidation state variability and consequent chemical behavior of coordination complexes, for example, in the catalytic use highlighted among recent studies

(Cheng et al., 2021). Complex formation and ligand bonding are significantly influenced by the oxidation states of the coordination center, particularly among metal-ligand interactions. The capacity of a ligand to donate and the character of its interaction with an oxidation state that is metal-dependent can determine the stability of the coordination complex that results. Increased oxidation states in transition metals will result in stronger metal-ligand interactions, which may cause a colored complex due to d-d transitions. For instance, altering the oxidation state in chromium and manganese complexes alters electronic transitions that lead to impacts on spectral behavior. Further, an understanding of how a change in different oxidation states influences bonding interactions helps one design new coordination compounds with desired properties for certain usage, e.g., catalysis or drug delivery. This bond flexibility in bonding circumstances underscores the importance of oxidation states in getting metal complexes ready for functional use (Tang et al., 2020).

2.4.3. Metal-Ligand Interactions

Metal-ligand interactions are at the core of coordination compound research, defining their stability, reactivity, and properties. At the molecular level, these interactions entail the formation of coordinate covalent bonds in which a metal center accepts electron pairs donated by ligands, such as uncharged molecules or charged ion-containing lone pairs of electrons. Electron donation is essential in order to form stable adducts, with specific geometries and coordination numbers. The intensity and nature of such interactions are influenced by various factors, like electronic character of the metal and ligand, sterics, and coordinating potential of the ligand. It should be mentioned that the counterion and the solvent can also affect the overall bonding environment, and this will create differences in the electronic and structural properties of the formed coordination complex, thus reflecting the dynamic aspect of metal-ligand interactions. Moreover, selectivity in metal-ligand interactions plays a very important role in fields such as materials science and biochemistry. For instance, in biological systems, the coordination of metal ions to biomolecules like enzymes and hemoglobin shows the importance of such interactions towards functional processes. The exotic electron configurations of the transition metals are often used as catalytic centers for assisting biochemical reactions, a demonstration of the importance of their coordination chemistry. This biological relevance is translated into artificial systems, where specially tailored metal-ligand complexes are achieved for drug delivery to molecular sensors. Strategic ligand design in order to develop high selectivity and binding affinity has been a focal area of synthetic coordination chemistry, ultimately paving the way to technological advancements and providing insights into the underlying principles of coordination interactions (Lucas et al., 2022). Metal-ligand interactions also involve concepts of supramolecular chemistry, where self-assembly phenomena and molecular recognition are made apparent. Supramolecular chemistry features the significance of noncovalent forces such as hydrogen bonding, π-π stacking, and electrostatic interactions in higher-order structure assembly. Coordination compounds can behave with complex properties due to these ancillary interactions, leading to sophisticated structures utilized as molecular machines or sensors. For example, the invention of mechanically interlocked objects. For example, the exploitation of mechanically interlocked compounds like catenanes and rotaxanes shows how ligand chemistry not just shapes metal coordination, but also increases the functionality of coordination compounds (Stoddart, 2017). By the use of these multi-faceted interactions, researchers can invent advanced materials with tailored properties that exceed conventional applications, ultimately revolutionizing our understanding of bonding in coordination compounds (Figure 2.9).

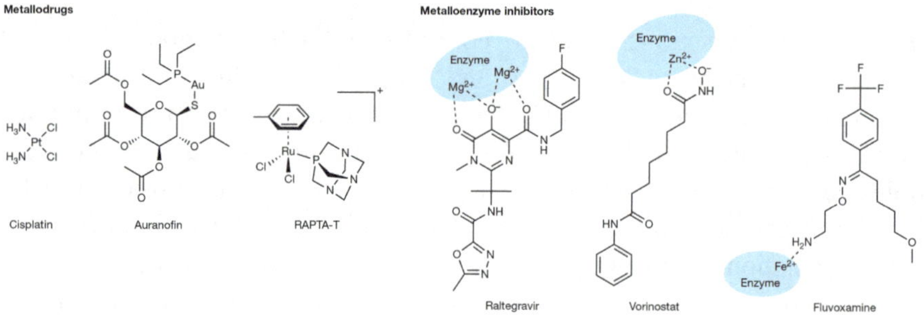

Figure 2.9. *Metal-ligand interactions.*

Source: https://media.springernature.com/m685/springer-static/image/art%3A10.1038%2Fs41570-018-0018-6/MediaObjects/41570_2018_18_Fig1_HTML.jpg.

2.5. THEORIES OF BONDING

Bonding theories are significant in the description of the structure and reactivity of coordination compounds, which have prominent appearances in most chemistry domains. Coordination compounds consist of a central metal ion coordinated with molecules or anions known as ligands. How these species interact has been extensively investigated, mainly through crystal field theory (CFT) and ligand field theory (LFT). CFT emphasizes electrostatic metal ion-ligand interactions responsible for the d-orbital splitting in coordination geometries. LFT extends CFT by combining with molecular orbital theory, which allows a more realistic description of covalency in metal-ligand bonds. Such a combination of bonding picture is able to describe electronic properties and stability of coordination complexes and allows rational design of new materials with specific properties. Furthermore, the extended tight-binding approach, known as xTB, has broadened the horizon of bonding theories within the context of coordination compounds. The semiempirical approach is founded on density functional theory (DFT) perturbation expansions and indicates a systematic method for predicting properties across the periodic table. The advantage of the xTB family is its fair cost-accuracy ratio while treating thousands of atoms in a system, a feature which is particularly useful within transition metal research. The approach provides a guideline for reproducible parameterization based on accurate gas-phase theoretical reference data, further elucidating the role of geometrical structures and noncovalent interactions in coordination chemistry (Bannwarth et al., 2020). By addressing these aspects, xTB is now an efficient tool in the study of bonding in complex molecular systems. Experimental methods in the last few years have also provided more insight into the bonding of coordination complexes. By combining various approaches, researchers have developed knowledge of nanoparticles' properties and their uses in numerous fields, ranging from catalysis to material science. The advent of very powerful analytic tools now allows probing molecular interactions at resolutions hitherto unsuspected, to enable the refinement of theoretical models for describing bonding processes. Optimal models combine classical bonding theories with modern computational methods, to enable the determination of optimal ligand structures that maximize metal center interactions. Therefore, recent research looks to employ both experimental and theoretical means to disassemble the complexities of coordination complex bonding to further drive material formation and optimization techniques across disciplines (Mourdikoudis et al., 2018).

2.5.1. Crystal Field Theory (CFT)

CFT is a fundamental theoretical model of the electronic structure and bonding in coordination compounds. CFT supposes that the ligand arrangement around the central metal ion significantly influences the metals d-orbital energies. As ligands approach the metal ion, they produce an electrostatic field that distorts the d-orbital energy, causing splitting of the orbitals. The extent to which this splitting happens depends upon the complex's coordination geometry and the character of the ligand involved. It is in this context that coordination complexes' character like magnetism, color, and stability are forecastable by chemists depending on how ligands interactions shape the metal ion d-electron configurations. The spatial arrangement of ligands matters in CFT, leading to various electronic structures for distinct coordination complexes. Octahedral complexes, for instance, have distinct orbital splitting patterns compared to tetrahedral complexes. In octahedral complexes, d-orbitals are divided into two sets: the lower-energy t2g orbitals and higher-energy, e.g., orbitals. This alignment enhances unique electronic transitions, which can be utilized directly to explain coordinating compounds' observed color. Compounds' stability and electronic distribution may also be influenced by the extent of splitting, depending on the strength of ligands, which are either strong or weak field ligands. Understanding these differences is crucial in justifying the behavior of coordination compounds in varying conditions. CFT not only accounts for the electronic structures of coordination compounds but provides insight into noncovalent interactions, which are a key factor in most areas of chemistry. As coordination chemistry continues to develop, concepts from CFT are combined with supramolecular chemistry concepts to further advance our understanding of complex molecular systems. For example, noncovalent interactions play a crucial function in biological functions, with metal ions being key components in catalysts and enzymes. Background in the history of development in such theories highlights the importance of having multi-disciplinary research to understand the complications involved in bonding and interaction in coordination compounds. With the evolution of the field, further studies in the dynamics of such interaction would be vital to further development in applications ranging from material science to biochemistry, to stay relevant for CFT in modern research activities (der Heggen et al., 2022; Hans-Jörg Schneider, 2022) (Figure 2.10).

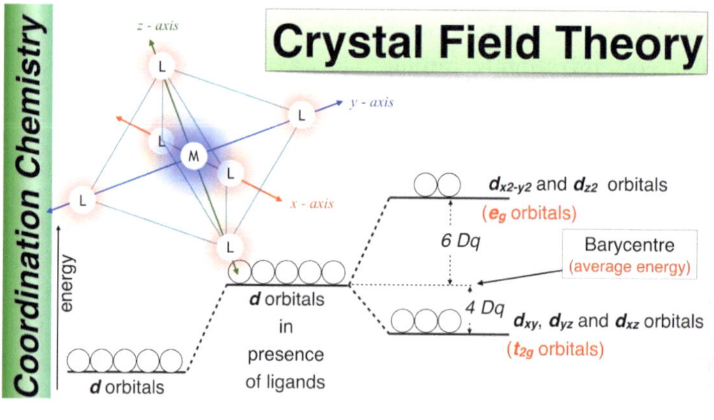

Figure 2.10. *Crystal field theory (CFT).*

Source: https://i.ytimg.com/vi/7d6xt0JlY3Y/maxresdefault.jpg.

2.5.2. Ligand Field Theory (LFT)

LFT is a simplified model for the electronic structure and bonding chemistry of coordination compounds. Through consideration of metal center-ligand interactions, LFT abstracts the underlying principles established by Crystal Field Theory (CFT) and adds a very important wave-function concept to describe bonding. Unlike the electrostatic quality of ligand-ligand interactions in CFT, LFT emphasizes the covalent nature of metal-ligand bonds. This understanding allows for a greater understanding of the geometric and electronic features that control the stability and reactivity of coordination complexes. Therefore, LFT not only assists us in predicting observable properties such as color and magnetism but also provides us with a sound theoretical basis to investigate intricate coordination geometries, enhancing our coordination chemistry knowledge in real-world applications.

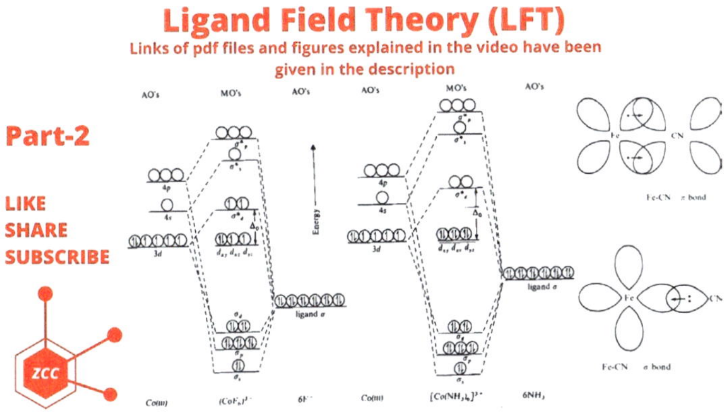

Figure 2.11. *Ligand field theory (LFT).*

Source: https://i.ytimg.com/vi/T2yzQzh6MGw/hq720.jpg?sqp=-oaymwEhCK-4FEIIDSFryq4qpAxMIARUAAAAAGAElAADIQj0AgKJD&rs=AOn4CLBZd3exlsSQqyqCH8tgB9M3170QAA.

The insights provided by LFT also extend towards the description of many metal-ligand complexes, particularly transition metals with their varied coordination surroundings. As far as ligand effects are involved, LFT indicates that the electronic environment of the central metal ion is greatly controlled by the nature of the ligands, which are either strong or weak field ligands. For instance, strong field ligands such as CN^- induce great d-orbital splitting favoring low-spin geometry, while weak field ligands such as Cl^- provide the possibility of increased spin states with less d-orbital splitting. These energetic differences play critical roles in explaining reactivity patterns and stability in some coordination complexes. Moreover, LFT has been at the forefront of rationalizing such phenomena as the spectrochemical series, providing a methodical way of predicting how different ligands influence d-orbital splitting parameters. Application of LFT is not limited to merely explaining static structures but instead goes on to explain dynamic interactions and transformations in coordination complexes. For example, LFT can guide our understanding of ligand exchange reactions at the heart of most catalytic and biological processes. Ligand exchange kinetic studies, such as on functionalized carbon nanoparticles, illustrate how LFT principles map onto material design in devices such as thermoelectric devices (Golovenko et al., 2024). Moreover, coinage metal cyclic trinuclear compound research has metal-ligand bonding subtleties at odds with conventional approaches, suggesting increased electronic interactions governed by inverted ligand field effects (Galassi et al., 2022). Such observations accentuate the versatility of LFT in bridging theoretical computations to pragmatic applications, demonstrating its applicability across a wide range of disciplines from molecular machine construction to optoelectronic devices (Figure 2.11).

2.5.3. Molecular Orbital Theory (MOT)

MOT is a sophisticated theory of coordination compound bonding emphasizing the role played by molecular orbitals formed by overlap of atomic orbitals. MOT is in sharp contrast to classical valence bond theory, which is localized bond based, and provides a description of delocalization in which electrons are distributed between more than a single nucleus. This theoretical adjustment works best in coordination chemistry, where the alignment of ligands with respect to the center metal ion may bring about complicated bonding scenarios. The mixing of metal d-orbitals and the ligand orbitals enables bonding, non-bonding, and antibonding molecular orbitals. This delocalization is crucial in enhancing bond stability and predicting the compounds electronic nature, which are significant in reactivity and functionality in various chemical situations. The significance of bonding interactions in coordination complexes can be also rationalized by using molecular orbital diagrams graphically illustrating the molecular orbital energies and ordering. For instance, in the coordination of a ligand with a transition metal, the various d-orbital energy levels are such that they allow for various electron configurations. The ligands may be contacting the center metal atom with different orbital types, i.e., σ and π interactions, resulting in various types of bonding situations which can profoundly affect the physicochemical properties of the formed complex. The creation of these special orbital interactions has been established in studies of compounds with one-dimensional uranyl fluoride chains, wherein molecular-level assembly occurs by virtue of second-coordination sphere noncovalent interactions, thereby exerting a critical influence on molecular behavior and properties (Decoteau et al., 2024). This methodology demonstrates how MOTs depiction of molecular interaction is key to predicting reactivity and stability in coordination complexes accurately. Furthermore, the predictive ability of MOT is also limited to determining structures but also to explaining advanced electronic behaviors expressed in coordination compounds. For instance, the electronic structure of bis(silylene) complexes illustrated through DFT calculations describes how bond interactions can be examined at the molecular orbital level. The highest filled molecular orbital (HOMO) in such complexes closely corresponds to the lone pair of electrons, indicating effective σ-bonding interaction but also leaving room for potential reactivity with other transition metals. The results above show how molecular orbital viewpoints inform the synthesis and characterization of new coordination compounds, making it possible to prepare materials with specific desirable properties (Fan et al., 2024). In conclusion, MOT not only offers a more profound understanding of bonding in coordination compounds but is also an extremely useful tool for elucidating and predicting their rich chemical diversity (Figure 2.12).

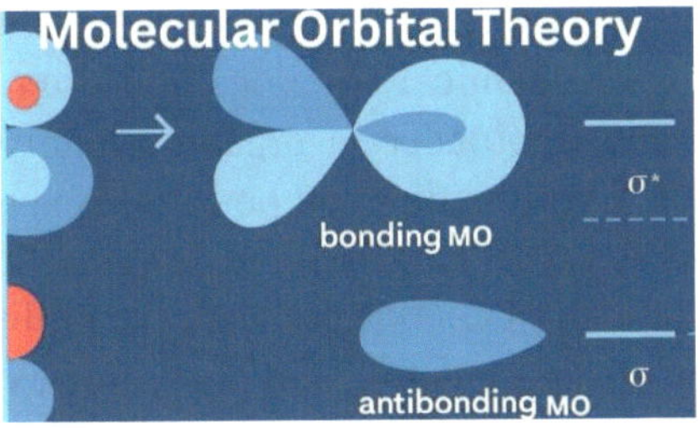

Figure 2.12. *Molecular orbital theory (MOT).*

Source: https://images.shiksha.com/mediadata/images/articles/1753788642phpA6 LqxI_480x360.jpeg.

2.6. APPLICATIONS OF COORDINATION COMPOUNDS

Applications of coordination compounds vary across the different scientific fields, especially in medicine and biotechnology. Their unique bonding properties enable them to form complex structures that can be modified for specific functions and render them beneficial in drug design and delivery systems. Supramolecular coordination complexes (SCCs), for instance, are found to be promising candidates for new anticancer agents. These have distinctive modes of action that have a tendency to surpass those of traditional small-molecule drugs, particularly due to their abilities for molecular recognition. Besides, the inherent flexibility of SCCs allows the addition of more than one function, adding further value as a platform for developing multimodal theranostic agents in medicine. The research in progress along this area is promising the potential for using such structures in effective therapies to be just beginning, and an intriguing new frontier is unfolding in drug research (Alexander Pöthig et al., 2019). Apart from their biomedical applications, coordination complexes have a pivotal role in chromatographic techniques, essentially facilitating analytical chemistry. Coordination complexes are able to form coordination bonds with transition metals, which allow for subtle interaction with stationary phases in chromatographic systems. This property is particularly beneficial in liquid chromatography that yields quantitative details of stoichiometry and mechanisms of ligand-exchange reactions. All these advancements have made the effective separation of complex mixtures, like optical isomers and drugs, possible. Besides, the versatility provided by coordination chemistry enables scientists to design separation conditions according to specific requirements. Therefore, these methods not only facilitate the study of coordination compounds in and of themselves but also provide a better understanding of biochemical reactions and drug interactions and thus the scope of applications is broader in organic chemistry and biochemistry (Даванков et al., 2018). Coordination compounds also significantly contribute to environmental chemistry, particularly for fighting pollution and remediation problems. Certain coordination complexes have the ability to absorb toxicants and thus clean them up from contaminated sites. For example, metal-organic frameworks that are derived from coordination compounds have proven useful in the adsorption of greenhouse gases and other poisonous materials. Their highly porous nature renders them suitable for the synthesis of new materials used in cleaning air and water. Additionally, the ability to tune coordination compounds means it is possible to create specific interactions that can be engineered to be able to target many various contaminants. Not only does this make it easier to develop novel pollution control methods but also serves to highlight the role coordination compounds will play in green environmental practice. With progressive advances in research, there remains a vast potential for the creation of more effective remediation processes based on advances in coordination chemistry.

2.6.1. Role in Biological Systems

The biological role of coordination compounds is wide-ranging, with particular importance in oxygen transport and enzymatic processes. Hemoglobin, a ubiquitous coordination complex, exemplifies this by being based on iron. This structure enables oxygen binding and release in a reversible process necessary for cellular respiration. The hemoglobin's complex bonding interactions with a chain of ligands that stabilizes its transition states enable efficient operations under varying physiological conditions.

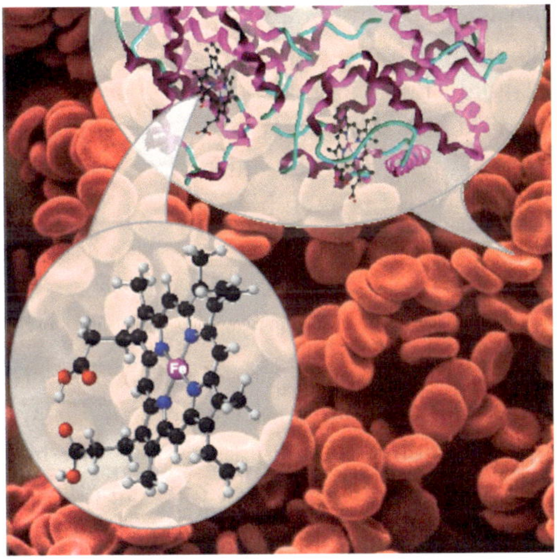

Figure 2.13. *The role of coordination compounds in biological systems.*

Source: https://chemlaba.wordpress.com/wp-content/uploads/2011/11/picture_05.png.

Additionally, changes in the iron coordination sphere can have a profound effect on the allosteric behavior of hemoglobin, thus modulating oxygen affinity according to environmental signaling like pH and carbon dioxide concentration, emphasizing the dynamic character of biomolecular interactions typical of biological systems (Sharpee et al., 2016). Aside from transportations functions, coordination complexes play a very important role in catalyzing biochemical transformations. Enzymes often have metal ions in their active sites, in which they achieve coordination complexes that facilitate substrate binding and subsequent catalysis. Metalloenzymes, for instance, use transition metals such as cobalt, zinc, and manganese, which leverage their particular electronic properties in stabilizing transition states and minimizing activation energy requirements. These processes acquire particular relevance in the context of metabolic pathways, in which enzymes should maintain efficiency and selectivity. The coordination capabilities of the metal-ligand interactions within these complexes are essential to knowledge of the stability and reactivity of the enzymes active site and influencing overall metabolic rates and cellular viability (Sharpee et al., 2016). Lastly, coordination compound functions extend into signaling and regulation in biological systems. Metal ions serve as secondary messengers or cofactors in signal cascades and influence cellular responses to external signals. For example, calcium ions serve as critical signaling molecules that mediate processes such as muscle contraction and neurotransmitter release. Binding of calcium to specific coordination sites alters protein conformations, initiating a sequence of intracellular signals. This regulatory potential lends credence to the significance of coordination compounds as structural and catalytic entities but even more so in communication in and between cells, demonstrating their pivotal role in attaining homeostasis in biologic systems (Sharpee et al., 2016) (Figure 2.13).

2.6.2. Use in Industrial Processes

In industrial processes, coordination compounds play a pivotal role because of their unique patterns of bonding and patterns of reactivity. Coordination compounds are highly specific and stable, which are highly desirable for catalysis, materials science, and pharmaceutical applications. For instance, metal complexes are ideal catalysts for chemical reactions, enhancing reaction rates and selectivity, thereby rendering industrial processes more efficient. The coordination capability of metal ions makes it possible to synthesize catalysts that can be tailored for precise reactions, maximizing yield and lowering side products. That specificity is a result of the complex interaction of bonding involved in such a compound, which could be modified for desired outcomes in synthetic conversion, something that has broad implications for productivity within chemical manufacturing sectors. Furthermore, coordination compounds are applied in environmental chemistry and waste management. The unique bonding characteristics render the compounds capable of extracting heavy metals from industrial effluent streams, thus contributing significantly to the prevention of environmental toxicity. Coordination complexes are able to complex with metal ions, rendering them less toxic or easier to eliminate. This process not only reduces the toxic effect of heavy metals on the environment but also reduces the pressing problem of metal accumulation in the food chain, as indicated in previous studies on heavy metal toxicity. The effectiveness of such processes in detoxifying waste shows the importance of research into the bonding in coordination compounds because it not only impacts industrial sustainability but also public health interventions (Jaishankar et al., 2014). The flexibility of coordination compounds is further utilized in material science, where their inherent properties aid in the production of superior materials. For example, encapsulation of metal ions within polymer matrices can lead to superior thermal and electrical material properties. These superior materials have direct practical applications in industry, particularly in electronics and energy industries, where superior performance can be converted to higher device efficiencies. Moreover, the ability to engineer the bond interactions in these coordination compounds allows for the creation of multifunctional materials that will be responsive to environmental stimuli, such as temperature or pH. Such responsiveness to environmental change defines the role of coordination chemistry in the creation of next-generation materials to meet current industrial requirements by filling the gap between theoretical design and useful applications across industries (Dorothée Liebschner et al., 2019).

2.6.3. Importance in Analytical Chemistry

The contribution of analytical chemistry toward the comprehension of coordination compound bonding cannot be overestimated. Analytical tools form a vital tool for comprehending the structure and properties of coordination complexes, illuminating their intricate bond formation and reactivity. Comprehension of the bonding interactions within the compounds is essential to their applications as diverse as catalysis and drug development. Thanks to the advanced analytical tools such as spectroscopy and chromatography, scientists are able to extract data regarding electronic metal center frameworks and ligand effects. In addition, these analytical tools allow scientists to study chemical space and tailor compounds for a specific function or design new materials. The growing reliance on such methods indicates the necessity for analytical chemistry in enhancing our knowledge about coordination compounds and realizing their application in all sciences. The usage of coordination compounds in various fields also indicates the necessity of analytical methods for effective exploration and identification. For example, fluorescent chemosensors provided by coordination chemistry are asserting their influence in a variety of applications ranging from biology and pharmacology to environmental science. These metal ion-ligand interaction-based chemosensors are highly specific with high affinity to produce measurable fluorescent signals, which provide meaningful information on the target ion concentration in advanced matrices. Their analytical chemical foundation guarantees accuracy and reliability in target analyte detection, as this is essential for their deployment in real situations. Moreover, with the continued evolution of analytical methods comes the potential for even greater capacity to monitor and quantify the coordination complex behavior within dynamic systems and, thus, expand their use throughout scientific disciplines, proving the interactive nature of analytical chemistry and the coordination compound science. In addition, with still more computational tools evolving, their integration with traditional analytical methods offers unprecedented opportunities for coordination chemistry explanation. The arrival of machine learning models, as elucidated by recent research work, allows researchers to predict open-shell transition-metal complex properties by analyzing their coordination environments. This synergy allows for the study of gigantic chemical spaces that were previously inaccessible with traditional analytical tools due to their limitations. Through accurate prediction of structural properties such as spin states and redox potentials, researchers can identify potential candidates for catalytic use or sensor development. The ability to rapidly assess the performance of novel compounds in these contexts underscores the fundamental role of analytical chemistry in basic research and applied sciences and its leadership position in the development of innovation in coordination compounds, as the need for efficient, high-throughput discovery continues to grow (Janet et al., 2021; Wu et al., 2017) (Figure 2.14).

Figure 2.14. *Importance in analytical chemistry.*

Source: https://www.azolifesciences.com/images/Article_Images/ImageForArticle_635_17395415973197374.jpg.

SUMMARY

Briefly, coordination compound bonding is a reflection of intricate interactions more than simple covalent paradigms. Interactions in coordination complexes are governed by a number of determinants like coordination number, metal-ligand type, and geometric shape. These determinants are significant in establishing the stability and reactivity of the final compounds. The results from more recent breakthroughs, namely in structures like the metal-organic APF-80 structure, illustrate the strength that new approaches have to deconstruct formerly challenging nucleophilic molecules at a molecular level. This expansion of analytical capabilities allows further understanding of structural dynamics in coordination chemistry, ultimately improving the overall field of materials science and pharmacology via precise characterization techniques that can provide innovative applications (Nakagawa et al., 2025). Also, theoretical models in the basis of analysis for coordination compounds emphasize the value of an integrated approach towards the understanding of their bonding characteristics. With the help of advanced computational methods, such as Density Functional Theory, one gains profound understanding of the non-covalent forces that dictate the bonding situation in such systems. For instance, in the study of the $[Fe(bpca)(\mu\text{-}bpca)Gd(NO_3)_4]$ complex, one discovers that the compounds cohesion and stability are attributable mainly to non-covalent forces, such as dispersion forces. This research not only validates the need for theoretical models to describe coordination bonding but also indicates the complexity of metal center-ligand interaction, which is crucial in the creation of materials for everyday applications (Toader et al., 2024). These findings ultimately have relevance beyond the realm of coordination chemistry, influencing new material design and the preparation of biologically relevant compounds. The intricate bonding mechanisms present in coordination compounds both define them and pave the way for new insight in catalysis and bioactivity. By using multimodal analysis tools, as illustrated in the APF-80 structure, scientists are able to effectively capture and characterize a wide variety of bioactive molecules. This sophisticated perception facilitates drug design innovation and materials science, as the extremely specific structural information gained from the study of coordination compounds can lead to the optimization of new functional materials across many fields. Thus, ongoing research in coordination compound bonding remains a significant aspect of theoretical and industrial chemistry.

Coordination compounds constitute a very fascinating branch of chemistry owing to their unique bonding characteristics. The central defining feature of coordination compounds is the coordination bond, a form of interaction between the central metal ion or atom with the enveloping molecules or ions that are known as ligands. This bond formation is primarily governed by Lewis acid-base theory concepts, where the metal acts as a Lewis acid and the ligands as Lewis bases that donate electron pairs. The ligand geometric arrangement around the metal center is primarily responsible for the properties of coordination compounds like color, magnetic nature, and reactivity. Understanding of this bonding framework is required in order to look at broader applications of coordination chemistry in areas such as biochemistry, catalysis, and materials science. Of particular significance to coordination compound bonding is the role played by ligand field theory, with its ability to describe electronic structure and color. According to this theory, the energy of d-orbitals is affected by the presence of ligands around a metal center, leading to the splitting of the orbitals into different levels of energy. The splitting directly depends on the character of the ligands and their coordination numbers, which stabilize or destabilize electronic structures. Observation of color in coordination complexes is predicated on electronic transitions between such split d-orbitals, and thus ligand field theory is a precious tool in the prediction of coordination compound spectral characteristics. Of course, improvement in computational methods, e.g., as offered through xTB methods (Bannwarth et al., 2020), has enhanced our understanding of these interactions and allowed more effective investigation of complex coordination chemistry. Aside from simple ligand interaction, bonding in coordination compounds is also influenced by the properties of the central metal cation, including oxidation state and coordination number. Different metals possess different affinities for ligands, which make the coordination complexes reactive and stable. For example, transition metals form a vast number of coordination compounds because they can exist in multiple oxidation states and also have different coordination numbers. This versatility is the foundation of their industrial catalytic application as well as the significance of transition metals in biology, such as in metalloproteins and metalloenzymes. The most recent reports have also referred to the potential for beyond Li-ion technologies in mass markets, projecting that coordination chemistry-inspired approaches would be pivotal in designing innovative energy storage devices, hence bridging the gap between theoretical chemistry and practical application (Tian et al., 2020).

REFERENCES

1. Alvarez, N., Kramer, M., Ellena, J., Filho, A., Torre, M. H., & Facchin, G. (2018). Copper-diimine coordination compounds as potential new tools in the treatment of cancer. *[Journal unspecified]*, *2*, 1–5. https://www.semanticscholar.org/paper/0b6cea9acc17233aa4f3939ffaa3f24aca7786c7 (accessed on 13th September 2025).

2. Bannwarth, C., Caldeweyher, E., Ehlert, S., Hansen, A., Pracht, P., Seibert, J., Spicher, S., et al. (2020). Extended tight-binding quantum chemistry methods. *Wiley Interdisciplinary Reviews: Computational Molecular Science*, *11*. https://doi.org/10.1002/wcms.1493 (accessed on 13th September 2025).

3. Cheng, X., Lei, A., Mei, T. S., Xu, H. C., Xu, K., & Zeng, C. C. (2021). Recent applications of homogeneous catalysis in electrochemical organic synthesis. *CCS Chemistry*, *4*, 1120–1152. https://doi.org/10.31635/ccschem.021.202101451 (accessed on 13th September 2025).

4. Cui, Y., Zhang, J. X., Xu, C. Q., Jiang, X., & Li, J. Y. (2022). Bonding nature of "ionic carbenes" in $[M_3(\mu_3\text{-}CH_2)]$-containing compounds: The covalent interaction. *Inorganic Chemistry*. https://www.semanticscholar.org/paper/6e5206073dfa24b752d8d2101d80bf26b026cc1f (accessed on 13th September 2025).

5. Decoteau, E. A., Raghavan, A., & Cahill, C. L. (2024). Structural, spectroscopic, and computational analysis of halogen- and hydrogen-bonding effects within a series of uranyl fluorides with 4-halopyridinium. *Inorganic Chemistry*. https://www.semanticscholar.org/paper/ad8d5a467111546a9fb188a99d521a013bac3230 (accessed on 13th September 2025).

6. Fan, J., Wang, Y., Shi, T., Yang, P., Zhou, G., Xu, J., & Su, B. (2024). Isolation and diverse reactivity of an unsymmetrical 1,2-bis(silylene)-stabilized pentacarbonyl chromium (0) species. *Inorganic Chemistry*. https://www.semanticscholar.org/paper/8580d5850499e3315df372abc1ca6696847442ed (accessed on 13th September 2025).

7. Galassi, R., Luciani, L., Graiff, C., & Manca, G. (2022). A reinterpretation of the imidazolate Au(I) cyclic trinuclear compounds reactivity with iodine and methyl iodide with the perspective of the inverted ligand field theory. *Inorganic Chemistry*, *61*, 3527–3539. https://www.semanticscholar.org/paper/21d84d9f29fd57d769a5dc25467f1421780f606f (accessed on 13th September 2025).

8. Gensch, T., Hopkinson, M. N., Glorius, F., & Wencel-Delord, J. (2016). Mild metal-catalyzed C–H activation: Examples and concepts. *Chemical*

Society Reviews, *45*, 2900–2936. https://doi.org/10.1039/c6cs00075d (accessed on 13th September 2025).

9. Ghanghas, M., Rom, T., & Paul, A. (2025). Featuring the role of coordination geometry in effective CO_2 fixation in lower-dimensional hybrid copper phosphonate frameworks. *Inorganic Chemistry*. https://www.semanticscholar.org/paper/1b844b5f44b8c79230bd9686fb00823574fbbcbb (accessed on 13th September 2025).
10. Golovenko, E. A., Pankin, D. V., Deriabin, K. V., Volkov, A., Kirichenko, S. O., Levin, O. V., & Islamova, R. (2024). Ligand exchange reaction between ferrocene and multiwalled carbon nanotubes: A contemporary approach. *Langmuir: The ACS Journal of Surfaces and Colloids*. https://www.semanticscholar.org/paper/fa3c305a307f242232ba5db9733b94c25100ed81 (accessed on 13th September 2025).
11. Guo, F. S., Bar, A. K., & Layfield, R. A. (2019). Main group chemistry at the interface with molecular magnetism. *Chemical Reviews*, *119*, 8479–8505. https://doi.org/10.1021/acs.chemrev.9b00103 (accessed on 13th September 2025).
12. Guo, Q. H., Jiao, Y., Feng, Y., & Stoddart, J. F. (2021). The rise and promise of molecular nanotopology. *CCS Chemistry*, *3*, 1542–1572. https://doi.org/10.31635/ccschem.021.202100975 (accessed on 13th September 2025).
13. Hamedpour, V., Sasaki, Y., Zhang, Z., Kubota, R., & Minami, T. (2019). Simple colorimetric chemosensor array for oxyanions: Quantitative assay for herbicide glyphosate. *Analytical Chemistry*, *91*, 13627–13632. https://doi.org/10.1021/acs.analchem.9b02822 (accessed on 13th September 2025).
14. Hwang, J. Y., Myung, S. T., & Sun, Y. K. (2017). Sodium-ion batteries: Present and future. *Chemical Society Reviews*, *46*, 3529–3614. https://doi.org/10.1039/c6cs00776g (accessed on 13th September 2025).
15. Jaishankar, M., Tseten, T., Anbalagan, N., Mathew, B. B., & Beeregowda, K. N. (2014). Toxicity, mechanism and health effects of some heavy metals. *Interdisciplinary Toxicology*, *7*, 60–72. https://doi.org/10.2478/intox-2014-0009 (accessed on 13th September 2025).
16. Janet, J. P., Duan, C., Nandy, A., Liu, F., & Kulik, H. J. (2021). Navigating transition-metal chemical space: Artificial intelligence for first-principles design. *Accounts of Chemical Research*, *54*, 532–545. https://doi.org/10.1021/acs.accounts.0c00686 (accessed on 13th September 2025).

17. Johnstone, T. C., Suntharalingam, K., & Lippard, S. J. (2016). The next generation of platinum drugs: Targeted Pt(II) agents, nanoparticle delivery, and Pt(IV) prodrugs. *Chemical Reviews*, *116*, 3436–3486. https://doi.org/10.1021/acs.chemrev.5b00597 (accessed on 13th September 2025).

18. Joshi, B., & Shivashankar, M. (2023). Recent advancement in the synthesis of Ir-based complexes. *ACS Omega*, *8*, 43408–43432. https://www.semanticscholar.org/paper/e276efa665d56c68350fd24ee13f1ab70e0b7548 (accessed on 13th September 2025).

19. Li, W., Feng, J., Zhang, X., Li, C., Dong, H., Deng, W., Liu, J., et al. (2021). Metallization and superconductivity in the van der Waals compound CuP_2Se through pressure-tuning of the interlayer coupling. *Journal of the American Chemical Society*, *143*, 20343–20355. https://doi.org/10.1021/jacs.1c09735 (accessed on 13th September 2025).

20. Liebschner, D., Afonine, P. V., Baker, M. L., Bunkòczi, G., Chen, V. B., Croll, T. I., Hintze, B. J., et al. (2019). Macromolecular structure determination using X-rays, neutrons and electrons: Recent developments in *Phenix*. *Acta Crystallographic Section D: Structural Biology*, *75*, 861–877. https://doi.org/10.1107/s2059798319011471 (accessed on 13th September 2025).

21. Liu, J., Xiao, H., Zhao, X. K., Zhang, N., Liu, Y., Xing, D., Yu, X., et al. (2022). Computational prediction of graphdiyne-supported three-atom single-cluster catalysts. *CCS Chemistry*, *5*, 152–163. https://doi.org/10.31635/ccschem.022.202201796 (accessed on 13th September 2025).

22. Lu, T., & Chen, F. (2011). Multiwfn: A multifunctional wavefunction analyzer. *Journal of Computational Chemistry*, *33*, 580–592. https://doi.org/10.1002/jcc.22885 (accessed on 13th September 2025).

23. Lucas, E. L., Lam, N. Y. S., Zhuang, Z., Chan, H. S. S., Strassfeld, D. A., & Yu, J. Q. (2022). Palladium-catalyzed enantioselective β-C(sp^3)–H activation reactions of aliphatic acids: A retrosynthetic surrogate for enolate alkylation and conjugate addition. *Accounts of Chemical Research*, *55*, 537–550. https://doi.org/10.1021/acs.accounts.1c00672 (accessed on 13th September 2025).

24. Luo, J., Montag, M., & Milstein, D. (2024). Metal–ligand cooperation with thiols as transient cooperative ligands: Acceleration and inhibition effects in (de)hydrogenation reactions. *Accounts of Chemical Research*, *57*, 1709–1721. https://www.semanticscholar.org/paper/040589561788e2bd97d68a0376aabc89e6d8087d (accessed on 13th September 2025).

25. Meng, Y. S., & Liu, T. (2019). Manipulating spin transition to achieve switchable multifunction's. *Accounts of Chemical Research*, *52*, 1369–1379. https://doi.org/10.1021/acs.accounts.9b00049 (accessed on 13th September 2025).

26. Mourdikoudis, S., Pallares, R. M., & Nguyễn, T. K. T. (2018). Characterization techniques for nanoparticles: Comparison and complementarity upon studying nanoparticle properties. *Nanoscale*, *10*, 12871–12934. https://doi.org/10.1039/c8nr02278j (accessed on 13th September 2025).

27. Nakagawa, T., Wada, Y., Chan, B., Baba, T., Hanaya, K., Koseki, Y., Asano, R., et al. (2025). Structural elucidation of nucleophilic compounds through synergistic coordination and hydrogen bonding in a metal–organic framework. *Journal of the American Chemical Society*, *147*, 29013–29025. https://www.semanticscholar.org/paper/16ec431e3dd41e2a90908f597452422e23e3d94a (accessed on 13th September 2025).

28. Oloyede, U. N., & Flowers, R. (2024). Coordination-induced bond weakening and small molecule activation by low-valent titanium complexes. *Dalton Transactions*. https://www.semanticscholar.org/paper/3a72d862a30e423199fbb218171ae0ac3906edbe (accessed on 13th September 2025).

29. Peng, T., Zhang, F. L., & Wang, Y. F. (2022). Lewis's base–boryl radicals enabled borylation reactions and selective activation of carbon–heteroatom bonds. *Accounts of Chemical Research*, *56*, 169–186. https://doi.org/10.1021/acs.accounts.2c00752 (accessed on 13th September 2025).

30. Pöthig, A., & Casini, A. (2019). Recent developments of supramolecular metal-based structures for applications in cancer therapy and imaging. *Theragnostic*, *9*, 3150–3169. https://doi.org/10.7150/thno.31828 (accessed on 13th September 2025).

31. Schneider, H. J. (2022). Noncovalent interactions: A brief account of a long history. *Journal of Physical Organic Chemistry*, *35*. https://doi.org/10.1002/poc.4340 (accessed on 13th September 2025).

32. Sharma, P., LaRosa, C., Antwi, J., Govindarajan, R., & Werbovetz, K. A. (2021). Imidazoles as potential anticancer agents: An update on recent studies. *Molecules*, *26*, 4213–4213. https://doi.org/10.3390/molecules26144213 (accessed on 13th September 2025).

33. Sharpee, T., Destexhe, A., Kawato, M., Sekulić, V., Skinner, F., Wójcik, D., Chintaluri, C., et al. (2016). 25th Annual Computational Neuroscience Meeting: CNS-2016. *BMC Neuroscience*, *17*. https://www.

semanticscholar.org/paper/23a30ca55aef3e307c9e761b22a34080ebf6985f (accessed on 13th September 2025).

34. Sies, H., Berndt, C., & Jones, D. P. (2017). Oxidative stress. *Annual Review of Biochemistry*, *86*, 715–748. https://doi.org/10.1146/annurev-biochem-061516-045037 (accessed on 13th September 2025).
35. Sigel, H., Sigel, R., & Sigel, A. (2022). Coordination chemistry of nucleotides and antivirally active acyclic nucleoside phosphonates, including mechanistic considerations. *Molecules*, *27*. https://www.semanticscholar.org/paper/0163b161efc3e6ab869104eca5b38b7f5437dd2f (accessed on 13th September 2025).
36. Stoddart, J. F. (2017). Mechanically interlocked molecules (MIMs)—Molecular shuttles, switches, and machines (Nobel Lecture). *Angewandte Chemie International Edition*, *56*, 11094–11125. https://doi.org/10.1002/anie.201703216 (accessed on 13th September 2025).
37. Strelnikova, V., Ovsyannikov, A., Agarkov, A., Iova, A. A., Vandyukov, A., Gubaidullin, A., Kleshnina, S., et al. (2024). Interplay between conformational flexibility, intermolecular H-bonding and 3d-metal cation extraction ability in a series of thiacalix[4]arene lower rim disubstituted Schiff base derivatives. *Physical Chemistry Chemical Physics: PCCP*. https://www.semanticscholar.org/paper/1239967e29ea3e030d99f532f2c99f69e2c166ce (accessed on 13th September 2025).
38. Sun, J., Xu, Y., Lv, Y., Zhang, Q., & Zhou, X. (2023). Recent advances in covalent organic framework electrode materials for alkali metal-ion batteries. *CCS Chemistry*, *5*, 1259–1276. https://doi.org/10.31635/ccschem.023.202302808 (accessed on 13th September 2025).
39. Tang, D., Chen, X., Kang, R., & Kroemer, G. (2020). Ferroptosis: Molecular mechanisms and health implications. *Cell Research*, *31*, 107–125. https://doi.org/10.1038/s41422-020-00441-1 (accessed on 13th September 2025).
40. Tian, Y., Zeng, G., Rutt, A., Shi, T., Kim, H., Wang, J., Koettgen, J., et al. (2020). Promises and challenges of next-generation "beyond Li-ion" batteries for electric vehicles and grid decarbonization. *Chemical Reviews*, *121*, 1623–1669. https://doi.org/10.1021/acs.chemrev.0c00767 (accessed on 13th September 2025).
41. Toader, M., Buta, C., Cimpoesu, F., & Ferbinteanu, M. (2024). Noncovalent bonding in d and f-type coordination compounds and lattices. A case study. *ChemPlusChem*. https://www.semanticscholar.org/paper/23105f159606f01f76870ed63542287c791b1c4d (accessed on 13th September 2025).

42. Tooke, C. L., Hinchliffe, P., Bragginton, É. C., Colenso, C. K., Hirvonen, V. H. A., Takebayashi, Y., & Spencer, J. (2019). β-Lactamases and β-lactamase inhibitors in the 21st century. *Journal of Molecular Biology, 431*, 3472–3500. https://doi.org/10.1016/j.jmb.2019.04.002 (accessed on 13th September 2025).
43. Van der Heggen, D., Joos, J., Feng, A., Fritz, V., Delgado, T., Gartmann, N., Walfort, B., et al. (2022). Persistent luminescence in strontium aluminate: A roadmap to a brighter future. *Advanced Functional Materials, 32*. https://doi.org/10.1002/adfm.202208809 (accessed on 13th September 2025).
44. Verma, C., Singh, A., Singh, P., Rhee, K. Y., & Alfantazi, A. (2024). Regioisomeric effect of heteroatoms and functional groups of organic ligands: Impacts on coordination bonding and corrosion protection performance. *Coordination Chemistry Reviews, 515*, 215966–215966. https://doi.org/10.1016/j.ccr.2024.215966 (accessed on 13th September 2025).
45. Woodruff, D. N., Winpenny, R. E. P., & Layfield, R. A. (2013). Lanthanide single-molecule magnets. *Chemical Reviews, 113*, 5110–5148. https://doi.org/10.1021/cr400018q (accessed on 13th September 2025).
46. Wu, D., Sedgwick, A. C., Gunnlaugsson, T., Akkaya, E. U., Yoon, J., & James, T. D. (2017). Fluorescent chemo sensors: The past, present and future. *Chemical Society Reviews, 46*, 7105–7123. https://doi.org/10.1039/c7cs00240h (accessed on 13th September 2025).
47. Zhu, C., Ji, X., You, D., Chen, T. L., Mu, A. U., Barker, K. P., Klivansky, L. M., et al. (2018). Extraordinary redox activities in ladder-type conjugated molecules enabled by B ← N coordination-promoted delocalization and hyperconjugation. *Journal of the American Chemical Society, 140*, 18173–18182. https://doi.org/10.1021/jacs.8b11337 (accessed on 13th September 2025).
48. Zhu, J., Hu, L., Zhao, P., Lee, L. Y. S., & Wong, K. Y. (2019). Recent advances in electrocatalytic hydrogen evolution using nanoparticles. *Chemical Reviews, 120*, 851–918. https://doi.org/10.1021/acs.chemrev.9b00248 (accessed on 13th September 2025).

CHAPTER 3

CRYSTAL FIELD THEORY (CFT)

Contents

INTRODUCTION

Crystal Field Theory (CFT) provides a valuable tool for understanding the electronic structure and properties of transition metal complexes in coordination chemistry. CFT deals specifically with the interaction of electric fields generated by ligands with the d-orbitals of the metal ion. Upon the coordination of ligands to a metal ion, their negative charge affects the d-orbitals of the metal ion, causing degeneracy to be lifted, leading to a rearrangement of the energy levels.

Such an effect is a determiner of stability, color, and magnetic properties of these complexes. While CFT investigates the significance of such interactions, it provides a framework for the prediction of various properties of transition metal complexes that are crucial to numerous fields including materials science, catalysis, and biological processes. The applicability of CFT is not confined to theoretical applications; it has practical applications in the design of top-quality materials as well. Based on understanding the influence of ligand field strength on d-orbital splitting, scientists can tailor material properties for a specific application. For example, in photonic crystal fibers, where theories of light propagation and management are applied, transition metal behavior can be engineered to enhance performance. These innovations can advance the coupling operations of photonic crystal fiber devices, in line with observations such as those shown in recent research that merges modal transmission-line theory into device design (Akbari et al., 2005). CFT adds not only to our theoretical insight but also to the engineering of new technologies that take advantage of these principles. Moreover, the study of CFT and its implications is even more moving into multidisciplinary fields, including optofluidics, which is the merger of microfluidics and optics for chemical analysis. CFT's comprehension of light-matter interaction is important in the development of analytical techniques, especially in miniaturized lab-on-a-chip systems. The latter is plagued with issues that are unique to smaller-scale devices, namely the reduced optical pathways (Sharkawy et al., 2007). However, using the postulates of CFT, these limitations can be overcome by maximizing light-ligand interaction, thereby increasing the efficiency and effectiveness of chemical analysis. Not only does current research in CFT increase theoretical understanding but also stimulates innovation in applied applications, with CFT being an important area of research in contemporary scientific investigation.

Definition of Crystal Field Theory (CFT)

The concept of CFT fundamentally reshapes our understanding of how transition metal complexes interact with ligands through electrostatic forces. CFT posits that ligands, which can be negatively charged ions or neutral molecules, influence the arrangement and energy levels of the d-orbitals of transition metals when they form coordination complexes. This is due to the interaction of the electric fields generated by the ligands surrounding it with the metal ion's d-electrons. When ligands get close to the metal center, the d-orbital splitting symmetry develops, determining the complex's electronic properties. The theory's consequences reach into a variety of chemical behavior such as color, magnetism, and reactivity, and thus it is a fundamental part of coordination chemistry. In understanding the facts of CFT, there is a need to consider very crucial distinction between different geometries of complexes, such as octahedral and tetrahedral geometries. In an octahedral geometry, electrostatic ligand interaction results in a typical pattern of splitting where energy levels of d-orbitals are shifted due to their relative positions in space with respect to the ligand position. This arrangement leads to a higher energy set of orbitals (eg) and a lower energy set (t2g), which is responsible for the crystal field splitting energy (Δ). Tetrahedral complexes have the opposite— splitting of d-orbitals in a manner that leads to different patterns of energy, hence influencing the electronic structure and stability of the complexes. The understanding of these geometric variations is important in determining the electronic behavior of various metal complexes in various scenarios, like catalysis and materials science. Moreover, the application of Crystal Field Theory exceeds the theoretical underpinnings to practical applications in materials science and inorganic chemistry. For instance, recent studies have elaborated on the significance of CFT in deciding the electronic properties of two-dimensional transition metal dichalcogenides (TMDs), which take various polymorph forms. The varying crystal and ligand field parameters of such materials reflect their stability as well as other electronic properties. As emphasized, ligand and crystal fields possess colossal influences on charge transfer processes, and hence, the interaction must be realized in detail in practical applications. Based on the application of CFT principles, chemists are able to design and maximize materials of needed properties, and hence, the significance of the theory is established in sophisticated research and development in modern chemistry (Coudert et al., 2015; Pasquier et al., 2018) (Figure 3.1).

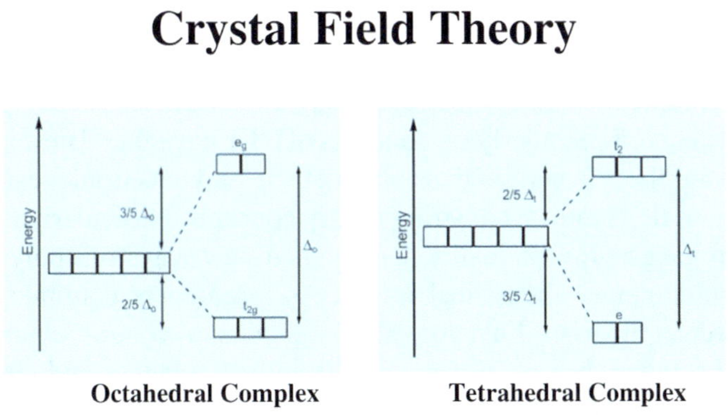

Figure 3.1. *The concept of crystal field theory (CFT).*

Source: https://scienceinfo.com/wp-content/uploads/2022/12/Octahedral-Complex.jpg.

Historical Background And Development

The origin of CFT is as old as the beginning of the 20th century due to the failure of the then dominant atomic models to explain the nature of transition metal complexes. CFT was an innovative theoretical model that sought to describe the electronic structures and properties of such complexes in terms of electrostatic metal ion-ligand interactions. In its early stages, CFT borrowed concepts from the earlier literature of valence bond theory and ionic/covalent bond theory to provide a more nuanced description of metal-ligand bond character.

By providing a methodical means of determining the energy of d-orbitals in a ligand field, CFT paved the way for further research in coordination chemistry and has since become an integral part of the discipline of inorganic chemistry, allowing chemists to predict the color, magnetism, and stability of metal complexes. The first formal presentation of CFT is attributed to the initial work of Hans Bethe and subsequent refinements by other scientists who contributed to the theoretical foundation of the model. The original formulation made a clear distinction between the different d-orbital types and how they were affected by ligand coordination geometries such as octahedral and tetrahedral. This separation was important in the standardization of procedures for examining intricate coordination complexes and determining their spectral characteristics. As progress continued through the second half of the 20th century, later developments like ligand field theory came about, offering an enhanced explanation by including molecular orbital theory. This was an important transition where CFT evolved from a stringent electrostatic issue to a richer approach that integrated the behavior of such complexes with elementary principles in solid-state physics and quantum mechanics with its broader applications in various scientific fields. More recently, applications of CFT have transcended traditional chemistry to condensed matter physics and materials science. The underlying concepts of CFT have found new applications in describing the effects such as magnetism and electronic excitations in various materials. Synthesis of string theory concepts, particularly the concepts included in gauge-gravity duality, has opened new ways of applying CFT in solving contemporary theoretical issues. As researchers continue to explore these interdisciplinarian linkages, CFT remains a central element of the scientific discourse, bringing disparate disciplines together and enriching our understanding of complex systems. As is seen from where advanced theories and approaches converge, the destiny of CFT and its applications in broader scientific scenarios lies in a further development, constantly searching for new uses while fortifying the traditional roots of CFT in chemistry and physics (Green et al., 2013; Mitra et al., 1999).

Importance of Crystal Field Theory (CFT) in Coordination Chemistry

In coordination chemistry, CFT is a basic theory used to describe the electronic structures of transition metal complexes. Taking into consideration the interactions between the ligands and the metal center, CFT provides a systematic approach to predict the stability, geometry, and electronic absorption spectra of the complexes. This theoretical framework allows chemists to account for color changes and magnetism, the hallmark in the practical use of these complexes in magnetic materials and dyes. CFT principles are understood to guide the synthesis of novel coordination compounds with tailored properties leading to advancements in fields like materials science and catalysis where electronic state control is necessary for enhanced performance.

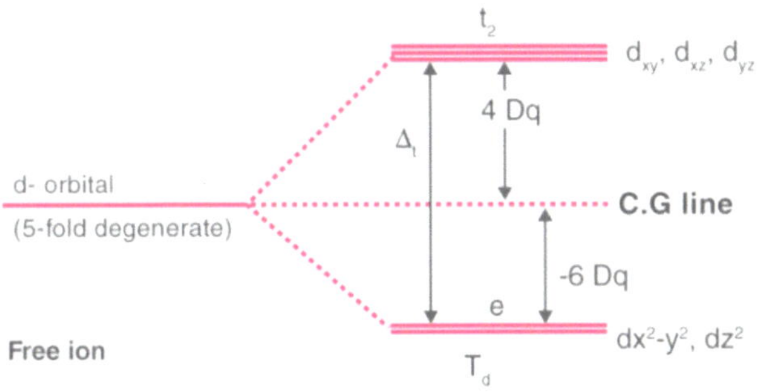

Figure 3.2. *Importance of crystal field theory (CFT) in coordination chemistry.*

Source: https://cdn1.byjus.com/wp-content/uploads/2016/01/crystal-field-theory-new-2-700x327.png.

The computational basis of CFT also adds to its usefulness in coordination chemistry. Advanced computational techniques have arisen to predict structural and electronic material properties of metal-organic frameworks (MOFs), providing insights into their potential applications in gas storage and separation technologies. These techniques encompass a range of quantum and classical simulation techniques for the evaluation of stability and performance criteria under diverse conditions. The ability to model and simulate such systems is strong enough that researchers are able to anticipate how variation in the ligand design or metal selection can affect the net behavior of a complex. This assists significantly in the creation of targeted materials that can be designed to react to specific energy storage, heterogeneous catalysis, and environmental remediation demands, thus demonstrating the far-reaching consequences of CFT uses (Coudert et al., 2015). Also, the compatibility of CFT with the metalloenzyme active sites exemplifies the profound relevance of this theory in biochemical process research. For instance, research conducted using non-heme iron complexes to model thiol dioxygenase activity exemplifies information derived from CFT on mechanisms and reactivity of enzymes. These complexes not only replicate the catalytic activity of their biological counterparts but also demonstrate how variations in ligand coordination can affect patterns of reactivity, thus informing synthetic mimicry and enzyme design. The relationship of electronic structure and function, as exposed by CFT, underscores the importance of precise molecular engineering in designing useful catalysts. Therefore, the application of CFT in the study of coordination compounds and biological systems highlights its central role in the formulation of both theoretical understanding and applied use in chemistry (Brunold et al., 2016) (Figure 3.2).

3.1. BASIC PRINCIPLES OF CRYSTAL FIELD THEORY (CFT)

The foundation of CFT lies in its ability to describe the transition metal ion behavior in coordination complexes. It deals with the effect of the ligands around the metal on the energy of the metal's d-orbitals. As ligands approach a metal ion, their negative charge generates an electrostatic field that alters the degeneracy of the d-orbitals. The d-orbitals in a single transition metal atom are degenerate, but the ligand breaks the degeneracy and creates differences in the energy levels. The splitting depends on the geometry of the complex—octahedral, tetrahedral, or square planar—and also on the type of ligand, which is either a strong field or a weak field ligand. Thus, understanding these interactions is crucial to the prediction of magnetic and spectral properties of such complexes. One significant advance in the application of CFT is the incorporation of realistic modeling methods that adhere to first-principle frameworks. These methods have been most significant in the study of transition metal oxides and correlated materials, where standard methods are likely to fail. For instance, electronic activity at the Fermi level plays a crucial role in dictating material properties like vanadates. It has been established in research that with the strict derivation of parameters through computational methods, scientists are able to analyze magnetic states and orbital interactions induced by these crystal-field effects. Upon looking at complex structures such as the C-type and G-type antiferromagnetic states, the role of crystal distortion and lifting of orbital degeneracy becomes relevant. These results point toward the significance of including meticulous modeling so as to discern the mechanisms of electron correlation and spatial configuration. Furthermore, significance of

CFT does not lie just in the explanation of electronic configurations but also in helping understand the color and magnetic behavior of transition metal complexes. The color of these complexes is largely due to d-d transitions, which are extensively controlled by differential splitting of d-orbitals caused by the ligand-ligand interactions. Strong-field ligands, for example, cause a larger splitting effect, often resulting in some electronic transitions within the visible region, producing intense color features. This ligand field strength and electronic transition interdependence forms the basis for optical feature prediction in metal complexes. Thus, CFT is a reliable theoretical basis that not only unscrambles the intrinsic fundamentals of coordination chemistry but also serves as a basis for advancements in fields like material science and solid-state physics, thus providing better insights into complex materials and their uses (Baianu, 2011; Liechtenstein et al., 2010).

3.1.1. Concept of Ligands and Metal Ions

The concept of metal ions and ligands is key to the understanding of coordination complexes, particularly under the framework of Crystal Field Theory (CFT). Ligands, being molecules or ions capable of donating electron pairs towards the metal ion, are key determinants in controlling the electronic environment in the vicinity of the metal center. This, in fact, has an immediate effect on the electronic structure, shape, and stability of the resultant complexes. Transition metal ions with their partially filled d-orbitals possess properties that allow them to adopt various coordination geometries. These are consequences of the coordination geometry of the ligands and the inherent properties of the metal ion, thus dictating the electronic transitions which occur within the complex. Here, one must appreciate the function of ligands for the forecast of the reactivity and behavior of metal complexes important to numerous applications, ranging from bioinorganic chemistry to catalysis. The nature of the interaction between metal ions and ligand is governed by the physics of Crystal Field Theory itself, which foretells the way ligands bend the degenerate d-orbitals of the transition metals. The different types of ligands can form different levels of electrostatic interactions that lead to d-orbital energy splitting. The strong field ligands will create extensive splitting, hence, larger separations in energies between the orbitals, while the weak field ligands can form less splitting. The electronic transitions are influenced by the splitting which consequently influences the absorption spectra and magnetism of the complexes. For instance, some metal-ligand interactions can result in some intriguing magnetic properties that are observed in complexes. Notably, hexanuclear complex $[Mn_6(\mu4\text{-}H_2L_2)2(\mu\text{-}HL_3)2(\mu3\text{-}OH)2(\mu1,3\text{-}O_2CC_2H_5)_4](ClO_4)2\cdot 2H_2O$ is a clear demonstration of the relationship between ligand structure and magnetism, showing how intramolecular interactions can lead to anti-ferromagnetic behavior, as outlined in recent studies (Chattopadhyay et al., 2017). Other than their influence on electronic structure, the type of ligands in a complex can influence the functional functionality of metal complexes in most applications, from material science to bioimaging. Transition metal sulfides, for example, possess different functional properties that cannot be related with those of their oxide counterparts, and therefore are emerging as an area of increasing interest in nanotechnology and catalysis (Dengo et al., 2017). The stability and reactivity of such complexes are often a function of good ligand design and functionalization principles, which may regulate the surface chemistry and general material behavior. Stable ligands, for instance, can immobilize nanoparticles in colloidal suspensions, allowing them to be applied practically in applications such as bioimaging and environmental remediation. Hence, a comprehensive understanding of the ligand-metal ion relationship not only accounts for fundamental chemical principles but also contributes to advancements in applied sciences, underscoring the important point of intersection between theoretical and applied applications in coordination chemistry (Figure 3.3).

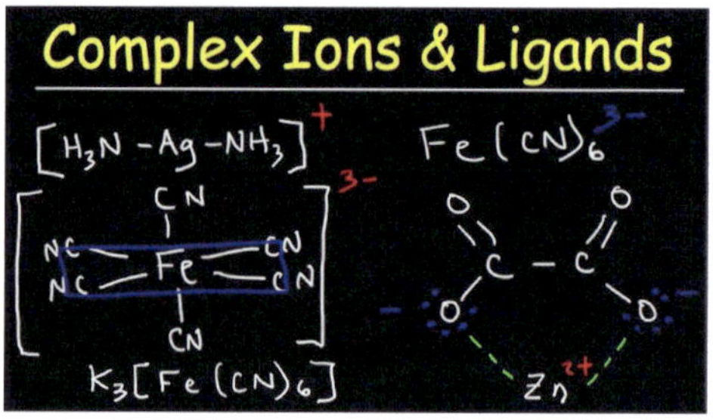

Figure 3.3. *The concept of ligands and metal ions.*

Source: https://i.ytimg.com/vi/r4H5XjJPn58/hq720.jpg?sqp=-oaymwEhCK4FEI-IDSFryq4qpAxMIARUAAAAAGAElAADIQj0AgKJD&rs=AOn4CLBziRUBn0HIQh8r6eFDXIHJNpZHXA.

3.1.2. Interaction Between Ligands and d-orbitals

The d-orbital ligation is a basic characteristic of Crystal Field Theory (CFT), having a significant influence on the electronic structure and properties of transition metal complexes. Charged or uncharged molecules around a metal ion, acting as ligands, cause the crystal field condition and alter the degeneracy of d-orbitals. This mutual interaction is responsible for varying degrees of rotational splitting of the d-orbitals, which is of tremendous importance in understanding the stability, color, and chemical reactivity of the complexes. The extent of splitting of the d-orbitals relies primarily on the nature and positions of ligands around the central metal ion, which can be studied using spectroscopic techniques. Studies of numerous transition-metal compounds, e.g., TiOCl and $RTiO_3$, demonstrate these principles, demonstrating the manner in which ligand structures can affect orbital excitations and crystal-field splitting energies (Freimuth et al., 2005). Investigation of ligand field interactions with d-orbital splitting further clarifies electronic and magnetic interaction functions within compounds such as iridates. In oxides like Sr_2IrO_4 and Ba_2IrO_4, ligand-d-orbital interaction not only modifies the energy range of various d-d excitations but also determines to a great extent the magnetic properties of these compounds. The strong hybridization of Ir-O with d-orbitals yields a characteristic electronic structure with strong spin-orbit coupling. In this regard, such interactions increase low energy d-d excitations as well as well-defined charge-transfer excitation thresholds that are a core aspect of the materials' superconductivity potential (Stoll et al., 2012). The understanding of these interactions facilitates the development of good Hamiltonians explaining the nature of such intricate systems. To sum up, ligand-d-orbital interaction as offered by CFT is very vital in the explanation of some chemical and physical transitions of transition metal complexes. The uniform trend of d-orbital splitting with ligand field effects is significant, emphasizing the ligand and coordination geometry. Examining the overall properties of these interactions through experimental and theoretical methodologies is significant since it enhances our understanding of their contribution to improved orbital excitations and magnetic properties in transition-metal compounds. By integration of data from studies on different compounds under different conditions, researchers are still able to advance the fundamental theories that describe electronic structures, energetics, and potential applications in catalysis and material science. Hence, ligand and d-orbital interaction remains a relevant area of investigation in transition metal chemistry, providing insight that cuts across numerous disciplines.

3.1.3. Splitting of d-orbitals in Different Geometries

Transition metal complex electronic structure is very crucial to interpret with the help of d-orbital splitting. According to Crystal Field Theory (CFT), the spatial arrangement of ligands surrounding a central metal ion directly influences the d-orbital energy levels. In octahedral geometries, for instance, interaction between the metal ion and ligands results in the degenerate d-orbitals being split into two energy levels: lower-energy t2g orbitals and higher-energy eg orbitals. This is due to varying amounts of electrostatic interaction that the orbitals undergo based on their orientation in relation to the incoming ligand field. Thus, the degree of splitting can predict the magnetic and optical behavior of the complex, solidifying the position of geometry in dictating the general behavior of d-block elements. Tetrahedral geometries present a different situation of d-orbital splitting where the d-orbitals split into one set of higher energy e orbitals and lower energy t2 orbitals.

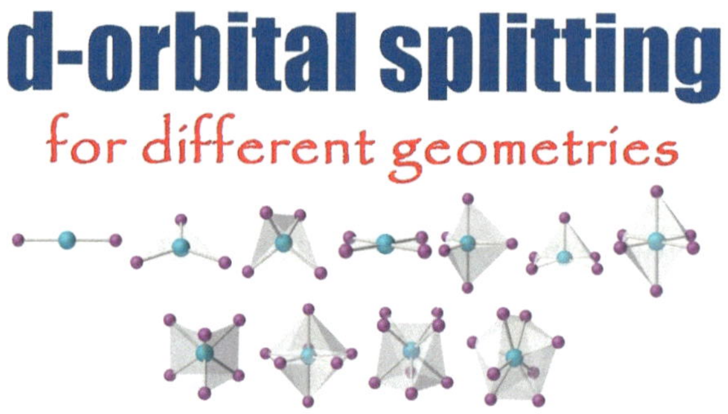

Figure 3.4. *Splitting of d-orbitals in different geometries.*

Source: https://i.ytimg.com/vi/mlI1hnJZleo/maxresdefault.jpg.

This condition is characterized by a smaller degree of splitting compared to octahedral complexes, mostly because of the ligand–metal interaction angle, which in tetrahedra approaches 109.5°, reducing significantly the direct overlap between metal-centered orbitals and ligands. Moreover, the unique characteristics shown by tetrahedral complexes, such as greater susceptibility to distortion and low-spin configuration, can be attributed to this altered splitting behavior. Identification of these geometrical differences enlarges the scope of CFT in explaining the various chemical behaviors exhibited by transition metal coordination compounds. Study of d-orbital splitting is also made difficult by the presence of other geometries, such as square planar and trigonal bipyramidal geometries. These differences lead to different splitting patterns, influencing the strength of the ligand field and the resulting electronic configurations. For instance, in square planar complexes, the d-orbitals are split similarly to octahedral complexes but with a typical energy order due to the varying ligand environments at the equatorial and axial sites surrounding the metal center. Local geometry becomes important with findings that show the spin and orbital interactions in such complexes can lead to effects like ferro-orbital ordering that influence their magnetic and reactive properties. Moreover, the study of exotic geometries, such as neighboring octahedra sharing faces, emphasizes subtleties overlooked in typical CFT analysis, placing geometry centrally in understanding transition metal chemistry ligand-ligand interaction nuances (Arita et al., 2011; Khomskii et al., 2015) (Figure 3.4).

3.2. TYPES OF CRYSTAL FIELD SPLITTING

Crystal field splitting is at the heart of the explanation of electronic structure in transition metal complexes according to Crystal Field Theory (CFT). The two important geometries that dictate the types of crystal field splitting to be observed are octahedral and tetrahedral environments. In an octahedral setting, the ligand coordination to the central metal ion results in the splitting of the d-orbitals into separate energy levels—viz., lower-energy t2g orbitals and higher-energy eg orbitals. These energy level differences are necessary since they influence the oxidation state of the metal ions, their magnetic moments, and how they respond to light approaching them, and hence their color and reactivity in the local chemical environment. Tetrahedral coordination gives another splitting pattern of the d-orbitals. Here, the energies of the orbitals are inverted compared to octahedral complexes. Here, the e orbitals are of lower energy than the t2 orbitals, and the energy gap is therefore greater than those of octahedral. This is a significant consideration when predicting the activity of metal complexes under tetrahedral conditions, as it has direct consequences on their stability and coordination chemistry. The discrepancies in splitting patterns between these two geometries have very much influenced the ligand field theory, especially in relation to the stability and spectral features of metal-ligand complexes. There are further complexities if one considers other geometries and how the strength of the ligand affects crystal field splitting. Square planar and trigonal bipyramidal arrangements, for instance, have unique splitting patterns and energy considerations to offer. Also, ligand type and the electrostatic environment around the metal ion will influence the degree of splitting, which is usually measured as the parameter Δ (delta). Study of these interactions is important in order to understand the quantum mechanical nature of transition metal chemistry. For instance, the configurations of a single divalent rare earth ion in cubic sites may drastically modify their propensity for electron hopping between states, for example, for La, Ce, Gd, and others, reflecting on the fine dependency of crystal field splitting and electronic transitions (Egranov et al., 2015; Schupper et al., 2005). These findings underscore the importance of understanding various crystal field environments in high-technology applications ranging from catalysis to materials research.

3.2.1. Octahedral Splitting

Octahedral splitting is a focal phenomenon in the Crystal Field Theory (CFT) setting, as it answers the most pressing question regarding how the energy levels of electrons in transition metal ions become altered when these ions are ligand-surrounded with an octahedral configuration. In this arrangement, the ligands, which are located at the vertices of an octahedron, exert an electrostatic field on the central metal cation's d-orbitals, bringing about a clear-cut separation of energy between the orbits. This is mainly due to alterations in the spatial orientation of the d-orbitals relative to the incoming ligand fields. The energy difference observed, typically symbolized as Δ, not only influences the electronic structure of the resultant complexes but also their magnetic and optical properties, highlighting the importance of ligand field effects in molecular chemistry.

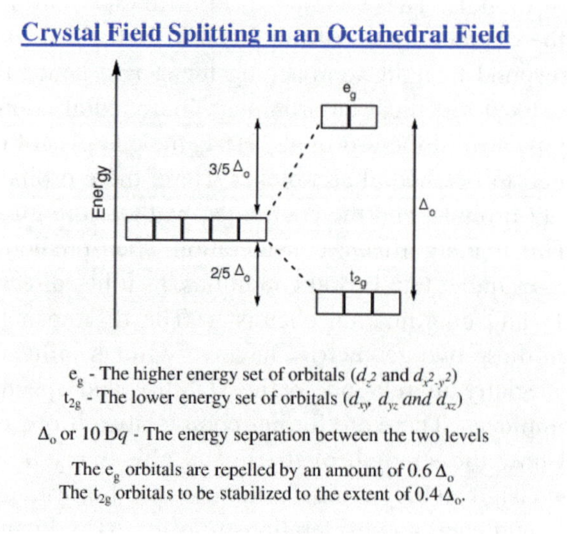

Figure 3.5. *The phenomenon of octahedral splitting.*

Source: https://0.academia-photos.com/attachment_thumbnails/43588125/mini_magick20180817-26294-fo40sy.png?1534548056.

Beyond its influence on electronic structure, octahedral splitting is also important in comprehending chemical reactivity and stability of coordination compounds. For instance, the splitting varies depending on the ligands' nature that enclose the transition metal. Weak field ligands result in less splitting energy, but strong field ligands result in more splitting between the t2g (lower energy) and eg (higher energy) orbitals. This, in turn, results in different electron configurations, which finally determine the magnetic properties of the compound—whether it appears as high-spin or low-spin configurations. Thus, the idea of octahedral splitting is a key factor in predicting transition metal complexes' properties and applications, revealing the delicate interplay between electronic structure and ligand interaction. Moreover, the effect of octahedral splitting is also parallel to the more general ideas observed in other crystal field systems such as the observed degeneracy and interaction in bigger crystal systems. As quoted in recent studies, for instance, the role of the $GdFeO_3$-type distortion in regulating orbital-spin structures in perovskite compounds, splitting within octahedra is a foundation of explaining complex behaviors in materialized functional systems (Akimitsu et al., 2003). The other dopants' interaction, as seen in the case of quantum dots, further elucidates how octahedral coordination dynamically tunes energy levels and transition probabilities and how this uncovers an enriching paradigm for subsequent studies in the area of material science (Bol et al., 2003). These interactions have profound consequences, emphasizing the role of octahedral splitting in regulating not only the electronic structure but also the emergent properties of most materials in new applications (Figure 3.5).

3.2.2. Tetrahedral Splitting

Tetrahedral splitting is a significant Crystal Field Theory (CFT) theory that explains the metal d-orbital energy level changes due to the spatial arrangement of the ligands in a metal ion center. When there is a tetrahedral coordination environment, the ligands are at the corners of a tetrahedron, resulting in a certain orientation that affects the degeneracy of the d-orbitals. Unlike the octahedral coordination which involves two sets of orbitals, tetrahedral coordination involves a lower-energy set which includes the d_xz, d_yz, and d_xy orbitals, but has the d_x2-y2 and d_z2 orbitals at higher energy levels. This splitting is primarily caused by the electrostatic interactions between the negative charge on the ligands and the positive charge on the metal ion, and therefore it is pivotal in comprehending the color and magnetic nature of transition metal complexes. The implications of tetrahedral splitting extend beyond theoretical models; they are of high significance to the material property of many applications. For example, manipulation of tetrahedral coordination in semiconductors can yield novel electronic properties. Recent studies involving Fe-doped $ZnGa_2O_4$ demonstrate how modifications in Fe octahedral and tetrahedral sites may stabilize ferromagnetic phases critical to future semiconductor applications. Electronic conduction-crystal symmetry interactions within these materials demonstrate how tetrahedral splitting might be able to impact material properties and ultimately technological advancement. Hence, knowledge about these concepts not only contributes to academic knowledge but also sets the stage for new applications in materials science and engineering, with evidence of the practical utility of CFT to real-world problems (Schwarz et al., 2005). Additionally, research into materials exhibiting tetrahedral coordination has revealed new behavior under various physical conditions. For instance, KOs_2O_6 has shown information on non-centrosymmetric structures, opening the avenue for the exploration of superconductivity and tetrahedral splitting effect-induced magnetism. In the material, anisotropic interactions between Os tetrahedra and O octahedra generate unique physical phenomena, such as a superconducting transition with a clean onset and some magnetic behaviors under an external field. The information in such publications illustrate the complex relationship between tetrahedral splitting and other physical properties, revealing extreme changes via different stages of matter. This investigation of such relationships improves our understanding of material science and suggests further research on the quantum nature of tetrahedral complexes, expanding the dimensions of theoretical and applied chemistry (Lebed et al., 2006).

3.2.3. Square Planar Splitting

In the Crystal Field Theory (CFT) model, the geometry of a metal complex plays a fundamental role in determining its electronic structure and consequent properties. Square planar geometry, characteristic for transition metal complexes, has a direct influence on d-orbital ordering and energy. In square planar complexes, such as those for d8 metal ions like nickel(II) or platinum(II), the dx2-y2 orbital is lifted to a higher energy by direct ligand-ligand interactions on the square plane corners. The other d-orbitals, i.e., dxy, dz2, and dxz/dyz, however, receive a relative energy lowering, which creates a resultant electronic distribution pattern that can have a significant influence on reactivity and catalytic behavior. Above all, the splitting of energy in square planar complexes has the immediate result of being a result of ligand field effects. In CFT, as a result of electrostatic interactions between the central metal cation and the ligands it is surrounded by, the d-orbitals are reoriented so that the specific symmetry of the ligands determines the size and type of this splitting. Square planar complex investigations show that if ligands coordinate to the metal center in a planar fashion, d-orbitals are stereoelectronically and electrostatically differentially stabilized. Though traditional CFT provides a simple comprehension of these processes, newer computational methods are expanding our comprehension of more complex square planar systems, leading to more advanced models that include thoughts on ligand-field splitting and overall electronic structure and therefore are forging new paths to comprehend coordination chemistry (Eisenberg et al., 2011). In the case of the dithiolene complexes, defined by their square planar shapes, both ancient and recent studies have shed light on the nuances of electronic interaction in such structures. The pioneering work done on dithiolenes in the early 1960s served as a stepping stone for an appreciation of how metal-ligand interaction can lead to intriguing redox properties and intense coloration in solution. This is partly because of the extensive degree of ligand-metal mixing that occurs within frontier orbitals of such complexes, which means that oxidation and reduction processes could nearly completely be attributed to the ligand rather than the metal center itself. Besides, recent advances in computational techniques, such as DFT+U methods, are beginning to bridge the gap between calculation and experimental observations, particularly for hydrated metal-ligand complexes wherein abnormal behaviors, such as ligand displacement, may play a significant role in the overall chemical environment (Medforth et al., 2007) (Figure 3.6).

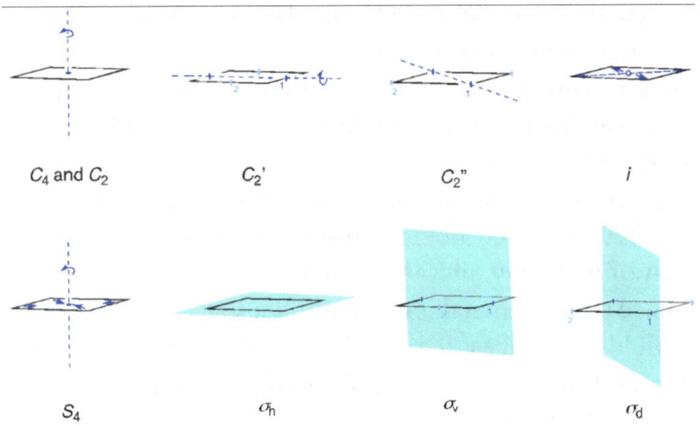

Figure 3.6. *Square planar splitting.*

Source: https://files.mtstatic.com/site_4334/405060/0?Expires=1756729484&Signature=Y-0YoSLAQwWkX4367W0bvQmsxbKkakXts2TELSI5ypf1LmT3YjSTcjPbEc2d~6itF278osalCsKJCI3fN4Xz6ZxTFdx49ctPONO~kGztIF-xcy6FcY4Ce-NEWr59SgsP9bgt1DJbfjlFasdtUOf7qseSQ2Sy7Ao-tW32Dho7g20c_&Key-Pair-Id=APKAJ5Y6AV4GI7A555NA.

3.3. FACTORS AFFECTING CRYSTAL FIELD SPLITTING

The phenomenon of crystal field splitting is greatly influenced by several important parameters, each contributing to the overall energy distribution of electrons in a transition metal complex. One of the major determinants is the nature of ligands that encircle the metal center, varying according to their ability to generate an electric field. Strong-field ligands like CO and CN^- tend to split more by having an increased influence on the metal cation's d-orbitals. Weak-field ligands like H_2O and F^- lead to less splitting. This variation in ligand strength dictates whether or not a complex will be high-spin or low-spin and thus affect its magnetic and spectroscopic features. Additionally, the geometry of ligand arrangement—octahedral, tetrahedral, or square planar—also has an effect on the magnitude and nature of crystal field splitting. Besides, the metal cations' identity plays a great role in crystal field splitting. Transition metals possess more than one oxidation state and different electron configurations, thereby affecting how d-orbitals respond to ligands. For instance, the presence of erbium ions within the crystal lattice in $SrEr_2O_4$ shows that different lattice sites create different crystal field parameters. From reports, neutron spectroscopy results show that the erbium ions occupy magnetically inequivalent sites, which influence their resulting crystal field levels. This inequivalence gives rise to the complexity of interaction among split d-orbitals and magnetic moments, which is indicative of the central role that the specificity of the metal has in establishing the properties that are realized in the material. Such inherent features have a basic function to play in the description of magnetic behavior and electronic states. Furthermore, environmental conditions, like temperature and applied fields, have a large effect on crystal field splitting. When the temperature is variable, thermal motion of ions may alter the ligand field strength and geometry of the ligands and hence alter the metal ions' electronic states. For instance, employing the external fields produces detectable variations in crystalline materials' bandgap width and defect modes localization, as shown in recent studies on metallic nanoball-filled cholesteric liquid crystals. Such manipulations not only confirm the theoretical explanation of crystal field theory but also provide a practical way to design material properties for specific applications. In effect, awareness of these interactive dynamics allows the designing of materials with particular applications in magnetism and electronics and sheds light on the multifaceted nature of crystal field splitting.

3.3.1. Nature of the Metal Ion

The electronic nature of the metal ion plays a crucial role in establishing the application and understanding of Crystal Field Theory (CFT). Metal ions, particularly in transition metals, possess characteristic electronic structures that play a major role in their chemical reactivity and interactions with ligands. The ionic size and ionic charge, along with oxidation states, decide how they modify the surrounding electric field produced by coordinated ligands. For instance, a divalent metal ion such as copper(II) has different coordination geometry and interaction strength in comparison to the trivalent one, copper(III). The difference affects the crystal field splitting, which is responsible for an instant influence on d-orbital energy levels. Thus, being aware of the metal ions nature is of key significance in order to foretell electronic transitions and consequently the optical and thermal properties of complexes established within coordination chemistry. Moreover, the variable metal ion oxidation states in different complexes add another dimension of complexity to their behavior in CFT. Transition metals readily display several alternate oxidation states, altering their d-electron configurations as well as their magnetic as well as spectral characteristics. This is where the concept of small polarons comes into action, as self-trapping of polarons depends on the unique character of these metal ions, which affects their interaction with the lattice and other ions (Shluger et al., 2002). For example, a transformation from a large polaron towards a more localized small polaron can control charge transport characteristics in semiconducting materials. The ability of transition metals to form different coordination complexes can produce a range of physical properties, highlighting the central nature of the metal ion in describing the principles behind CFT. In addition, the coordination between metal ions and the coordinating ligands results in drastic changes in physical properties, such as conductivity and magnetism. CFT predicts that the metal ion geometry will control the ligand geometrical pattern, and this is also upheld in research exploring semiconductor effects. Only recently has it been shown that it is possible to convert materials between insulators and metals with record thermoelectric activity by changing metal ion configurations within porous systems, such as $SrTiO_3$ (Brinkman et al., 2010). This observation points to the importance of the properties of metal ions in the present usage from energy storage to electronics. The intrinsic relationship between metal ions and their crystalline environment not only enriches our understanding of CFT but also opens new avenues for innovation in material science (Figure 3.7).

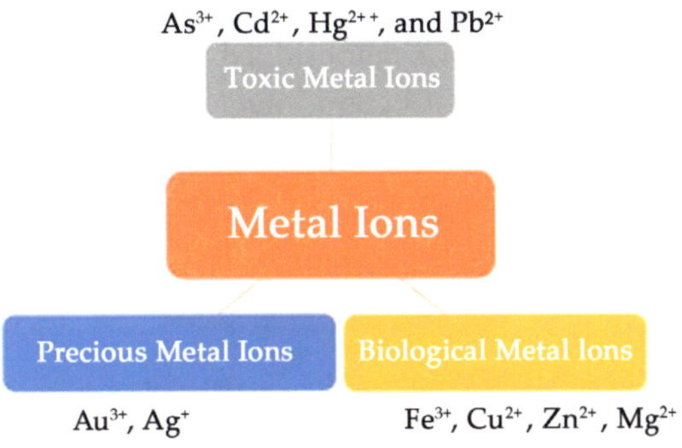

Figure 3.7. *The nature of the metal ion.*

Source: https://ars.els-cdn.com/content/image/3-s2.0-B9780323983501000256-f08-02-9780323983501.jpg.

3.3.2. Type of Ligands Involved

In Crystal Field Theory (CFT), the nature of ligands involved plays a significant role in defining electronic arrangements and properties within transition metal complexes. Ligands are classified as strong field and weak field based on their ability to cause splitting of d-orbitals in transition metals. Good field ligands such as CN^- and CO produce significant splitting among lower and higher energy d-orbitals, typically due to low-spin configurations. Weak field ligands, such as halides Cl^- and Br^-, produce smaller splitting and are likely to induce high-spin configurations. This is key to determining magnetic properties and color exhibited by complexes and hence their applications in a range of areas from catalysis to material science. The ligand's chemical structure determines its ability to coordinate at metal centers directly. Polydentate and bidentate ligands like those in coordination complexes coordinate at more than one point at a central metal atom, enhancing stability and affecting geometric configurations. For instance, in manganese complicated studies, ligands such as $\{H_2L_1\}-$ and $\{H_2L_2\}3-$ have been employed to create coordination assemblies that exhibit unique magnetic properties depending on their constituent binding organizations. These ligands facilitate the generation of complex architectures, as evident from the generation of complexes with slow magnetic relaxation and high intramolecular interaction, demonstrating how ligand selection can affect complex behavior and activity (Chattopadhyay et al., 2017). In addition, ligand development has enabled heterobimetallic complexes that involve different metal ions, further enhancing coordination chemistry flexibility. The low-dimensional forms of the ligand self-assembly are capable of giving rise to sophisticated magnetic systems. For example, introduction of both octacyanotungstate(V) and anisotropic Mn(III) within a cyanobridged compound resulted in various magnetic behaviors depending on ligand interactions and organization. These developments point to the importance of being aware of ligand species in CFT. By arranging ligand structures, chemists can now design materials with specified magnetic properties, which creates new opportunities in technology and materials science (Majcher et al., 2017).

3.3.3. Coordination Number and Geometry

The coordination number and geometry idea is one of the things that must be enjoyed in the way ligands interact metal ions under Crystal Field Theory (CFT). Coordination number refers to the coordination number of the donor ligand atoms around a central metal ion, an influence that greatly determines the spatial arrangement of the ligands around the ion. The resulting geometries from varying coordination numbers are octahedral, tetrahedral, and square planar, all of which possess varying spatial orientations that exert varying influences on the electronic properties of the complexes. These geometrical configurations are not a decoration; they play a pivotal role in determining the magnetic and spectroscopic properties of the complexes, highlighting the role of molecular structure in predicting chemical reactivity and stability. In CFT, the geometry taken up by the coordination number leads to the splitting of the d-orbitals of the metal ions, and this significantly impacts the electronic transitions that are able to occur within the complex. For instance, in an octahedral field, the d-orbitals experience a specific energy level change due to ligand orientation, leading to a characteristic crystal field splitting pattern. This splitting is essential to explain the colors displayed by transition metal complexes, as well as their magnetic properties, which can be highly different from one geometrically distinct configuration to another. This coordination geometry effect is also seen in compounds like $BiFeO_3$, where ordering effects and phase transitions are well-related to the stereochemical activity of lone pairs within the ligands, pointing out how coordination number, geometry, and electronic behavior are interdependent in CFT (Atanasov et al., 2011). Besides, coordination environment also controls metal center interactions among themselves in complex systems such as Ni(II) ion-containing complexes. Coordination geometry here results in intermolecular magnetic coupling that can alter collective magnetic behavior of the material. As is the case for isostructural Ni(II) complexes, square planar geometrical arrangement leads to characteristic magnetic properties that are an outcome of intramolecular and intermolecular interactions. The effective spin Hamiltonian approach used to describe such systems suggests the manner in which geometry not only specifies coordination number but also anisotropic magnetic properties arising from such geometries (Cornia et al., 2003). Such insight enhances the depth of understanding of how molecular structure is connected to property, substantiating once more the relevance of incorporating coordination number and geometry within the phenomenology of Crystal Field Theory (Figure 3.8).

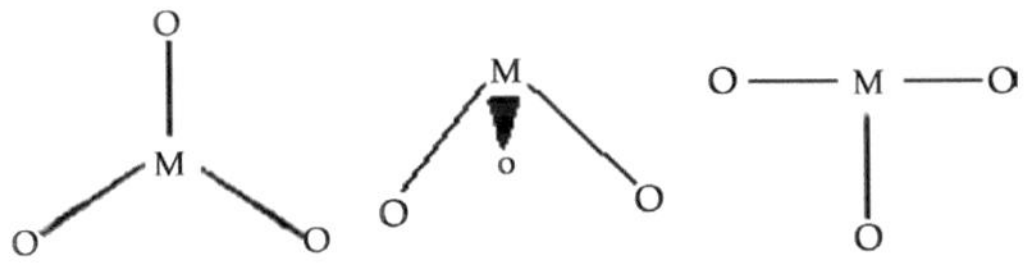

Figure 3.8. *The concept of coordination number and geometry.*

Source: https://encrypted-tbn0.gstatic.com/images?q=tbn:ANd9GcR-iafrmU-lDmusSz61UbQGCD8ZIqw9uhpNAg&s.

3.4. CRYSTAL FIELD STABILIZATION ENERGY (CFSE)

Definition of CFSE is crucial in explaining the electronic structure of transition metal complexes based on Crystal Field Theory (CFT). CFSE is employed to measure the energy stabilization that occurs when d-orbitals are split by ligands. When ligands approach a central metal ion, they generate an electric field which acts upon degenerate d-orbitals to produce a large splitting based on the ligand geometric structure. In octahedral complexes, for example, the d-orbitals split up into two groups of orbitals: the lower t2g and higher eg orbitals. The energy difference between these states governs the overall stability of the complex and is pivotal in the prediction of the electronic arrangements and magnetic properties of the resultant compounds, hence making CFSE an essential application in inorganic chemistry.

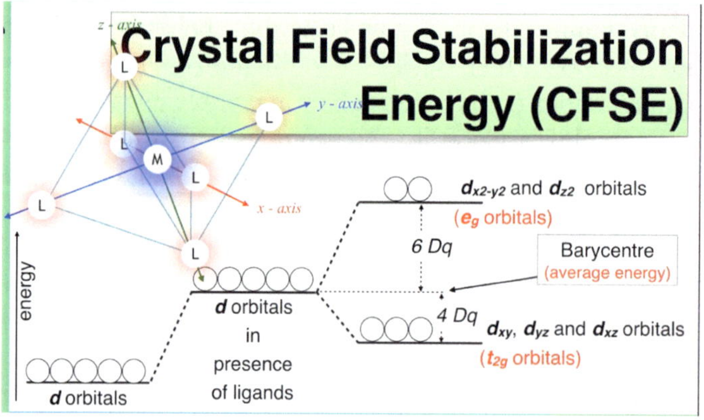

Figure 3.9. *The concept of crystal field stabilization energy (CFSE).*

Source: https://i.ytimg.com/vi/Q8COZ8hC0vE/maxresdefault.jpg.

CFSE not only influences the stability of transition metal complexes but also has significant impacts on their reactivity and color. Because ligands vary in their ability to produce electrostatic interactions—depending on where they are positioned on the spectrochemical series—different ligands yield differing CFSE values. Strong field ligands such as CN^- and CO produce greater splitting and higher CFSE than weaker field ligands such as H_2O and Cl^-. This variation actually affects the absorption spectrum of these complexes, thus their color; i.e., a complex with significant CFSE would absorb certain wavelengths of light to reflect or transmit, thereby imparting them with distinct colors. The connection between CFSE and these observable properties illustrates the application of CFT in predicting and explaining chemical behavior. Recent advances show how computational methods develop our understanding of CFSE in complex systems. As a case in point, studies on layered compounds like [CuCl]LaNb(2)O(7) show multifaceted connections between electronic states and crystal structure, determining the way in which Jahn-Teller distortion can affect CFSE. In such cases, d-orbital interactions are further complicated by structural factors and multi-dimensional spin models, demonstrating the dynamic nature of transition metal complexes (Rosner et al., 2009). Such high-level computational simulations allow for deeper interpretation of CFSE by means of modeling these interactions and stabilization effects prediction based upon various ligand environments. As our knowledge of CFSE continues to evolve through both theoretical and experimental approaches, it continues to remain at the center of CFT, tying the mathematical idealization of electronic theory to actual chemical reality, such as redox behavior and ligand field effects (Figure 3.9).

3.4.1. Definition and Significance of Crystal Field Stabilization Energy (CFSE)

The concept of CFSE is at the heart of explaining electronic interaction in transition metal complexes as developed by Crystal Field Theory (CFT). CFSE is the energy that comes about with d-orbital electron splitting whenever there is an external electric field supplied by ligands. This energy difference is quantitatively defined in terms of the relative spatial disposition of the ligands about the inner metal ion that dictates the relative d-orbital energies. As ligands approach a metal ion, they can cause splitting of the degenerate d-orbitals into separate energy levels. The size of this splitting, and thus the CFSE, depends on factors such as ligand identity, geometry, and oxidation state of the metal sites. CFSE is crucial in predicting the stability and reactivity of coordination compounds, and acquiring knowledge of their magnetic, spectroscopic, and chemical properties. CFSE is not just theoretical but makes its way into practical uses in various areas such as catalysis, materials science, and bioinorganic chemistry. Increased values of CFSE are usually linked with increased stability of specific geometries and oxidation states in metal complexes and thus affect reaction pathways as well as catalytic efficiency. For instance, in catalysis by transition metals, optimal CFSE can enhance electron transfer reactions and thus support elevated reaction rates and selectivity. Secondly, as demonstrated in the research of heteropoly structure, knowledge of CFSE can be employed to synthesize and design complicated materials with target properties, which can be used in such fields as sensors and energy storage devices (Simmons et al., 1963). Thus, knowledge of CFSE is not merely crucial for theoretical purposes but also for developing new chemical technology and materials engineering solutions. In recent advancements, the interplay between CFSE and compositional and environmental complexity has gained more significance, particularly in high-entropy ceramics. The long-range effects of CFSE have been viewed in the stabilization of distinct phases of complex oxides, where structural configuration competition exemplifies the need to balance CFSE with entropy effects (Gaboardi et al., 2024). As scientists prepare compositionally sophisticated materials, the ability to forecast and manipulate CFSE becomes increasingly vital. CFSE control allows chemists to take advantage of properties that can enhance thermal stability, catalysis efficacy, and overall performance in practical applications. The definition and significance of CFSE thus become closely linked with the present advancements in chemical sciences, with an opportunity for subsequent generations of materials and technologies to emerge.

3.4.2. Calculation of Crystal Field Stabilization Energy (CFSE) for Different Geometries

The calculation of CFSE is important in the determination of electronic properties of transition metal complexes using the Crystal Field Theory (CFT) platform. The CFSE results due to the interaction of d-orbitals of transition metal cations with electric fields generated by ligand surroundings. Different geometries, such as octahedral, tetrahedral, and square planar arrangements, produce typical patterns of d-orbital splitting, which have a particular influence on the energy calculations of CFSE. In the case of octahedral complexes, for instance, the d-orbital energies are split into two levels: the lower t2g energy and the higher eg orbitals, so it is straightforward to calculate using the number of electrons occupying these orbitals. Tetrahedral complexes exhibit the reverse splitting pattern, with a lower CFSE than octahedral complexes for a metal ion and ligand field strength. The method adopted in calculating CFSE for different geometries demands thorough knowledge of ligand field theory and symmetry. For octahedral complexes, CFSE can be computed with the help of the formula CFSE = (number of t2g electrons $\times$ -0.4 Δo) + (number of eg electrons $\times$ 0.6 Δo), where Δo stands for the crystal field splitting energy. Conversely, tetrahedral complexes formula is an electron contribution reversal due to opposed energy levels, so it is CFSE = (number of e electrons $\times$ -0.4 Δt) + (number of t2 electrons $\times$ 0.6 Δt), where Δt is tetrahedral splitting. Understanding these calculations, more than knowing electronic structures, allows for predictions of chemical behavior across various milieus, illustrating the utilitarian application of CFT in synthetic and biological milieus. More than calculation in principle, CFSE implications extend to actual metalloprotein dynamics and functional roles of metal ions. For example, metalloprotein coordination geometry, for example, octahedral or tetrahedral geometry, determines metal ion participation in enzymatic processes and other biological functions. The local spatiotemporal geometry of the ligands surrounding the metal site can have a profound effect on electron distribution, which further determines CFSE. Adopting the principle of novelty, more recent work has highlighted the importance of accurate determination of coordination geometry to gain further understanding on how metalloproteins perform their function. This demand matches constraints in existing models, pointing to the importance of geometric ideas as established in literature to metal ion coordination and its implication to biological systems (Bowen et al., 2012; Yao et al., 2016). Lastly, correct CFSE calculations for different geometries provide insightful findings on the delicate relationship between structure and function in transition metal chemistry (Figure 3.10).

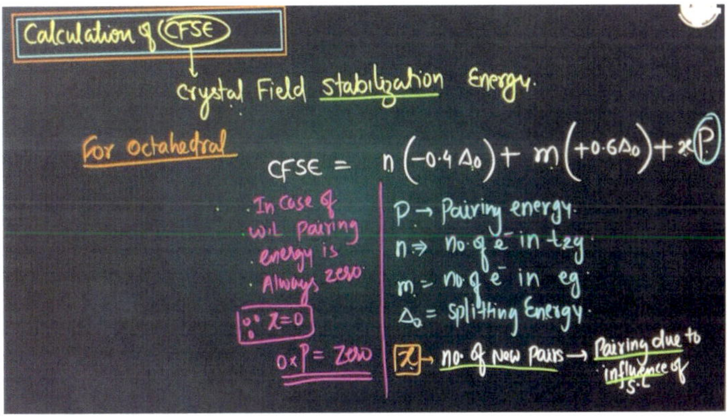

Figure 3.10. *The calculation of crystal field stabilization energy (CFSE).*

Source: https://classroom-images.cdn.askfilo.com/classroom/1727358560865_lcyxfpci_4664533.jpg.

3.4.3. Role of Crystal Field Stabilization Energy (CFSE) in Determining Stability of Complexes

The CFSE theory is central to the investigation of the stability of metal complexes within the context of Crystal Field Theory (CFT). CFSE arises as a result of interactions between the d-electrons of transition metal ions with the electric field created by the surrounding ligands. Since the coordination geometry in a center metal ion changes d-orbitals energy levels, the alteration in CFSE directly influences the thermodynamic stability of these complexes. Complexes with a higher CFSE are generally favored as they denote greater energy stabilization. Therefore, the level of this stabilization not only determines the architecture of the complex but also its reactivity and overall behavior in various environments that are relevant to biological and industrial applications. Stability in metal-ligand complexes is further dependent on the geometry of the coordination environment, which is intrinsically related to CFSE. For example, octahedral and tetrahedral complexes will possess unequal CFSE values due to the variation in the spatial structure of their constituent ligands. In octahedral complexes, splitting of d-orbitals results in a greater stabilization effect compared to the case of tetrahedral complexes, where the orbital energy differences are less. This variation is essential in defining the preferred geometry of coordination favored by various ligands, as certain ligands opt for one geometry over another owing to their natural electronic nature. For instance, the chelating ligand behavior of methimazole for some zinc complexes demonstrates the impact of changes in ligand coordination on CFSE, which in turn enhances stability and leads to complexes like [Zn(MeImS)(2)] with new properties and reactivity patterns (Aragoni et al., 2015). Furthermore, the role of CFSE in determining complex stability also applies to the prediction of catalytic activity and selectivity in chemical reactions. In the context of transition metal catalysis, the electronic structure at metal centers as determined by CFSE plays a significant role in determining the adsorption bond strength of reactants and the catalytic efficiency in general. Electronic factors governing the process, such as bond lengths and bond angles, are deeply rooted in CFSE of the complexs geometry and interactions between ligands. This is a critical insight in material development for applications like oxygen evolution and reduction reactions, where catalysts consisting of transition metals have to weigh reactivity against stability (Grimaud et al., 2014). Tracking CFSE allows researchers to better design and tailor catalysts and eventually to further improve sustainable energy technology.

3.5. SPECTROCHEMICAL SERIES

Spectrochemical Series is a critical component of Crystal Field Theory (CFT) that depicts the variations in the abilities of ligands to control the electronic structure of the transition metal complexes. It orders ligands according to the energy gaps they create in the metal ion's d-orbitals. The series depends from strong field ligands such as CN^- and CO, which cause extensive d-orbital splitting, to weak field ligands such as I^-, which cause zero splitting. The series not only determines the transition metal complex colors based on changes in the absorption spectrum but also determines the metal ion spin states. It is important to understand the Spectrochemical Series in order to predict the stability, geometry, and reactivity of precipitated complexes using various ligands because these properties are intrinsically linked with metal-ligand electronic transitions. It is possible for scientists to make predictions about the behavior of coordination compounds under certain conditions using the Spectrochemical Series. For instance, the influence of ligand identity on electronic properties of transition metal complexes affects practical applications in a variety of fields ranging from catalysis to material science. Ligands that induce strong field conditions enhance low spin configuration stability, while weak field ligands prefer high spin states. The phenomenon is not only the case for static systems but also under dynamic conditions. For example, external stimuli such as pressure can influence the spin state of a transition metal complex depending on the position of the ligand within the spectrochemical series. It is shown in a study that high pressure is required to induce a spin transition in octahedral complexes, which signifies the importance of the ligand strength in determining electronic structures under various conditions (Chakraborty et al., 2019). With the help of sophisticated theoretical methods, such as the Partial Density of States (PDOS) analysis, a more precise understanding of the Energetics as observed in the Spectrochemical Series can be obtained. The PDOS analysis provides essential details about the interaction between metal orbitals and ligand orbitals, producing an effective description of metal orbital energies and coordinating ligand energies. Success of such models is also observed in the determination of the Spectrochemical Series itself, which can now be employed to describe the light absorption characteristics of various complexes with accuracy. Furthermore, insights derived from ligand-d-orbital energy relationships may be employed to predict luminescence lifetimes, hence bridging theory and laboratory practice in the realm of coordination chemistry (Amolo et al., 2017) (Figure 3.11).

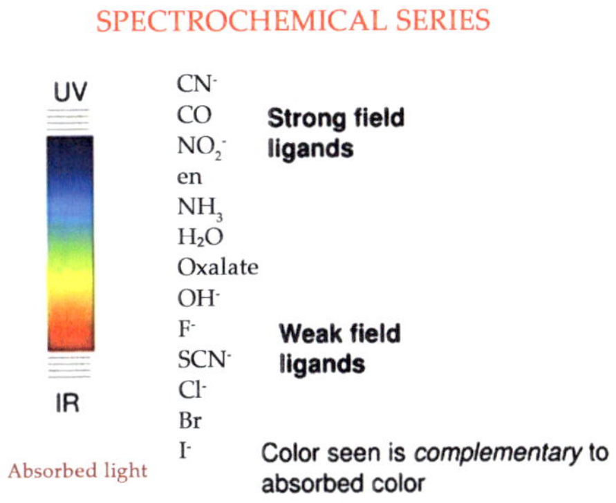

Figure 3.11. *The spectrochemical series.*

Source: https://d13loartjoclyn.cloudfront.net/article/1690460247_What%20is%20a%20Spectrochemical%20Series.jpg.

3.5.1. Explanation of the Spectrochemical Series

The spectrochemical series is a highly significant term in the Crystal Field Theory (CFT) model, as it categorizes ligands based on their ability to split the d-orbitals of transition metal complexes. This is due to the electrostatic interactions occurring between the ligand and metal cation, leading to the d-orbitals possessing varying energy levels. Ligands which create a powerful field environment, i.e., CO and CN^-, are at the top of the series, leading to larger energy separations between t_2g and e_g orbitals. Weak field ligands such as Br^- and I^- produce minimal splitting, leading to smaller energy gaps. Hence, ligand order in the spectrochemical series influences the electronic, optical, and magnetic properties of coordination compounds directly and has a significant role to play in determining whether a complex will be paramagnetic or diamagnetic. The relationship between the spectrochemical series and the features of transition metal complexes is underscored by knowledge of Ligand Field Theory (LFT) and its development into more complex approaches such as Partial Density of States (PDOS) analyses. Researchers utilize PDOS to modify parameters used in constructing the spectrochemical series through an examination of the electronic structures of various permutations of metal and ligand. This analysis provides insight into metal t_2g and e_g orbital energies and ligand orbital contributions, allowing for an advanced understanding of light absorption and emission behavior. Furthermore, the energy differences which are implied by PDOS-LFT not only create the spectrochemical series but also allow for correlation with luminescence lifetimes, allowing a bridge to be formed between theoretical expectation and experimental observation (Amolo et al., 2017). As coordination chemistry goes on evolving, ongoing research of the spectrochemical series allows chemists to anticipate and design new complexes for specific purposes, i.e., catalysis or photonic materials. Refining computational techniques has complemented our understanding toward the behavior of ligands, as described in recent reviews toward the mechanistic exploration of ligand substitution reactions and the rules governing them. This knowledge is useful because chemists try to access the unique properties brought about by diversified ligand environments, which lead to the development of new synthetic paths. This growing body of literature emphasizes the significance of including molecular orbital theory in CFT considerations since it describes a full picture for understanding the dynamics of ligand interactions, rate processes, and the resulting effects on the overall properties of transition metal complexes (Gray et al., 1966).

3.5.2. Relationship Between Ligand Strength and Splitting

The fundamental concept in Crystal Field Theory (CFT) directly relates to ligand strength and the subsequent splitting of the d-orbitals in transition metal complexes. Ligands are either weak or strong and play a significant role in affecting the energy levels of the d-orbitals. They are good field ligands that lead to more splitting of the d-orbitals by being able to exhibit strong σ-donation and π-acceptance interactions. Weak field ligands such as I^- and Br^- are responsible for weaker splitting effects, and this can be largely attributed to the reduced electronegativity and lesser bonding capacities of these ligands. This variation in ligand strength controls the electronic structure, color, and magnetic properties of the resulting complexes, a paramount aspect to consider in transition metal chemistry.

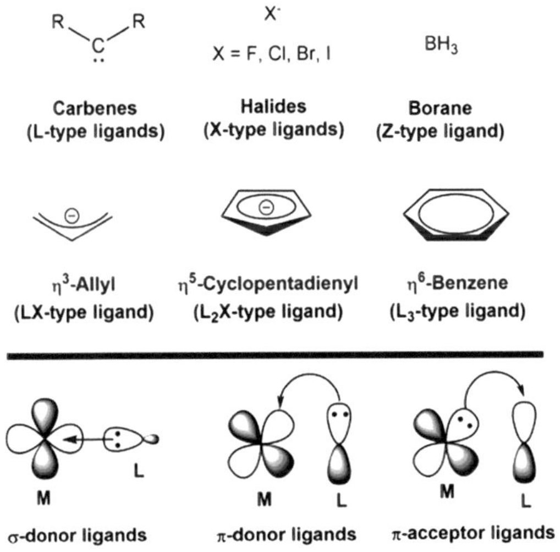

Figure 3.12. *Relationship between ligand strength and splitting.*

Source: https://ars.els-cdn.com/content/image/3-s2.0-B9780128202067000883-f00088-02-9780128202067.jpg.

The consideration of ligand strength also involves consideration of the spectrochemical series, which ranks ligands on the basis of their field strengths. This series is a prime indicator of the magnitude of splitting experienced by d-orbitals in various complexes. For example, the series tells us that ligands such as NH_3 and en (ethylenediamine) are positioned above water, suggesting a stronger influence on orbital splitting than aqueous ligands. Research demonstrates that the varying strengths of ligands are represented in the resulting energy gaps between split d-orbitals, which in turn impact characteristics such as color and magnetism. These interactions highlight the complex interplay of ligand interactions and geometric metal cation organization to give rise to characteristic metal complex properties in coordination chemistry. Additional probing of the specific mechanisms of ligand influence on orbital splitting reveals additional complexities, particularly with respect to the nature of the metal center and coordination geometry. For instance, d-orbital splitting is quite varied in octahedral complexes compared to tetrahedral complexes, highlighting the importance of ligand as well as the site of coordination. Such interactions have been clarified through recent application of density-functional theory (DFT) to transition metal complexes. In studies of ruthenium(II) polypyridine complexes, it was found that while traditional ligand field theory is a simplified explanation, partial density-of-states analysis enables a better comprehension of the variations in splitting energies, consistent with observations of ligand behavior in the spectrochemical series (Amolo et al., 2017). This combined approach highlights both the significance of ligand identity and geometry in electronic transition prediction and in absorption spectra of coordination compounds (Figure 3.12).

3.5.3. Implications for Color and Electronic Transitions

Implications of color and electronic transitions according to Crystal Field Theory (CFT) are closely linked with the d-orbital arrangement in transition metal complexes. The theory describes how the geometrical arrangement of the ligands around a central metal ion is the reason for altering the energy levels of d-orbitals, leading to special electronic configurations. The different amounts of splitting of the d-orbitals result from this reorganization and also influence the absorption of provided wavelengths of light. When light is encountered by such complexes, some wavelengths are absorbed, while others are reflected or emitted, thus forming the color observed. Ligand strength and geometry differences can result in the varied electronic transition probabilities, which play an important role in controlling the color the substance looks. This phenomenon is crucial to understanding the color characteristics of many transition metal complexes that are of utility in industries ranging from pigments to catalysis. The electronic transitions even expand the understanding of the color-changing metal complex, particularly under varying external conditions such as pressure or temperature. Studies have shown that changes in these physical properties can lead to significant changes in the crystal field environment, thereby influencing the d-orbital splitting and electronic interactions. For example, in manganese monoxide, pressure increases the bare Coulomb interaction of certain orbitals, a non-intuitive effect caused by the delocalization of Wannier functions. This change in electronic structure predicts that the color emission of such materials changes under altered physical conditions, opening up new possibilities for functional material design. These pressure-induced electronic transitions demonstrate the dynamic interaction between optical characteristics and structural integrity, confirming the relevance of knowing these interactions in the CFT model (Aryasetiawan et al., 2010). Furthermore, information concerning crystal structures, such as presented in $(CuCl)LaNb_2O_7$, clarifies the atomic order/electronic transitions relationship that is responsible for coloration. The (CuCl)$LaNb_2O_7$ polymorphs with their richness demonstrate the way structural changes bring about order-disorder phase transitions, which in turn influence magnetic and electronic properties. Methods like high-resolution synchrotron powder x-ray diffraction, among others, inform us about the precise crystal structures that control d-orbital activity in such materials. As ion movement and displacement, tilting of NbO_6 octahedra, are influenced when the structural environment shifts, electronic transition patterns may be radically changed. Therefore, understanding these transitions and how they translate with respect to color is not only expanding on the foundational knowledge of CFT but also bridging the way for altering materials with unique optical properties for the purposes of applications in electronics and photonics (Tsirlin et al., 2010).

3.6. APPLICATIONS OF CRYSTAL FIELD THEORY (CFT)

Applications of CFT cross many fields, but perhaps most significantly with respect to understanding the electronic properties of transition metal complexes. CFT assists in explaining how ligand arrangements around a metal center can influence its electronic structures and properties. For instance, the shape of a complex—octahedral, tetrahedral, or square planar—can have a very significant effect on the d-orbital energy levels involved. This is particularly useful in fields such as coordination chemistry, where CFT is utilized to predict stability and reactivity of metal complexes. By calculating crystal field splitting energies, chemists are able to make informed decisions in ligand selection and eventually control the synthesis of complex materials with programmed functions in catalysis and electronics. CFT is also very important in solid-state chemistry, especially in the study of metal-organic frameworks (MOFs) and their properties. As the chemistry of the d-electrons and ligands controls the electronic and structural aspects of these materials, CFT provides the theoretical framework for predicting and understanding it. The methodologies based on CFT are found to be useful to predict the material's crystal structures and understand their thermal and mechanical behavior. For example, accurate structural predictions of crystals from ligand binding geometries can lead to better porous material designs used in gas storage or gas separation that are critical for energy and environmental applications. The power to drive structural prediction is a significant step towards harnessing CFT for modern materials science. Additionally, among the uses of CFT are the study of strongly correlated systems, where the electron-electron interactions prevail. Studying the electrostatic interactions stabilized by strong d-electron correlations, CFT elucidates the stability and phase transitions of such systems. This is valuable for developing systems with their own unique physical properties, such as superconductivity or exotic magnetic behavior. For example, the charge transfer processes due to the energetic reconstruction of d-orbitals under varying conditions are essential in the development of new materials with desirable functionalities. Thus, the inclusion of CFT in existing computational methods provides possibilities for new approaches to material design, and one can achieve a total framework for predicting and ascertaining novel phases in strongly correlated systems (Coudert et al., 2015; Dobrosavljević et al., 2018) (Figure 3.13).

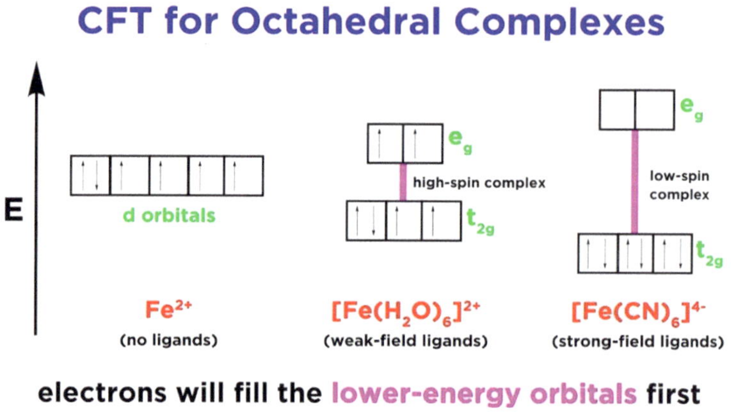

Figure 3.13. *The applications of crystal field theory (CFT).*

Source: https://galaxy.ai/_next/image?url=https%3A%2F%2Fimg.youtube.com%2Fvi%2F8lT21wKoXyQ%2Fmaxresdefault.jpg&w=3840&q=75.

3.6.1. Predicting Magnetic Properties of Complexes

Knowing how to forecast magnetic properties in coordination complexes is necessary in the use of Crystal Field Theory (CFT). CFT provides a framework explaining how ligand geometry around a metal ion determines its d-orbital splitting, and therefore magnetic characteristics. In transition metal complexes, the configuration of electrons in d-orbitals is crucial in determining whether a complex is paramagnetic or diamagnetic. A magnetic moment, which is a measure of unpaired electrons, can be estimated from quantum numbers from the specific coordination sphere of the metal. To be able to properly predict such properties computationally, computational methods are nowadays a part of ongoing research to better understand the electronic processes that result in magnetism in complexes through the simulation of the various ligand geometries and how these correspond to d-orbital population. Diverse computation techniques are now utilized in enhancing predictions of magnetic properties, namely the use of Density Functional Theory (DFT). DFT has been a key factor in the modeling of transition metal complex electronic structure, allowing researchers to investigate refined differences in magnetic moments due to varying ligand field strengths and geometries. Of special importance has been the DFT+U approach, which has been successful in reproducing the complex interactions between transition metal and ligand, particularly in spin state interpretation. For instance, detailed investigations of manganese porphine complexes have demonstrated how binding interactions with substrates like Au(111) can dramatically alter the spin state of the metal ion, and revealed that even those subtle alterations are vital for magnetic property predictions in practical applications. By combining theoretical expectation and experimental verification, DFT methods provide an increasingly accurate picture of the magnetic landscape of coordination compounds, constructing a more nuanced understanding of how molecular organization maps to magnetic behavior. Research on coordination complexes' magnetism goes beyond theoretical models and enters into applications, commanding significant impact across as varied a range as materials science and biomedicine. For example, accurate prediction of the magnetic behavior of complex systems is essential in the synthesis of catalysts and magnetic materials for technological applications. In addition, in the medical sector, understanding the magnetic properties of metal complexes facilitates the design of diagnostic and therapeutic agents where networks of biomolecule interaction are simulated and characterized in order to predict how the agents affect biological processes. Combining computational methods with experimental strategies, the ultimate value of CFT in the exploration of magnetic behaviors highlights the role of interdisciplinary strategies. Such developments will lay the foundation for optimizing drug design and personalized medicine protocols, thereby making the forecasting of the magnetic properties of complexes germane to the advances of modern scientific endeavors.

3.6.2. Understanding Color in Transition Metal Complexes

The colorful appearance of transition metal complexes is primarily a result of electronic transitions in d-orbitals, and this phenomenon is basically accounted for by Crystal Field Theory (CFT). As per the theory, ligands cause a perturbation in d-orbital energy levels such that d-orbital degeneracy is split. In octahedral complexes, for example, the interaction of metal ions with surrounding ligands produces two distinct levels of energy, the t2g low-energy and eg high-energy orbitals. The actual wavelengths of light being absorbed during electronic transitions between the two levels determine the perceived color. Because ligands vary in strength—from weak field ligands that cause minimal splitting to strong field ligands that create more splitting travel—a color variation that can be quite dramatic can occur, illustrating the direct relationship between the environment of the ligand and the hue perceived by the eye in transition metal complexes. The geometry of the orientations of the ligands around the metal center also determines the color characteristics of these complexes. Different geometries, such as octahedral, tetrahedral, or square planar, cause different patterns of d-orbital splitting, and it is this pattern of splitting that ultimately controls what specific wavelengths of light are absorbed.

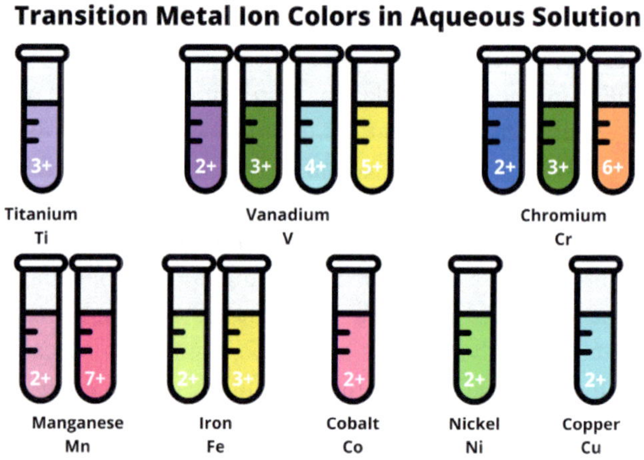

Figure 3.14. *Understanding color in transition metal complexes.*

Source: https://sciencenotes.org/wp-content/uploads/2020/09/Transition-Metal-Ion-Colors-in-Aqueous-Solution.png.

For example, a tetrahedral complex will have a distinct pattern of splitting compared to an octahedral complex depending on the position of the ligands with respect to the d-orbitals. Such diversity of geometric structure not only contributes diversity in color in transition metal complexes but also influences their electronic and magnetic properties. Hence, understanding the influence of coordination geometry and properties of ligands on color can provide an insight into the broader application of these complexes in catalysis, material science, etc. Moreover, magnetic coupling among transition metal complexes is also of utmost significance to their optical characteristics. Zero-field splitting (ZFS) parameters are at the core of the explanation of the magnetic behaviors of the high-spin states of such complex ions. Belvin et al. (2017) describes how one can directly measure ZFS, thus acquiring crucial information regarding the geometries and electronic surroundings of the metal centers concerned. The ability to directly monitor ZFS parameters sheds further light on how magnetism and color are intertwined in these systems. Such knowledge is not only important to advancing our knowledge of transition metal complex chemistry but also to maximizing their useful applications in areas such as molecular electronics and advanced materials, where control of color and magnetism can result in innovative advancements (Figure 3.14).

3.6.3. Role in Catalysis and Material Science

The connection between crystal field theory (CFT) and its application in catalysis and material science puts electronic structure center stage as a controller of metal ion reactivity. CFT assists in describing how the geometry of ligands surrounding a central metal ion adjusts energies of d-orbitals, thereby influencing catalytic activity. For instance, transition metals normally serve as a catalytic center where the electronic character can be delicately tuned by modifying the ligand environments. This tunable character allows for the creation of highly selective and active catalysts for numerous chemical reactions. Moreover, through the utilization of MOFs that have been designed based on CFT principles, researchers have succeeded in developing more efficient catalysts involving metals in processes like hydrogenation and oxidation, enabling more sustainable industrial processes. Material science also benefits significantly from the know-how acquired through CFT, particularly regarding the stability and performance of materials at the atomic level. The predictive precision of density functional theory (DFT) makes computational material design tailored to a specific application possible, marking a paradigm shift in material discovery. Improvements have been made that can precisely model energy materials, such as lithium-ion battery materials, (Jain et al., 2016). By understanding the role played by crystal field effects in stabilizing materials, researchers can choose promising candidates that have the potential to exceed traditional materials. For example, studies have confirmed that doping certain elements in semiconductor materials has proved to stabilize desired phases and maximize electronic properties, thereby enhancing their usability in various fields of technology. The work of DFT in illuminating the fundamental mechanism of crystal field theory has profound implications in catalysis and higher-order material design. In an example involving research on TiO_2, DFT computations have revealed key information about the stability of different polymorphs, wherein specific dopants played a significant role affecting the material's characteristics (Assadi et al., 2012). These kinds of observations indicate that careful dopant selection can enhance material reactivity and stability, which are crucial for use in sensor technology and photocatalysis. As computational methods continue to evolve, the integration of CFT and DFT will further ease the process of material discovery so that researchers will be able to examine an enormous variety of compositions and structures. Lastly, such a concordance between the theoretical models and practical applications presents prospects for groundbreaking advancements in both catalysis and material science.

SUMMARY

In Conclusion, CFT provides an elementary framework to explain the electronic structure and behavior of transition metal complexes. By emphasizing the role of ligand fields in determining the d-orbital energies, CFT explains the geometrical shapes and electronic patterns of these complexes. This model is also useful in the prediction of magnetic as well as optical properties of numerous coordination compounds as the interaction of the electrostatic forces of the ligands with d-electrons leads to phenomena like magnetism and color, which are observable. High-spin and low-spin work further emphasizes the role of the strength of ligands, further increasing our understanding of complex behavior in coordination chemistry. Overall, CFT is a powerful instrument both in research and application, stimulating advances in materials science and catalysis. Moreover, CFT implications extend beyond simple coordination compounds to more complex systems as exemplified in models for biological iron enzymes. More recent research focuses on the way that the coordination geometries of non-heme iron complexes can be engineered so as to model specific enzymic reactions, as exemplified by investigations involving isotopically labeled species of dioxygen and nitric oxide. This approach encompasses spectroscopic and computational techniques for determining the electronic structures of such models, yielding valida evidence for the role of CFT in bioinorganic chemistry. The results of these models reaffirm the ability of CFT to foretell, particularly in bridging the gap between theoretical studies and chemical behavior in actuality, and better understanding enzymic pathways and reactions. Finally, the contribution of CFT in teaching us about correlations between structure-function in transition metal complexes cannot be more aptly put. Remarkably, the relevance of CFT is also underscored by its generalization to iridium fluorides that have been of current concern via potential superconductivity. Investigations of the proximity of iridium fluorides to the SU(2) symmetric limit provide a prime example of how ideas from CFT can be transferred to new materials in condensed matter physics. These findings point to the richness and diversity of CFT, as it remains a central theme for ongoing research in both materials science and coordination chemistry. The context established through CFT continues to drive diverse explorations, deepening knowledge of electronic interactions and their implications within scientific disciplines.

CFT clarifies the electronic structure of transition metal complexes by explaining how ligand coordination with a metal ion influences its energy levels. It is crystal field splitting-based, wherein the degenerate d-orbitals of the transition metals experience varying energy levels when exposed to an electric field created by ligands surrounding them. This split has a bearing not only on the electronic structure of the metal ions but also on the color, magnetism, and reactivity of the complexes. An understanding of such interactions is helpful in the prediction of the stability of some coordination geometries and the nature of bonding between the ligands and the metal and is a precept for a number of applications in material science and catalysis. Crystal field interactions become also more important in high-finesse materials, such as 5d transition metal oxides iridates. Density functional theory and LDA+U calculations predict a complex electronic structure governed by electron correlations and spin-orbit coupling and resulting in various ground states for distinct parameter ranges. For instance, in $Y_2Ir_2O_7$, the variation of the strength of correlation (U) tunes the system from magnetic metal to Mott insulator, and impressively, to Dirac semi-metal for intermediate U. This crossover demonstrates the non-trivial interplay between electronic phases and magnetic ordering with the richness of the phenomena that can be observed from the underlying principles of CFT, particularly in spin-orbit coupled materials, as detailed in Turner et al. (2010). Apart from the electronic and magnetic features, the usability of CFT to practical applications is also concerned with understanding behavior in liquid crystal devices. Material manipulation in such devices is informed by the theoretical framework of CFT, which is capable of predicting liquid crystal state response to electric field application. As an illustration, bistable liquid crystal behaviors are described using dynamic Landau de Gennes theory to investigate latching mechanisms and their dependence on material parameters and surface topologies. These findings are helpful in optimizing device performance and functions, proving the generalizability of CFT principles beyond the traditional coordination chemistry to fields like optoelectronics. Ongoing researches and advances of these theories continue to stretch the boundaries of materials science and require more studies as in Amos et al. (2010).

REFERENCES

1. Adroja, D. T., Batulin, R. G., Gabbasov, B. F., Gilmutdinov, K., et al. (2015). Magnetic and spectral properties of multi-sublattice oxides $SrY_2O_4:Er^{3+}$ and $SrEr_2O_4$. https://core.ac.uk/download/43597081.pdf (accessed on 13th September 2025).
2. Akbari, M., Dabirian, A., Birks, T. A., Coldren, L. A., Desoer, C. A., Eggleton, B. J., et al. (2005). Propagation of light in photonic crystal fiber devices. http://arxiv.org/abs/physics/0508230 (accessed on 13th September 2025).
3. Akimitsu, J., Amow, G., Arao, M., Bocquet, A. E., Brandow, B. H., Cwik, M., Cyrot, M., et al. (2003). G-type antiferromagnetism and orbital ordering due to the crystal field from the rare-earth ions induced by the $GdFeO_3$-type distortion in $RTiO_3$ with $R = La, Pr, Nd$ and Sm. http://arxiv.org/abs/cond-mat/0312244 (accessed on 13th September 2025).
4. Amolo, G., Casida, M. E., Kituyi, L., Magero, J., et al. (2017). Partial density of states ligand field theory (PDOS-LFT): Recovering an LFT-like picture and application to photo properties of ruthenium (II) polypyridine complexes. http://arxiv.org/abs/1707.03665 (accessed on 13th September 2025).
5. Amos, R. M., Care, C. M., Jones, J. C., Spencer, M., et al. (2010). Zenithal bistable device: Comparison of modeling and experiment. https://core.ac.uk/download/112366.pdf (accessed on 13th September 2025).
6. Aragoni, M. C., Arca, M., Bettoschi, A., Caltagirone, C., Castellano, C., Demartin, F., Isaia, F., et al. (2015). Zinc (II)-methimazole complexes: Synthesis and reactivity. https://core.ac.uk/download/pdf/54592659.pdf (accessed on 13th September 2025).
7. Arita, R., Nakamura, K., & Nohara, Y. (2011). *Ab initio* derivation of correlated superatom model for potassium loaded zeolite A. http://arxiv.org/abs/1111.4815 (accessed on 13th September 2025).
8. Aryasetiawan, F., Zaanen, J., Tomczak, J. M., Methfessel, M., Miyake, T., & Syono, Y. (2010). Realistic many-body models for manganese monoxide under pressure. http://arxiv.org/abs/1006.0565 (accessed on 13th September 2025).
9. Assadi, M. H. N., Hanaor, D. A. H., Li, S., & Sorrell, C., et al. (2012). *Ab initio* study of phase stability in doped TiO_2. http://arxiv.org/abs/1210.7555 (accessed on 13th September 2025).
10. Atanasov, A., Palewicz, A., Sani, A., Udovenko, A. A., Harris, A. B., Gavriliuk, A. G., Gavriliuk, A. G., et al. (2011). Magnetoelectric ordering

of *$BiFeO_3$* from the perspective of crystal chemistry. http://arxiv.org/abs/1005.3102 (accessed on 13th September 2025).

11. Baianu, I. C. (2011). *Food Physical Chemistry and Biophysical Chemistry*. https://core.ac.uk/download/pdf/290434.pdf (accessed on 13th September 2025).
12. Barradas-Bautista, D., Fernández-Recio, J., Pallara, C., Rosell, D., et al. (2018). Structural prediction of protein–protein interactions by docking: Application to biomedical problems. https://core.ac.uk/download/157811597.pdf (accessed on 13th September 2025).
13. Batista, V. S., Chandross, M. E., Leung, K., Medforth, C., et al. (2006). Density functional theory and DFT+U study of transition metal porphines adsorbed on Au (111) surfaces and effects of applied electric fields. http://arxiv.org/abs/cond-mat/0610742 (accessed on 13th September 2025).
14. Belvin, C., Dincă, M., Gedik, N., Li, Y., et al. (2017). Rapid and precise determination of zero-field splitting by terahertz time-domain electron paramagnetic resonance spectroscopy. http://arxiv.org/abs/1702.06613 (accessed on 13th September 2025).
15. Blaha, P., Schwartz, K., Madsen, G. K. H., Kvasnicka, D., Luitz, J., Schwarz, K., Dietl, T., Filippetti, A., Spaldin, N. A., Sanvito, S., Hahn, Th., Itoh, S., Kugel, K. I., Lauer, M., et al. (2005). Ferromagnetism in Fe-substituted spinel semiconductor *$ZnGa_2O_4$*. http://arxiv.org/abs/cond-mat/0412530 (accessed on 13th September 2025).
16. Bol, A. A., Gallagher, D., Norris, D. J., Ito, H., Gumlich, H. E., Yu, I., et al. (2003). Theory of luminescent emission in nanocrystal *ZnS:Mn* with an extra electron. http://arxiv.org/abs/cond-mat/0202358 (accessed on 13th September 2025).
17. Bowen, K. H., Jr., Jena, P., Joseph, J., Jorly, J., et al. (2012). Evolution of superhalogen properties in PtCln clusters. https://core.ac.uk/download/51293307.pdf (accessed on 13th September 2025).
18. Brink, J. van den, Daghofer, M., Gegenwart, P., Singh, Y., et al. (2017). Models and materials for generalized Kitaev magnetism. http://arxiv.org/abs/1706.06113 (accessed on 13th September 2025).
19. Brinkman, A., Ohtomo, A., Zunger, A., Vining, C. B., Van de Walle, C. G., Kan, D., Reagor, D., et al. (2010). Field-induced water electrolysis switches an oxide semiconductor from an insulator to a metal. http://arxiv.org/abs/1011.3681 (accessed on 13th September 2025).
20. Brunold, T. C., Fiedler, A. T., Fischer, A. A., Lindeman, S. V., et al. (2016). Synthesis, X-ray structures, electronic properties, and O_2/NO reactivities of thiol dioxygenase active-site models. https://core.ac.uk/download/213076498.pdf (accessed on 13th September 2025).

21. Chakraborty, R., Head-Gordon, M., & Stauch, T. (2019). Quantum chemical modeling of pressure-induced spin crossover in octahedral metal-ligand complexes. https://core.ac.uk/download/237711153.pdf (accessed on 13th September 2025).
22. Chattopadhyay, K., Craig, G. A., Frontera, A., Heras Ojea, M. Á., et al. (2017). Dangling and hydrolyzed ligand arms in $[Mn_3]$ and $[Mn_6]$ coordination assemblies: Synthesis, characterization, and functional activity. https://core.ac.uk/download/80696931.pdf (accessed on 13th September 2025).
23. Cornia, A., Barra, A. L., Kersting, B., Kortus, J., Taft, K. L., Schwarz, M. P., Kahn, O., et al. (2003). Ferromagnetic coupling and magnetic anisotropy in molecular Ni (II) squares. http://arxiv.org/abs/cond-mat/0301373 (accessed on 13th September 2025).
24. Coudert, F. X., & Fuchs, A. H. (2015). Computational characterization and prediction of metal-organic framework properties. http://arxiv.org/abs/1506.08219 (accessed on 13th September 2025).
25. Dengo, N., Gross, S., & Vittadini, A. (2017). Functionalization of colloidal transition metal sulphides nanocrystals: A fascinating and challenging playground for the chemist. https://core.ac.uk/download/141699569.pdf (accessed on 13th September 2025).
26. Dobrosavljević, V., Lanatà, N., Lee, T. H., Stevanović, V., et al. (2018). Critical role of electronic correlations in determining crystal structure of transition metal compounds. http://arxiv.org/abs/1710.08586 (accessed on 13th September 2025).
27. Efimenko, A., Fumagalli, R., Giacobbe, C., Gubanov, A. I., et al. (2017). Possibility to realize spin-orbit-induced correlated physics in iridium fluorides. https://core.ac.uk/download/154336989.pdf (accessed on 13th September 2025).
28. Egranov, A. V., Shendrik, R. Y., Sizova, T. Y., Smirnova, N. L., et al. (2015). Instability of some divalent rare earth ions and photochromic effect. http://arxiv.org/abs/1507.04456 (accessed on 13th September 2025).
29. Eisenberg, R., & Gray, H. B. (2011). Noninnocence in metal complexes: A dithiolene dawn. https://core.ac.uk/download/4892476.pdf (accessed on 13th September 2025).
30. Freimuth, A., Möller, A., Revcolevschi, A., Abel, E., Auerbach, A., Büchner, B., Ballhausen, C. F., et al. (2005). Optical study of orbital excitations in transition-metal oxides. https://core.ac.uk/download/525037759.pdf (accessed on 13th September 2025).

31. Gaboardi, M., & Monteverde, F. (2024). Entropy-driven expansion of the thermodynamic stability of compositionally complex spinel oxides. https://core.ac.uk/download/656008294.pdf (accessed on 13th September 2025).
32. Gray, H. B., & Langford, C. H. (1966). *Ligand Substitution Processes.* https://core.ac.uk/download/4890295.pdf (accessed on 13th September 2025).
33. Green, A. G. (2013). An introduction to gauge gravity duality and its application in condensed matter. https://core.ac.uk/download/16263488.pdf (accessed on 13th September 2025).
34. Grimaud, A., Hong, W. T., Risch, M., Shao-Horn, Y., et al. (2014). Toward the rational design of non-precious transition metal oxides for oxygen electrocatalysis. https://core.ac.uk/download/83235357.pdf (accessed on 13th September 2025).
35. Jain, A., Persson, K. A., & Shin, Y. (2016). Computational predictions of energy materials using density functional theory. https://core.ac.uk/download/294625349.pdf (accessed on 13th September 2025).
36. Khomskii, D. I., Kugel, K. I., Sboychakov, A. O., Streltsov, S. V., et al. (2015). Role of local geometry in spin and orbital structure of transition metal compounds. http://arxiv.org/abs/1510.05371 (accessed on 13th September 2025).
37. Lebed, A. G., Larkin, A. I., Schuck, G., Karpinski, J., Yamaura, J., Maki, K., et al. (2006). Crystal growth, structural studies and superconducting properties of beta-pyrochlore KOs_2O_6. http://arxiv.org/abs/cond-mat/0601465 (accessed on 13th September 2025).
38. Liechtenstein, A. I., McMahan, A. K., Khomskii, D., Mazurenko, D. A., Aryasetiawan, F., Khaliullin, G., Wannier, G. H., et al. (2010). Modeling of complex oxide materials from the first principles: Systematic applications to vanadates RVO_3 with distorted perovskite structure. http://arxiv.org/abs/1007.2058 (accessed on 13th September 2025).
39. Majcher, A., Mironov, V. S., Pilet, G., Vostrikova, K. E., et al. (2017). Neutral low-dimensional assemblies of a Mn(III) Schiff base complex and octacyanotungstate(V): Synthesis, characterization, and magnetic properties. https://core.ac.uk/download/161840783.pdf (accessed on 13th September 2025).
40. Medforth, C. J., Kadish, K. M., Leung, K., Turner, P., & Zahran, Z. N. (2007). *Ab initio* molecular dynamics study of manganese porphin hydration and interaction with nitric oxide. http://arxiv.org/abs/cond-mat/0701544 (accessed on 13th September 2025).

41. Mitra, A. N. (1999). *The Dimensions of Field Theory: From Particles to Strings*. https://core.ac.uk/download/25272507.pdf (accessed on 13th September 2025).
42. Pasquier, D., & Yazyev, O. V. (2018). Crystal field, ligand field, and interorbital effects in two-dimensional transition metal dichalcogenides across the periodic table. http://arxiv.org/abs/1810.01302 (accessed on 13th September 2025).
43. Pyatnov, M. V., Timofeev, I. V., & Vetrov, S. Y. (2014). Photonic defect modes in cholesteric liquid crystal with a resonant nanocomposite layer and a twist defect. http://arxiv.org/abs/1406.6855 (accessed on 13th September 2025).
44. Rosner, H., & Tsirlin, A. A. (2009). Structural distortion and frustrated magnetic interactions in the layered copper oxychloride $[CuCl]LaNb_2O_7$. http://arxiv.org/abs/0901.0154 (accessed on 13th September 2025).
45. Schupper, N., & Shnerb, N. M. (2005). Inverse melting and inverse freezing: A spin model. http://arxiv.org/abs/cond-mat/0502033 (accessed on 13th September 2025).
46. Sharkawy, A., Monat, C., Choi, C. J., Erickson, D., Janasek, D., Psaltis, D., Skoog, D. A., et al. (2007). Liquid-infiltrated photonic crystals–Enhanced light-matter interactions for lab-on-a-chip applications. http://arxiv.org/abs/0707.1194 (accessed on 13th September 2025).
47. Shluger, A. L., & Stoneham, A. M. (2002). Small polarons in real crystals – concepts and problems. https://core.ac.uk/download/1766355.pdf (accessed on 13th September 2025).
48. Simmons, V. E. (1963). *Heteropoly Tungstocobaltoates and Tungstocobaltiates Based on CoO_4 Tetrahedra: Magnetic Properties, Spectra, Chemistry, and Structures*. https://open.bu.edu/bitstream/2144/30177/1/Simmons_Violet_1963_web.pdf (accessed on 13th September 2025).
49. Stoll, H., van den Brink, J., Pierloot, K., Hozoi, L., Helgaker, T., & Katukuri, V. M. (2012). *Ab initio* determination of excitation energies and magnetic couplings in correlated, quasi two-dimensional iridates. http://arxiv.org/abs/1203.2054 (accessed on 13th September 2025).
50. Tsirlin, A. A., Abakumov, A. M., Van Tendeloo, G., Rosner, H., Goodenough, J. B., Arpe, R., & Petricek, V. (2010). Interplay of atomic displacements in the quantum magnet *$(CuCl)LaNb_2O_7$*. http://arxiv.org/abs/1005.4898 (accessed on 13th September 2025).
51. Turner, A. M., Vishwanath, A., Ashcroft, N. W., Savrasov, S. Y., & Wan, X. (2010). Electronic structure of pyrochlore iridates: From topological

Dirac metal to Mott insulator. http://arxiv.org/abs/1007.0016 (accessed on 13th September 2025).

52. Yao, S. (2016). *Structure-Function Analysis and Characterization of Metalloproteins.* https://core.ac.uk/download/143831396.pdf (accessed on 13th September 2025).

CHAPTER 4

LIGAND FIELD AND MOLECULAR ORBITAL THEORY

Contents

INTRODUCTION

Ligand field theory and molecular orbital theory are the cornerstone of understanding the electronic structures of transition metal complexes. These areas of research unite theory and experiment in describing the ways that interactions among ligands influence metal ion behavior. Key to this investigation is an awareness that the geometry of a coordination complex and the nature of the ligands present can completely transform the character of the metal center.

This is reinforced by the historical context under which such theories evolved, particularly in the 1960s when there was extensive research focusing on the electronic chemistry of dithiolene complexes. This research led to a paradigm shift because it showed that traditional oxidation-state interpretations were insufficient for explaining the unique electronic structures and properties of such complexes, as summarized in Eisenberg et al. (2011). As the discipline developed, it was apparent that greater sophistication of understanding electronic structure was necessary. Molecular orbital theory offers a robust context in which to explore the metal d-orbital and ligand orbital interactions, considerations of symmetry and energy being included. This theory accounts for results that have emerged from experimental data, such as for the overlap of metal and ligand orbitals in transition metal complexes. Further, modern computational methods such as density functional theory (DFT) allow more precise modeling of such interactions, although issues such as many-electron self-interaction errors remain prevalent. As cited in Gani et al. (2016), such errors are of vital importance and not only determine energetic calculations but even the metal to ligand electronic charge distribution. To sum up, the union of ligand field and molecular orbital theories provides a complete description for the understanding of the complexity of transition metal complexes. Historical evolution of the theory has given rise to a foundation for current studies, and the essential importance of ligands in defining the electronic, magnetic, and spectroscopic properties of metal ions has been emphasized. As further studies come to light, a multi-dimensional approach that incorporates both energy and density delocalization errors will be needed to advance the understanding of these intricate systems. With the intersection of theoretical models with empirical phenomena, ligand field theory and molecular orbital theory remain at the forefront of inorganic chemistry, informing further studies of metal complexes and their diverse applications.

Definition of Ligand Field Theory (LFT)

LFT is a fundamental model in coordination chemistry, defining the electronic structure of transition metal complexes. This theory extends the principles of Crystal Field Theory (CFT) to incorporate the effects of covalent bonding interactions between metal ions and ligands. Ligands, in LFT, are treated as sources of an electric field that influence the metal d-orbitals' energy levels and spatial locations. As ligands get closer to the central metal ion, they cause splitting of the degenerate d-orbitals by way of electrostatic interactions, thereby forming discrete energy levels. This differential energy allocation is important for the comprehension of the electronic structures and resultant magnetic and optical properties of coordination compounds since it dictates the disposition of electrons in the d-orbitals under different ligand surroundings. One significant aspect of LFT is the ability to explain the variations in electronic properties resulting from different ligand field strengths. The nature of the ligand as a strong or weak field significantly affects the extent of orbital splitting and thus the oxidation states of the metal and also the coordination geometry. Strong field ligands are responsible for causing larger splitting of the d-orbitals, which can lead to low-spin configurations of paired electrons. Weak field ligands are responsible for causing smaller energy gaps and high-spin configurations with unpaired electrons.

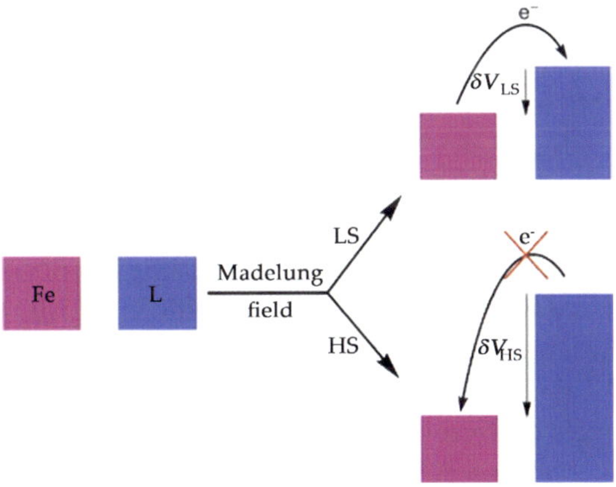

Figure 4.1. *Ligand field theory (LFT).*

Source: https://ars.els-cdn.com/content/image/3-s2.0-B9780080977744008196-f00819-14-9780080977744.jpg.

This fundamental axiom is seen in transition metal dichalcogenides (TMDs), where various coordination environments (such as disrupted octahedral and trigonal prismatic geometries) are responsible for variant electronic properties, showing the significance of ligand interactions in declaring the chemical behavior of a compound (Pasquier et al., 2018). LFT is also central to facilitating predictions regarding the stability and reactivity of transition metal complexes. By an examination of the crystal and ligand field parameters, researchers are capable of establishing how different ligands stabilize particular electronic structures, influencing directions of chemical reaction and novel complex formation. Subsequent studies employing advanced techniques, such as the Wannier-function approach, further clarified the influence of ligand field contributions on molecular-level systems (Pasquier et al., 2018). Besides, the progress of ligand field theory makes its way into applications in practice, including catalysis and material science, where the knowledge of this kind facilitates the preparation of more effective catalysts and materials with specific properties. Thus, the advancement of LFT continues to be important in our extension of knowledge concerning complex transition metal chemistry (Figure 4.1).

Overview of Molecular Orbital Theory (MOT)

MOT is one of the pillars of modern theoretical chemistry that provides a framework for understanding the electronic structures of molecules. Unlike Valence Bond Theory, where the emphasis is on localized electron pairs and sigma-pi bond interactions, MOT sees the distribution of electrons over entire molecules as delocalized systems in molecular orbitals. This approach is particularly valuable for the explanation of diatomic and polyatomic molecules with complex bonding situations, such as those found in transition metal complexes. Thus, molecular orbitals can be filled according to the principles of quantum mechanics in order to predict stability, reactivity, and spectral properties of molecules more precisely, encompassing a wider range of chemical systems and processes. One of the key applications of MOT is observed in the study of transition metal complexes where emphasis is laid on the theory of ligand field. As ligands approach a transition metal center, the interaction sets the energy levels of d-orbitals such that ligand field molecular orbitals are created. These interactions not only transfer information about the optimum geometric configurations of the complexes but also identify their electronic spectra and magnetic properties. Ligand field theory therefore modifies the d-orbital energies and assists in classification of the resultant molecular orbitals, allowing predictions to be made about electronic transitions and the resultant energetic implications. This understanding is crucial to create new materials and molecules with selective characteristics, a developing field with the advancement in computation powers and model theory (Reiher et al., 2014). Experimental studies of molecular orbital interaction are also represented by contemporary research such as bis(phthalocyaninato)lutetium(III) ($LuPc_2$) on highly oriented pyrolytic graphite surfaces. Here, molecular orbital theory converges with experimental techniques like scanning tunneling microscopy to give a window on the adsorption modes and electronic structuring of $LuPc_2$. The self-assembly and orientation of molecules at the interface with the substrate and the redshifting of the lowest unoccupied molecular orbital toward the Fermi level are concrete signs of the principles at work in MOT. These findings not only attest to the practicability of molecular orbital explanations in tangible scenarios but also highlight the sensitive relationship between electronic and molecular design, showing how MOT can be used to guide rational component construction in nanotechnological applications (Hietschold et al., 2012).

Importance of These Theories in Coordination Chemistry

The study of LFT and molecular orbital (MO) theory has immensely enhanced our understanding of coordination chemistry, particularly the role of ligand interactions in affecting metal ion properties. LFT, which is concerned with the spatial distribution and resultant splitting of the d-orbitals in the presence of ligands, helps predict electronic structures and magnetic properties of complexes. This theory explains color alterations in coordination complexes and provides an immediate correlation between molecular structure and observable properties.

By putting these issues into LFTs framework, chemists can gain a deeper knowledge of the behavior and reactivity of transition metals, which will facilitate the development of catalysis and materials science. In this way, the theoretical insight gained through the employment of LFT is priceless in scholarly study and applied applications in coordination chemistry. In the same manner, molecular orbital theory has a crucial contribution to making the bonding features of coordination complexes better understood. Unlike LFT, MO theory is more detailed in that it takes into account the overlap of the metal and ligand atomic orbitals. The overlap gives rise to bonding and anti-bonding molecular orbitals, which eventually determine the stability and reactivity of the overall complex. Through MO theory, chemists are able to describe the action of ligand substitution reactions—a reaction that is at the heart of many catalytic cycles—by examining how differences in ligand properties affect electronic environments.

The predictive and describable nature of these kinds of reactivity informs us further about fundamental mechanisms in coordination chemistry and prescribes lines of innovation towards new applications in designed materials. The significance of these theories extends beyond explanation alone; they result in a more refined, empirical understanding of reaction mechanism, particularly in ligand substitution reactions. Characterization of these reactions is an active field of research as new techniques and methodologies enable observation and control of these reactions. As illustrated in recent research, the mechanistic study of ligand substitution reactions with particular focus on monodentate ligands and transition metals such as cobalt(III) and platinum(II) has aroused more interest (Gray et al., 1966). The understanding obtained through the application of ligand field and molecular orbital theories provides the background required for such research. Moreover, as chemists develop more sophisticated models and computational techniques, the unification of these theories will continue to shape the direction of future coordination chemistry, requiring them to become part and parcel of future academic and industrial research (Gray et al., 1968).

4.1. HISTORICAL BACKGROUND

Historical development of ligand field theory and molecular orbital theory mainly transformed coordination chemistry knowledge. Early 20th-century research established foundational concepts of coordination compounds, emphasizing the geometric arrangements of ligands around metal centers. The pioneering work by scientists such as Alfred Werner in 1893, who introduced the concept of coordination number, set the stage for understanding how different ligand arrangements influence the electronic properties of metal complexes. His early contribution led to the opening of in-depth studies of the electronic structures of d-block metals, allowing for the inclusion of quantum mechanics in coordination chemistry. This came after molecular orbital theory was introduced in the mid-20th century, giving a deeper understanding by incorporating the overlap of atomic orbitals, hence providing a deeper analysis of metal-ligand complex bonding. The creation of high-level spectroscopy techniques throughout the latter part of the 20th century continued to propel research and development of ligand field and molecular orbital theory. Among the significant achievements was the use of X-ray absorption spectroscopy in analyzing electronic states in metal complexes. Experiments involving nickel(II) tetramesitylporphyrin (NiTMP) were some of the experiments that utilized femtosecond pulses of X-rays, enabling researchers to track electronically configuration transitions in real time. This improvement allowed for improved understanding of the short-lived excited states that exist before thermal relaxation, as outlined in recent articles, with an emphasis on the metal-centered electronic configurations' impact during catalytic reactions. These techniques enhanced the understanding of electronic structure and also highlighted the dynamic nature of metal-ligand interactions, outlining the delicacy of the interplay of geometry and electronic aspects in coordination compounds (Stickrath et al., 2016). In addition, as bioinorganic chemistry emerged, the significance of ligand field and molecular orbital theory was ever more well understood in both academic studies and applied science. Books entirely devoted to bioinorganic chemistry emphasized the importance of knowing these theories for a full understanding of metal ion function within biological systems. The theory system formulated by ligand field theory was also what helped to account for the metalloenzymes' actions, with the metal electronic structure playing a significant role in catalytic activity and substrate affinity. The evolutionary course of these theories is not merely of academic interest but also relates to practice in such fields as pharmaceuticals and materials science. Thus, the development of these theories is part of a broader movement in chemistry that aims to highlight both theoretical account and experimental confirmation, demonstrating their immense utility to the emerging science of chemistry (Bertini et al., 1994).

4.1.1. Development of Ligand Field Theory (LFT)

The development of LFT emerged as a major step towards explaining the interaction of ligands with transition metal ions, based on earlier models such as Crystal Field Theory (CFT). Initially, CFT addressed metal complex electronic configurations of d-orbitals in ligand environments but was not developed to explain phenomena in observed reaction kinetics and spectroscopy. LFT supplemented a more robust molecular orbital theory by introducing covalent interactions between metal and ligands and thus enabling a superior interpretation of electronic structure and its influence on complex behavior. This paradigm change at the theoretical level allowed chemists to observe delicate disparities in reactivity and colorimetric properties signifying diverse electronic environments at the metal centers, and which is vital for uses ranging from catalysis to biological systems. Beyond this, advances in experimental techniques have played a pivotal role in Ligand Field Theory development, particularly in kinetic studies of ligand substitution reactions. Emergence of new methodologies in the field of fast-reaction techniques has facilitated exploration of hitherto unexploited metal complexes, broadening the horizon of LFT and its applications.

Along with the further development of ligand substitution reaction study, researchers began to construct models on the basis of experimental data, i.e., those related to monodentate ligands. Such a correlation is stressed in recent studies with the aim of evaluating ligand behavior in different metal systems. Advances in LFT provide a foundation on which fresh knowledge can be deposited, so that theory is progressively refining with a growing appreciation of the mechanisms for ligand interaction within coordination chemistry (Gray et al., 1966). The convergence of experimental proof and theoretical conception has brought Ligand Field Theory to the limelight of chemical investigation, particularly in understanding electronic band structures in metal oxides. Recent work based on ab initio methods illustrates how electron correlation effects significantly influence band structures that ultimately decide optical and electronic properties of coordination compounds. As an example, investigation of interactions in oxide systems such as MgO involves rigorous investigation of the band gap and its evolution with respect to local Hamiltonians. These modern approaches enhance LFT by yielding information on polarization effects and charge relaxation processes that were hitherto hard to account for employing simpler models. With the advancement of LFT by researchers, its explanatory potential for inorganic materials and advanced chemical systems increasingly becomes important, bringing together the areas of experimental and theoretical chemistry (Mitrushchenkov et al., 2007) (Figure 4.2).

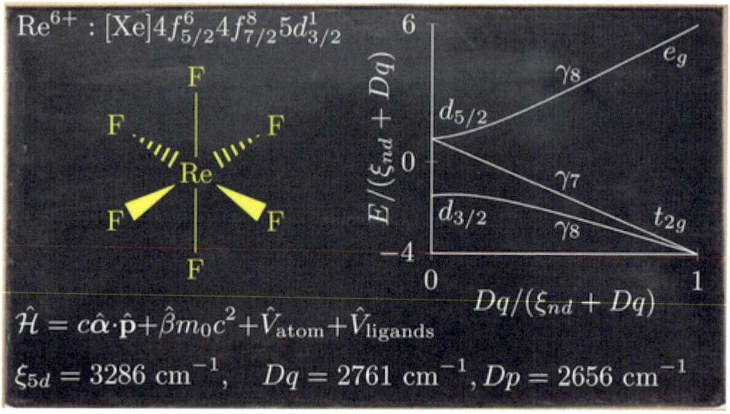

Figure 4.2. *The development of ligand field theory (LFT).*

Source: https://pubs.acs.org/cms/10.1021/acs.inorgchem.4c01771/asset/images/medium/ic4c01771_0006.gif.

4.1.2. Evolution of Molecular Orbital Theory (MOT)

Development of molecular orbital (MO) theory is one of the significant advancements in chemical bonding knowledge, particularly for transition metal complexes where ligand interactions contribute significantly. Early in the 20th century, the valence bond theory was developed, and it held sway over the domain of interpreting chemical bonding initially. But soon came to be realized that this model was short of accounting for the electronic character of coordination compounds. Molecular orbital theory turned out to be a better option, allowing the electronic structure to be more satisfactorily described by handling electrons in a delocalized manner rather than localized bonds. This transformation was pivotal, and even more important given that scientists began to focus on explaining the subtleties of metal-ligand interactions at the heart of ligand field theory and drove the advancement of MO theory further to its modern form. The role of ligand field theory to the molecular orbital framework cannot be overstated, particularly when one speaks of qualitative and quantitative descriptions of d-orbital splitting among transition metals. The formulation of ligand field theory was founded on the general MO theory principles by introducing the influence of ligands on d-orbital energy. This was particularly exemplified in two-dimensional transition metal dichalcogenides, which exist in unique polymorphs dictated by coordination chemistry. It has been established by research that the distorted octahedral and trigonal prismatic symmetries result in significant variations of crystal and ligand field splittings of the d-electron states, thereby influencing their electronic behavior (Pasquier et al., 2018). This is where the relationship between ligand field effects and molecular orbitals comes into play, emphasizing the fact that an integrated approach to molecular descriptions of transition metal complexes is necessary. As molecular orbital theory matured, it appeared in application in experimental and computational arenas. Methods such as electron spin resonance have been refined and enabled chemists to explore the geometries of d-levels in layered transition metal oxides, shedding light on ligand distortions and interlayer charge imbalances, essential in altering electronic properties. For example, data for Sr_2IrO_4 reveal inversion of ordering of the d-level from expectations based on conventional ligand-field theory, illustrating the versatile and subtle nature of the molecular orbitals in such systems (Bogdanov et al., 2015). These experimental works not only verify molecular orbital theory but also mark its evolution as a result of developing evidence, ultimately leading to a deeper understanding of the electronic processes involved with ligand interaction in complex materials (Figure 4.3).

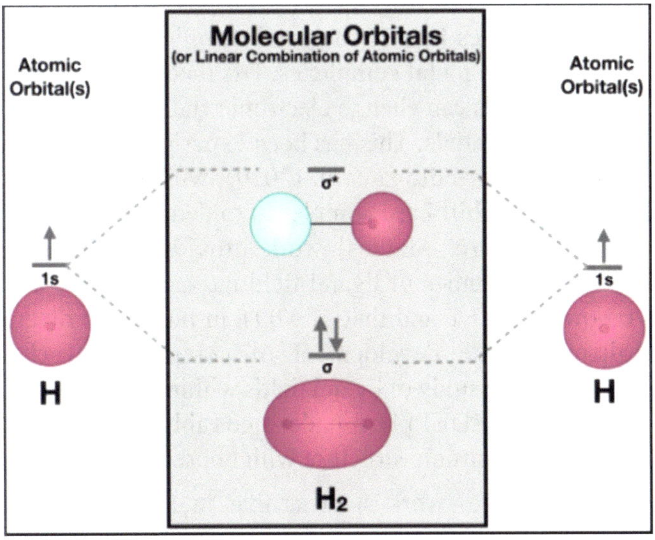

Figure 4.3. *The evolution of molecular orbital (MO).*

Source: https://files.mtstatic.com/site_4334/296664/0/webview?Expires=1756-731322&Signature=QGKsvDE6MLK1L4rbVbCkZOtU6C8aMRNvyGfOx0GRhW-K039rqeQXKk5-n-lu8hPiHyXP7PjRa9x3SYlLvalhjs~Soh5hZI64-EWbNU1LTOum-n8FjdAlRdTTA9oGK-Zxg5jOXAnSUqy0yngC3kwIHlkOPWih~h2Hv2ZLsD94JGcA_&Key-Pair-Id=APKAJ5Y6AV4GI7A555NA.

4.1.3. Key Contributors to the Theories

Ligand Field Theory (LFT) and Molecular Orbital Theory (MOT) owe their genesis primarily to contributions by some key figures in coordination chemistry. William H. Huggins is one such key contributor whose contribution focused primarily on the importance of electronic interaction between metal centers and ligands. His work explained how these interactions could alter d-orbital energies and, in turn, influence the electronic and geometric properties of metal complexes.

From here, this methodology became core to the understanding of ligand stabilization of certain electronic structures, something fundamental to predict reactivity and properties of transition metal complexes. Huggins' findings laid the ground for further investigations on the effect of various ligand environments on molecular orbits towards furthering an understanding of the functionality of coordination compounds in synthetic and biological reactions. Another important figure whose research furthering electron transfer processes brought LFT and MOT closer together is Robert A. Marcus. Marcus' theory of electron transfer kinetics has fundamental significance for the behavior of ligand-activated transition metal complexes. His theory describes how electric fields generated by ligands can change electronic states and induce measurable changes in reduction potentials. This has been experimentally verified by work like that described in Alexandrova et al. (2019), which investigates the effect of proximal cations on Schiff base complexes to determine how electric fields impact electronic structure. Marcus' work provided additional theoretical understanding of the dynamics of ligand-field interactions, thereby enhancing the predictive power of LFT and that of MOT in numerous diverse chemical systems. Furthermore, the development of computational chemistry has significantly enhanced the study of ligand fields within these theories. Advanced density functional theory (DFT) has made it possible to visualize and predict metal-ligand complex electronic structure with unprecedented accuracy.

Application of DFT to work such as that in Fuhrer et al. (2019) is a demonstration of how spin orbit interactions can produce new electronic structures in topological materials with a new complexity introduced by ligands. Such computational methods not only guide theoretical models but serve to make an assessment of theoretical ideas, including spin textures and how they influence material properties. The interaction between computational tools and theoretical advances is the very nature of the collaborative process of scientific advancement in ligand field and molecular orbital theories, which applies to materials science and elsewhere.

4.2. FUNDAMENTAL CONCEPTS OF LIGAND FIELD THEORY (LFT)

The Basic Concepts of LFT present a set of principles for describing the electronic structure of transition metal complexes that is really quite distinct from that of main-group compounds. One of the main principles of LFT is that coordination and structure of ligands around a metal ion can be employed to alter the d-orbital energy levels of the metal ion, leading to distinctions between the geometries and characteristics of complexes. LFT places central emphasis on the role of ligand interactions and symmetry in determining the electronic character and stability of metal complexes. By examining the influence of different ligands on d-orbital splitting, LFT classifies ligands as strong or weak field ligands, which influences color, magnetism, and reaction rate in coordination complexes. LFT, thus, is a fundamental principle in the broader field of coordination chemistry and has relevance for understanding transition metal molecular interactions. According to ligand field theory, the understanding of ligand substitution reactions is most crucial as they form the basis of dynamic nature of metal-ligand interactions. Substitution reactions are the simplest processes whereby ligands are interchanged in a coordination complex, and they have an enormous influence on the electronic environment and reactivity of the metal center. The mathematical analysis and mechanistic data derived by LFT allow one to categorize various substitution routes, accounting for the differences in terms of ligand class and metal identity. Not only does this analysis reveal the underlying principles determining these reactions, it also reflects observed characteristics in metal complexes, such as their stability and spectral properties. The marriage of kinetic observations and ligand field theory also contributes further to the exploration of these basic chemical processes, thus positioning substitution reactions at the forefront of modern coordination chemistry (Gray et al., 1968). Advances in analytical techniques have reignited passion for mechanistic exploration of ligand substitution reactions, which allow scientists to explore the theoretical foundations of ligand field theory more profoundly. With the help of sophisticated methodologies such as rapid reaction kinetics and spectroscopy, researchers are now able to investigate previously inaccessible reaction channels and illuminate less common metal-ligand systems. Such an advancement in methodology has the impact of confirming and enhancing LFTs applicability in various chemical milieux, thereby fueling the construction of globalized concepts about ligand interactions. Secondly, the dispute regarding inherent features of ligand substitution helps in ongoing refinement of theoretical models, particularly as molecular orbital (MO) theory is an alternative perspective on electron distribution and coordination complex bond formation. Synthesis of LFT with MO theory ultimately highlights the significance of electronic structure in regulating reactivity and properties in metal coordination complexes (Gray et al., 1966).

4.2.1. Crystal Field Splitting

Crystal field splitting is the underlying principle of explaining the electronic structure of transition metal complexes because it describes how ligands affect metal d-orbitals to produce energy differences. Ligands create a repulsion when coordinating to a central transition metal ion that causes the asymmetrical distribution of d-electron energy levels, a phenomenon known as splitting. In octahedral complexes, the five d-orbitals are separated into two sets: the lower-energy t2g and the higher-energy, e.g., The extent of splitting is determined by a range of factors, including the geometry and type of ligands involved, resulting in a large range of electronic properties. Crystal field theory, thus, is a basis concept in ligand field theory that illustrates how variations in ligand field strength may affect reactivity and stability of transition metal complexes to a significant degree. In more sophisticated crystal field splitting treatments, orbital mixing and ligand field strength are parameters that play significant roles in the electronic state of complex molecules. For instance, studies on layered oxides like Sr_2IrO_4 center on how distortion of ligands and unbalance of charge can reorganize electronic d-orbital levels, ultimately leading to a reorganization of the energy states. This kind of interaction is crucial for material comprehension in the case of materials that have huge spin-orbit coupling, where simple crystal field diagrams fail to adequately describe electronic configurations. Measurements of g-factors larger than two reveal an inversion of the anticipated d-level orderings, meaning that conventional ligand-field theory must be supplemented to explain these new electronic structures. These investigations highlight the strength of the ligand field paradigm, allowing chemists to design materials with targeted electronic characteristics for implementation in quantum devices and electronics (Bogdanov et al., 2015). The historical context to the development of ligand field theory lays open its roots in the exploration of metal complexes of dithiolate ligands, which revealed captivating electronic properties and anomalous oxidation effects. Early research revealed that ligand–metal interactions were central to the electronic behavior of these complexes, which often departed from traditional descriptions in terms of oxidation states. These experiments unveiled complexes with non-typical geometries and high orbital overlap, which indicated that the inclusion of ligand contributions is necessary in determining oxidation states. The developments in computational methods that came after have only ratified these observations, proving that it is most often the ligand that engages in the crucial processes during redox events and not merely the metal center. Thus, ligand field theory combined with molecular orbital theory has transformed our understanding of metal-ligand interaction, resulting in sophisticated interpretations in inorganic chemistry (Eisenberg et al., 2011) (Figure 4.4).

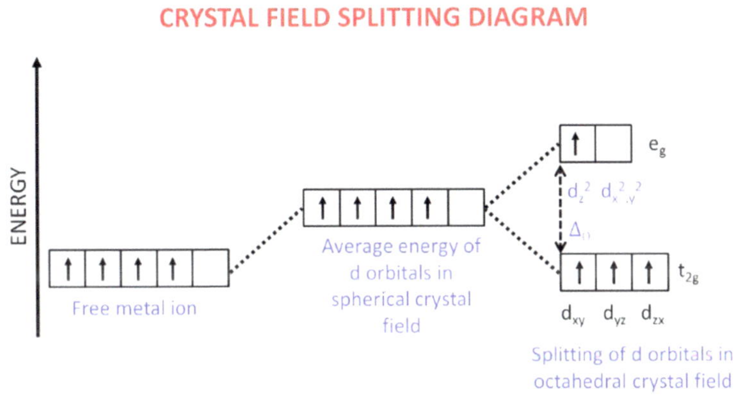

Figure 4.4. *The concept of crystal field splitting.*

Source: https://www.teachoo.com/16449/3730/Question-5--Choice-2----b/category/Chemistry-CBSE-Class-12-Sample-Paper-for-2022-Boards--Term-2-/.

4.2.2. Coordination Number and Geometry

Coordination number is a key idea in the description of the geometry of metal complexes, a fundamental building block of ligand field theory and molecular orbital theory. Coordination number is the number of ligand atoms that are directly bonded to the central metal ion. The value significantly influences the electronic character and the spatial geometry of the complex. Four and six are typical coordination numbers that typically correspond to square planar and octahedral geometries, respectively, and are extensively explored in coordination chemistry. These geometries dictate the ligand field splitting, producing variations in color, magnetism, and stability of the complexes. Secondly, the metal ion's electron configuration and the character of the ligands are the key factors in deciding proper geometry since some metal-ligand interactions prefer certain structural types, thus demonstrating the subtle interplay among coordination number and geometry. With the inclusion of ligand field theory, one can better understand that different geometries are a product of varying coordination numbers and that these geometries are an expression of the electronic structure of the involved complexes. For instance, square planar complexes are often found for the situation of d8 metal ions, such as platinum and palladium, owing to strong ligand interactions to maximize orbital overlap. By contrast, octahedral complexes can typically have d6 or d7 metal ions, which can allow for additional ligand interactions and lead to varied electron distributions across the metal-centered orbitals. The ligand field splitting thus created has a profound effect on d-orbital energies and dictates the electronic transitions governing the color and reactivity of the complexes. Coordination number and geometry thus constitute an entry into additional explorations of the electronic properties and functionalities of metal complexes. Novel advances in computational methods, such as density functional theory (DFT), accentuate the requirement for precise modeling of coordination geometries and their corresponding electronic structures. Careful consideration of electronic interactions in these metal-ligand frameworks is significant, particularly in addressing issues like self-interaction errors that may misbehave results within energy calculation approximations. As observed in studies, charge can be redistributed extensively between metal and ligand atoms, influencing magnetic moments and electronic structures (Gani et al., 2016). This points to the dynamic character of coordination chemistry studies, where the knowledge of coordination number and resulting geometry is not merely theoretical but is vital for the development of novel materials with distinct electronic properties, broadening the horizon of inorganic chemistry as a whole (Eisenberg et al., 2011).

4.2.3. Factors Affecting Ligand Field Strength

The ligand field strength of coordination complexes is generally influenced by many factors, predominantly the electronegativity and character of the involved ligands. Electron-withdrawing ligands possessing greater electronegativity induce a greater withdrawal of electron density from the metal center, thus raising the overall ligand field strength. This is responsible for understanding many transition metal complexes as strong-field ligands, for example, CN^- and CO, cause splitting of the d-orbitals that can surpass energy barriers to spin transitions. However, weak field ligands, for example, I^-, cause a lower splitting, which influences the electronic and magnetic characters of the complex. Thus, it is the properties of the ligand that determine the strength of the ligand field, which determines reactivity, color, and stability of coordination compounds. Geometry of the final complex is also an important factor in determining the strength of the ligand field. Arrangement of the ligand around the metal center can lead to substantial variation in d-orbital splitting patterns. As an example, octahedral complexes will generally exhibit a different splitting of d-orbitals from tetrahedral complexes based on their corresponding ligand geometries. The degree of deviation from ideal geometries—often influenced by steric and electronic factors—contributes even more to such diversity. Deviation from ideal geometries, for instance, to square planar or trigonal bipyramidal geometries, can also bring in special ligand field behaviors that are of central importance to applications in catalysis and material science. Awareness of these geometrical influences is a necessary prerequisite for consideration of the tailored design of complexes for specific applications in bioinorganic chemistry and materials science, for example. In addition, involvement of multiplet states and charge transfer interactions makes significant contributions to definition of the ligand field strength. Presence of excited states that facilitate electron transition between the metal and the ligand can cause alterations in effective ligand field experienced by the metal d-orbitals. Electronic structure calculations such as those provided in Gani et al. (2016) observe the need to correct many-electron errors in performing such analyses to accurately account for the actual charge distribution behavior. Proper accounting of such interaction not only illuminates the nature of bonding but also delivers an understanding of magnetic behavior in complexes. Such refinements are essential in the subtle calibration of theoretical prediction so that ligand field strength characterization is robust against the inherent complexities of molecular interaction intrinsic to transition metal complexes (Figure 4.5).

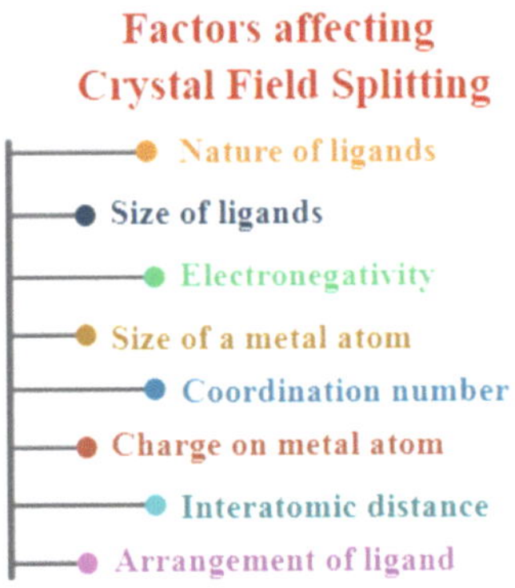

Figure 4.5. *Factors affecting ligand field strength.*

Source: https://psiberg.com/wp-content/uploads/2022/03/factor-splitting.svg.

4.3. MOLECULAR ORBITAL THEORY (MOT) BASICS

MOT is a theoretical approach to the study of chemical bonding and molecular structure. In contrast to Valence Bond Theory's explanation of bonds as resulting from atomic orbital interactions, MOT gives a wider view with the inclusion of electron distribution across entire molecules. Atomic orbitals here are assembled to provide molecular orbitals which may be classified as bonding, antibonding, or non-bonding. Electrons fill the molecular orbitals using principles essentially the same as those that govern atomic orbital filling. With this method, chemists can make molecular geometry, bond order, and magnetic molecular states predictive, and thus have improved insight into molecular stability and reactivity. The relevance of MOT extends well beyond simple diatomic molecules, impacting many areas, including bioinorganic chemistry, where complex metal-ligand interactions have a tendency to be studied from this viewpoint (Bertini et al., 1994). The way in which ligands engage metal centers within complexes can be profoundly understood with the lens of Molecular Orbital Theory. Transition metal complex ligands create a ligand field that may split the degenerate metal ion d-orbitals due to bimodal asymmetric approach of bonding ligands. This creates different energy levels for d-orbitals, which ultimately influence the electronic structure of the complex. How these molecular orbitals are filled will define physical and chemical properties such as color, magnetism, and reactivity of the complex. Hence, it becomes essential to understand ligand field theory from the MOT point of view in order to grasp the important facets of coordination compounds, particularly transition metal compounds where splitting of d-orbitals is maximum (Harden et al., 2023). Molecular orbital studies not only inform the metal complex bonding character but also catalytic activity and overall reactivity. The character of metal-ligand interactions can be an important determinant of how effective catalytic processes are, particularly industrial and environmental processes. It is possible for scientists to model these interactions by computation and choose which ligands would provide optimal reactivity or selectivity. For example, in green chemistry, catalyst synthesis that minimizes hazardous waste and energy use is based on understanding how molecular orbitals influence reaction pathways. This implies the application of Molecular Orbital Theory to more practical catalysis areas not only sheds light on the chemical fundamentals but also provides innovative possibilities for the sustainable processes of the chemical industry (Bertini et al., 1994; Harden et al., 2023).

4.3.1. Atomic Orbitals and Molecular Orbitals

Understanding atomic orbitals and molecular orbitals is central in the overall context of ligand field and molecular orbital theory. Atomic orbitals, described in terms of shape and energy, determine the probable positions of electrons around single atomic nuclei. They are a prerequisite to enabling chemical bonding to take place because they are the source from which molecular orbitals are derived by their combination. Molecular orbitals are a product of the overlap between atomic orbitals of bonding atoms, and electronic distribution spreads over more than one nucleus. Interaction significantly alters electron energy and organization, producing bonding and antibonding orbitals that direct molecular stability and reactivity. The knowledge of atomic orbitals, thus, gives a starting point to explore the complex nature of molecules by ligand field effects, particularly in transition metal complexes. Ligand field theory, in coordination chemistry of transition metals, complements the molecular orbital model by describing the effect of ligands on the metal ion d-orbitals. The presence of ligands around a transition metal ion produces a ligand field that distorts the energy of the d-orbitals to provide various electron configurations and thus various physical and chemical characteristics. This d-orbital energy distortion is something that can be investigated through molecular orbital calculations, which reflect the metal-ligand interaction. Recent advances, particularly those in gas phase density-functional theory (DFT), have provided chemists with a greater insight into the interactions. As for instance, in studies such as that addressed in Amolo et al. (2017), there is a comprehensive overview of ruthenium(II) complexes, demonstrating the measurable assessment of ligand interactions through orbital energy modeling. The integration of molecular orbital theory and ligand field theory further yields useful information on magnetism and electronic transitions of polynuclear transition metal complexes. The development of computational methods, including excitation-selected configuration interaction methods, enables researchers to examine magnetic interactions and predict the system behavior based on a molecule's electronic structure. For instance, the findings presented in Jiménez Calzado et al. (2011) illustrate that it is indeed possible to correlate the interaction between transition metals quite successfully with their magnetism, thereby validating the use of molecular orbital analysis in predicting complex phenomena such as magnetic susceptibility. This integration of molecular and atomic orbital theories not only explains the bonding phenomena but also enhances the predicting capacity as it relates to a material's property, cementing the coordination of electronic structure and chemical behavior within coordination complexes (Figure 4.6).

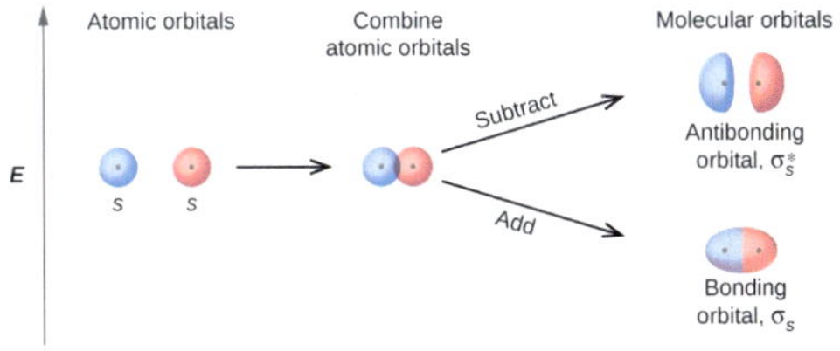

Figure 4.6. *The understanding of atomic orbitals and molecular orbitals.*

Source: https://s3-us-west-2.amazonaws.com/courses-images-archive-read-only/wp-content/uploads/sites/887/2015/04/23211933/CNX_Chem_08_04_ssigma1.jpg.

4.3.2. Bonding and Antibonding Orbitals

Bonding and antibonding orbitals is one of the most critical concepts in describing the electronic structures of molecules, particularly for ligand field and molecular orbital theory. Constructive addition of atomic orbitals results in bonding orbitals, which possess electron density that is larger between the nuclei, hence stabilizing the molecule. This stabilization occurs in metal-ligand interactions in coordination complexes, where efficient overlap of ligand atomic orbitals with metal's d-orbitals yields strong bonding interactions. However, antibonding orbitals arise from destructive interference between overlapping orbitals such that a condition of lower electron density is encountered between the atomic nuclei. Their occurrence can reverse the stabilization provided by the bonding orbitals and decrease the strength of bonds as a whole, as well as influence properties and reactivity of the resulting complexes (Gray et al., 1966). Moreover, relative energies of antibonding and bonding orbitals provide useful information regarding electronic transitions in coordination compounds. With ligands bonded to a central metal ion, d-orbitals are split into different energy levels, an effect that is strongly observed in transition metals due to the lack of symmetry in their coordination of ligands. The crystal field splitting influences the filling of the antibonding and bonding states by electrons and is also responsible for the optical and magnetic properties of the complexes. For those systems with strong field ligands, increased stability is typically observed as electrons favorably fill the bonding orbitals. Conversely, when electrons are in antibonding orbitals, this may decrease the bond strength and altered chemical behavior, demonstrating the critical role of electron structuring in controlling molecular properties. Understanding these interactions is integral to the formulation of theories of ligand substitution reactions, where changes in ligand coordination cause dramatic variations in electronic configurations (Azarbayejani et al., 2025). Furthermore, the intentional regulation of bonding and antibonding orbitals encourages enhanced comprehension of reactivity trends in coordination compounds, especially according to ligand field theory. For example, for the explanation of certain ligand substitution mechanisms, it is clear that ascertaining the character of metal-ligand interactions and their associated orbital arrangements encourages clarity according to interaction dynamics in the substitution. The molecular orbital theory provides a solid foundation for correlating experimental spectral and magnetic data with the electronic structure of the complexes. As the field grows, the inclusion of advanced molecular orbital techniques will be instrumental in accounting for phenomena such as cubic field splitting in group II-VI compound systems to further develop transition metal chemistry (Gray et al., 1966). Summary, thus, antibonding and bonding orbitals are central concepts that not only decide the structural stability of transition metal complexes but also enable an overall perception of their reactivity and properties in molecular orbital theory and ligand field.

4.3.3. Energy Level Diagrams

Energy level diagrams are a fundamental tool for the visualization of the complex interplay between metal centers and ligands in coordination complexes and for the explanation of the principles of both Ligand Field Theory and Molecular Orbital Theory. Such diagrams represent the relative energies of molecular orbitals resulting from the overlap of atomic orbitals of metal and ligand atoms. By depicting the energy levels and their corresponding electron occupation, energy level diagrams provide a better understanding of the electronic structure, bonding, and reactivity of transition metal complexes. Specifically, they enable chemists to predict how the structure and electronic character of ligands govern the metal ion's behavior, something that is valuable in applications for catalysis and materials science. These graphical representations help to render the abstract ideas involved in theoretical treatments more understandable for both students and researchers in the field. The significance of energy level diagrams when viewed within the context of coordination complexes extends beyond mere representation; it is a consideration of how different kinds of ligands affect the splitting of the metal d-orbitals. The extent of splitting of the orbitals in tetrahedral and octahedral complexes varies with the nature of the ligands, which are either strong field or weak field ligands based on their ability to split d-orbitals. Strong field ligands such as CN^- cause greater separation of energy levels, leading to pairing of electrons and showing preference for low-spin state. Conversely, weak field ligands like Cl^- provide low orbital splitting, allowing high-spin configurations. This is the distinction that is essential to explaining the magnetic and spectroscopic behavior of complexes because energy level diagrams concisely summarize the significance of ligand fields to electronic geometry. Furthermore, these diagrams form the basis of comprehending electron transfer processes that are pivotal in redox processes of catalysis. Recent computer chemistry developments made predictions from energy level diagrams more accurate, particularly through improved electronic structure calculations, e.g., on the basis of density functional theory (DFT) and related methods. The new methods provide information that cannot be obtained with previous methodologies, allowing investigation of complex interactions in coordination compounds with better accuracy. For instance, they make it possible to examine how varied geometric arrangements and ligand orientations affect the stabilization of specific electronic states and the associated ligand-to-metal charge transfer energies. This dynamic interaction also is explained by findings in redox-active ligand studies, which, like in the literature, discount conventional oxidation-state depictions and describe that electron transfer can be significantly localized on the ligands over the metal centers themselves, as with complications illustrated in energy level diagrams (Eisenberg et al., 2011; Held et al., 2017). The intersection of theoretical and computational advances is an exhilarating new front in the exploration of energy levels in coordination chemistry, enabling symbiotic understanding of ligand field and molecular orbital theory (Figure 4.7).

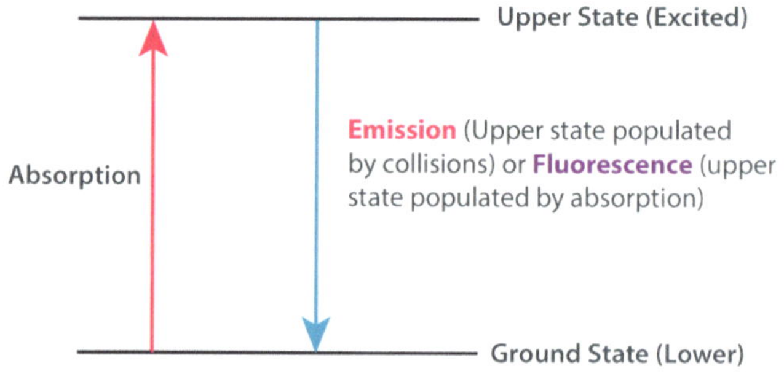

Figure 4.7. *Energy level diagrams.*

Source: https://cdn1.byjus.com/wp-content/uploads/2019/03/what-is-energy-level-diagram.png.

4.4. COMPARISON OF LIGAND FIELD THEORY (LFT) AND MOLECULAR ORBITAL THEORIES (MOT)

The discussion of LFT and MOT reveals tremendous differences in their approaches to understanding the electronic structures of transition metal complexes. LFT is primarily interested in the role played by ligand interaction in altering metal d-orbitals, the manner in which symmetry and spatial orientation of the ligands result in splitting the orbitals into various energy levels. This theory explains effects such as crystal field stabilization energy quite well and is invaluable for the prediction of magnetic and spectroscopic properties of coordination complexes. LFTs limitation comes in the fact that it does not take fully into consideration the molecular nature arising from orbital mixing and instead simplifies the interaction to point charges rather than considering the complexity of electron distribution across the system. Thus, while LFT is providing rudimentary understandings, it falls short of a full description when contrasted with MOT. Molecular orbital theory does, however, have a more refined picture by also considering the combining of the metal and ligand atomic orbitals to produce molecular orbitals distributed across the entire complex. This approach allows for a more comprehensive explanation of electron delocalization and interaction of bonding, with emphasis being given to the significant contribution ligand orbitals make to the stabilization of the electronic system. The addition of concepts such as sigma and pi bonding, and anti-bonding interactions, enables better prediction of spectroscopic and electronic behavior. For instance, using molecular orbital diagrams, one can describe the connection between the number of d-electrons and the resultant spin state of the compound, which is crucial for the interpretation of magnetic properties of transition metal complexes. Hence, MOTs detail makes it necessary for a proper explanation of metal-ligand interactions. To further shed light on the complexities in comparing LFT and MOT, the recent developments in computational methods enhances our understanding of these theories. For example, the development of approximate density functional theory (DFT) allows the removal of many-electron self-interaction errors that lead to improved prediction of electron density and magnetic moments of transition metal systems. With these sophisticated computational techniques, researchers have discovered characteristic charge transfer effects, most notably the selective loss of charge at metal sites, in accordance with molecular orbital theory predictions. These results further support the need to address electron delocalization along with traditional energy considerations while investigating advanced behavior in coordination compounds. Accordingly, these advancements provide a compelling justification for synergistically integrating LFT and MOT under a unified theoretical framework that would more effectively explain the intricate chemistry of transition metal complexes (Eisenberg et al., 2011; Gani et al., 2016).

4.4.1. Similarities in Predictive Power

Relative comparison between ligand field theory and molecular orbital theory shows wide similarities between their predictive ability towards explaining the electronic structure and properties of complex molecules. Both theories aim to describe ligand interaction with metal centers in coordination complexes, determining the spectral and magnetic characteristics of these compounds. By using frameworks that assess electron distribution and bonding interaction, both theories are able to make predictions of complex geometry and stability. This convergence enhances our understanding of ligand behaviors across different conditions, ultimately aiding the development of computation methods, e.g., quantitative structure-activity relationships (QSAR) and quantitative structure-property relationships (QSPR), from exact modeling of electronic properties to predict biological action and stability. Further, the synthesis of machine learning approaches with the traditional ligand-based screening protocols constitutes the revolutionary situation of predictive modeling in chemistry. Current advances in computer-aided drug design (CADD) are aimed at emphasizing the advantages of such methods in efficiently searching vast chemical databases. Algorithm construction, such as the Ligand Classifier of Adaptively Boosting Ensemble Decision Stumps (LiCABEDS), illustrates how machine learning can supplement ligand field and molecular orbital theories by enhancing the predictability of pharmacological properties. These computational methods optimize the rate of success of drug candidates in clinical trials by checking their likely interactions, selectivity, and permeability through biological barriers, thereby unifying predictive methods for the discipline of chemistry (MA et al., 2012). Additionally, the introduction of non-linear classifiers and machine learning models derived from standard quantum chemistry data demonstrates an increasing complementarity between traditional theoretical frameworks and advanced computational methodologies. By constructing models relating molecular and solid properties from unselected samples in chemical space, researchers are uncovering recurring correlations between reactivity and molecular structure. The convergence of these methodologies corroborates the notion that both ligand field theory and molecular orbital theory can provide valuable insights into mechanistic pathways governing ligand binding and reactivity in biological molecules. As the research continues, the combination of cutting-edge computational modeling with classical theories is likely to yield increasingly robust predictive capabilities in the explanation of intricate chemical interactions (Adamo et al., 2016).

4.4.2. Differences in Approach and Application

In exploring the complicated landscape of ligand field and molecular orbital theory, the approach and use variation gets more acute. Ligand field theory is centered mainly on the electrostatic interaction between the metal ions and ligands and provides a model for a classification of ligands based on their field strength—weak or strong field ligands. This distinction leads to varying electronic configurations and organization of energy levels, primarily observed in transition metal complexes. Molecular orbital theory overtakes the limitations of ligand field theory by considering both localized and delocalized electrons in calculations. Providing a wider context to the complex bonding nature, the method acknowledges the contribution of overlapping atomic orbitals, whose effects can elucidate nuances such as resonance and hybridization that are given short shrift in ligand field theory. The alternative methodologies thus offer complementary perspectives and not necessarily competing interpretations. Their working consequences are observable in application areas such as catalysis, spectroscopic analysis, and reaction mechanisms. For instance, ligand field theory correctly predicts color in coordination compounds on the basis of d-d transitions controlled by ligand-metal atom interactions at the center. Such predictability is important in analytical chemistry when colorimetry is applied. Molecular orbital theory, on the other hand, allows one to investigate the mechanism of ligand substitution reactions more intensively. By emphasizing electron density distribution and potential energy surfaces, molecular orbital theory allows for more subtle comprehension of reactivities and reaction routes, especially in multisligand systems. Such differences in application are demonstrated rather well in ligand studies within inner sphere complexes, relevant to metal ion substitution mechanisms (Gray et al., 1968). Besides, the speeding-up pace of experimental techniques positions both theories in action, testing their respective boundaries and evoking a constant progress in ligand substitution researches. As noted, the subject is currently at its developing stage, particularly with regard to the branch of quick-reaction techniques that allow scientists to measure reaction velocities previously unknown and investigate previously established complexes on the basis of less common metal atoms. This dynamic stimulates a re-examination of basic ideas and structures in ligand field and molecular orbital theories. As an illustration, whereas ligand field theory has simple models for predicting reactions, as can be seen with cobalt(III) and platinum(II) complexes, molecular orbital theory has the graphical and computational means by which to map over kinetic data against other variables such as spectral and magnetic characteristics. Thus, as scientific study goes on, the interplay among such beginning theories can lead to greater insight and prompt the generation of new models, which increase our knowledge of ligand substitution reactions (Gray et al., 1966).

4.4.3. Situations Favoring One Theory Over the Other

For transition metal complexes, ligand field theory will generally hold sway in circumstances where conditions are low-spin and strong-field ligands are present. This theory is quite capable of describing d-orbital splitting by the ligand environment surrounding them. For example, the application of strong-field ligands like CN^- in octahedral complexes leads to large crystal field splitting, creating distinct electronic transitions which can be simply characterized by ligand field theory. It is particularly helpful in the prediction of electronic state multiplicity and magnetism. The relative simplicity of ligand field theory is specifically interesting when these low-energy transitions are highlighted and their connection to the vibrational modes. These characteristics shed light on complex relationships between electronic configurations and ensuing spectroscopic ramifications, making this model a critical mediator in chemistry and solid-state physics under some conditions (Natelson et al., 2010). Molecular orbital theory is preferred when there are delocalized electrons, complex bonding conditions, and higher-order interactions.

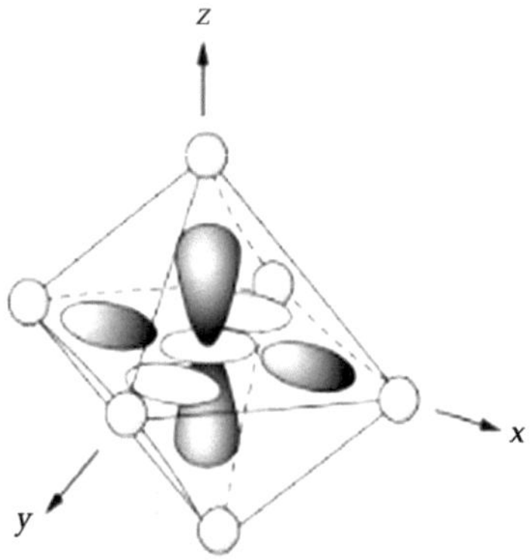

Figure 4.8. *The context of transition metal complexes, ligand field theory often prevails in scenarios characterized by low-spin conditions and strong-field ligands.*

Source: https://ars.els-cdn.com/content/image/3-s2.0-B9780128110485000079-u07-01-9780128110485.jpg.

The theory is best applied to reveal effects that arise from the nuances of molecular structure and electronic configuration not addressed within ligand field theory. Molecular orbital theory, for instance, provides a rigorous framework for the Kondo effect exhibited in nanoscale molecular devices in which correlations between charge and spin dynamics unspool in strange fashions due to intense electron interactions. This interplay of emergent behavior and molecular orbital interactions not only enhances the understanding of molecular electrical conductance but also enables investigation of intricate molecular systems outside the strictures of rudimentary ionic or covalent models (Natelson et al., 2010). Such flexibility of reasoning ensures molecular orbital theory's applicability in modern research and technological advances. In most instances, it is important to appreciate the fact that every theoretical framework has a scope by electronic localization and bonding interactions. Moments when density and electronegativity differences largely influence bond order and strength serve to demonstrate either theory's shortcomings. For example, comprehending covalent and multi-center bonding interactions transitions should involve an in-depth methodology that can incorporate understandings from both theories. The intrinsic relation between electronic localization changes and bond order modification means that these theories are not in competition with one another but rather parts of a consistent corpus of thought. As noted, the phenomenon observed in multi-center bonding shows how hybridizations between bond types can evolve under varying circumstances and highlights the significance of context in determining theoretical applicability. Hence, the situational, dynamic power of ligand field and molecular orbital theories parallels their interdependence in explaining complex chemical systems (Figure 4.8).

4.5. APPLICATIONS IN COORDINATION CHEMISTRY

Coordination chemistry is immensely helped by the advancements in molecular orbital theory and ligand field theory, which have been crucial in providing significant details regarding the electronic structures of metal complexes. With these theoretical outlooks, chemists can make predictions and explain the behavior of various coordination compounds, especially their electronic characteristics and geometrical shapes. Ligand field theory is specifically interested in how the presence of ligands influences the energy levels of the metal d-orbitals, leading to varying geometrical forms such as octahedral and tetrahedral structures. These forms are essential to understand because they directly lead to coordination complex reactivity and stability. Moreover, knowledge derived from these theories allows for the synthesis of complexes with properties designed to fulfill specific applications, ranging from catalysis to materials chemistry. One of the principal applications of these theories lies in the study of redox-active ligands, which has been groundbreaking in the demystification of the complexity of electron transfer processes in metal-ligand complexes. The large literature on metal complexes of dithiolate ligands has shown how conventional oxidation-state descriptions are not able to accommodate their new electronic structures. In a curious turn of events, the early work revealed the tendency for electron transfer reactions to occur to a significant extent on the ligand rather than the metal center, a discovery upheld by advanced computational methods (Eisenberg et al., 2011). This understanding has broad implications in coordination chemistry, where the synthesis of new complexes can take advantage of the unique redox properties of single ligands to synthesize more effective catalysts or materials with desirable electronic properties. Furthermore, new developments in computational methods, such as density functional theory (DFT), have made it easier for us to understand electronic delocalization in metal-ligand complexes, pushing the boundaries of coordination chemistry. This has been followed by recent advancements in the correction of delocalization and self-interaction errors in many-electron systems to yield more accurate predictions of electronic distribution and properties (Gani et al., 2016). The models have proved that charge transfer chemistry at the metal-ligand interface can be dramatic on magnetic properties and complex stability. While researchers move closer to achieving electronic modifications through various tuning methods, such research focuses on the intricate interaction between metal centers and ligands, paving the way for groundbreaking applications in enhancing the fields of sustainable energy, sensor science, and molecular electronics.

4.5.1. Color and Spectroscopy of Complexes

The interaction of the light with the transition metal complexes results in a highly fascinating array of colors, which is characterized by the electronic transitions occurring in these complexes. The observed color arises due to the ligand field effects that surround the metal ion, which determine the energies of the d-orbitals. Because different ligands create different extents of splitting between the d-orbitals, ground-to-excited-state energy difference decides the wavelengths of light absorbed and consequently color reflected. Predictability of such colors is not only relevant to the understanding of the beautifying nature of metal complexes but also as a valuable tool in the field of spectroscopy, enabling chemists to quantify electronic environments through UV-Vis absorption spectra. This interdependence illustrates how field strengths and geometries of ligand types govern the spectroscopic properties of complexes, which provide information regarding electronic structure. In addition, the analysis of excited states of transition metal complexes shows intricate behavior explainable by advanced methods of spectroscopy. The richness of possible electronic states, ranging from metal-centered to ligand-to-metal charge transfer states, indicates the intricacy of such systems. Techniques like charge transfer number analysis allow precise description of such excited states through quantification of electron transitions among atom groups, as in the techniques incorporating TheoDORE. This not only allows for comprehension of the photophysical and photochemical behavior of the complexes but also maximizes the accuracy of data involved in describing the excited states. By employing such methods, it is possible for researchers to conclude important information about the interaction dynamics in complexes, and how various states are linked to electronic property and stability effects by ligands. In addition to static properties acquired from absorption spectra, dynamic processes after excitation also play a vital role towards understanding functionality of transition metal complexes in catalysis and photochemistry applications. For instance, explanation of the nonradiative decay mechanisms—where the energy absorbed is not followed by the release of photons—yields information regarding photoreactive system efficiency in utilizing light. Spin-orbit coupling effects have the potential to complicate the landscape further in excited states, influencing pathways for energy transfer and conversion. By employing the synergy of state-of-the-art electronic structure theories and computational methodologies, for instance, density functional theory, it is feasible to represent these dynamics with good accuracy. The study of vibrational modes in combination with electronic transitions, as in the case of metal phthalocyanines, does a great job in bridging theoretical expectations and experimental findings, forming the bridge between ligand field theory and molecular orbital thinking in the context of complex spectroscopy (Arillo-Flores et al., 2013; Daniel et al., 2018) (Figure 4.9).

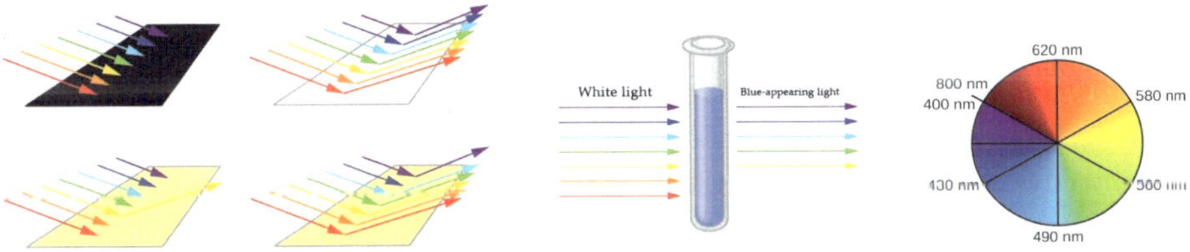

Figure 4.9. *Color and spectroscopy of complexes.*

Source: https://i.ytimg.com/vi/MuK-Kb_eBw4/maxresdefault.jpg.

4.5.2. Magnetic Properties of Transition Metal Complexes

The magnetic properties of transition metal complexes are significantly influenced by electronic structures dictated by ligand field and molecular orbital theory. Transition metals have inherently unpaired d-electrons, which are responsible for their magnetism. The d-orbital configuration and energy levels could be controlled by the character of ligands coordinating at the metal center. Strong-field ligands will induce more electron pairing due to the smaller difference in energy between the d-orbitals, leading ultimately to lower magnetic moments. Weak-field ligands, however, allow high-spin states, in which there are unpaired electrons, thus enhancing the magnetic character of the complex. It is necessary to comprehend these interactions, particularly while analyzing complex systems, since varying the type of ligands can result in varied magnetic properties pertinent to materials science and quantum computation. Furthermore, advanced experimental techniques have elucidated the magnetic properties of some transition metal complexes, improving theoretical predictions. For instance, the two-dimensional material Sr_2IrO_4 exhibits novel magnetic behavior due to its specific d-level geometry and ligand distortion. Studies reveal that in contrast to what ligand-field theory predicts, despite being canonical, for the magnetic properties of the system in terms of g-factors greater than two, indicates a reversal of the normal ordering of d-levels. This finding accentuates the necessity for comprehending the reason why charge imbalances and structural dispositions may affect the magnetic properties of transition metal complexes and establishing the inadequacies of classical descriptions. These intricacies provide avenues towards new applications in oxide electronics where precise control of magnetic behavior is important for the development of devices (Bogdanov et al., 2015). Additionally, control of electronic behavior through the adjustment of ligands in metal phthalocyanines (MPc) demonstrates the generality of transition metal complexes in controlling magnetic behavior. By varying various open-shell transition metals, scientists can tune the electronic gap and total magnetic moment of such compounds. Structural as well as vibrational properties have been shown by the research carried out using spin density functional theory to contribute significantly in determining the electronic and magnetic properties of MPc. For instance, fluorination of metal phthalocyanines illustrates how structural changes can have a direct correlation with experimental data, confirming the viability of the design of materials through predetermined magnetic properties. The observations are not only useful for theoretical ideas but also useful for applications in devices such as magnetic sensors and organic electronics, confirming the correlation between ligand interactions and transition metal complexes' magnetic behavior (Arillo-Flores et al., 2013) (Figure 4.10).

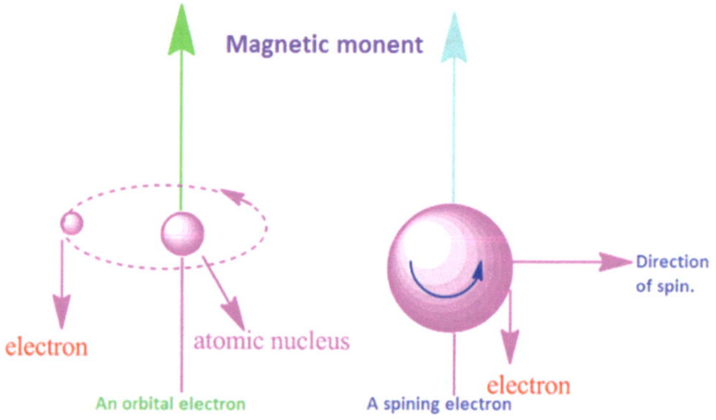

Figure 4.10. *The magnetic properties of transition metal complexes.*

Source: https://blogger.googleusercontent.com/img/b/R29vZ2xl/AVvXsEjE8ISL9w-FG3v9FV34nbhq4SprsNyqBkakg4oJo9TEPc-foTmsMfnmTBpiDqTFoBtNeK6b-Ax0xykKEwQ3MHS7cKfx9EFXc4Jp7TsUo5P0tyoEXLHOkgGGwU15UyYXqmg-wGsZ1sLn_LBvgU/s1600/mmmnoy.png.

4.5.3. Reactivity and Stability of Coordination Compounds

The stability and reactivity of coordination compounds are truly defined by the nature of the ligands present, along with the electronic and geometric nature of the metal core. In the study of these determinants, it can be observed that type of ligand—whether monodentate or bidentate—can drastically influence rate and reaction pathway. Monodentate ligands, because of having one site of coordination only, in most cases exhibit distinctive trends of reactivity relative to their bidentate analogues because the latter can form more stable chelate rings. Stability of complexes formed is generally linked with the chelate effect whereby the formation of multiple bonds between the metal ion and the ligand enhances overall stability. This phenomenon points out how fine the balance between stability and reactivity is, making the point that while coordination compounds may be extremely stable, they are nonetheless susceptible to substitution reactions, especially on the basis of ligand types (Gray et al., 1968). Advances in experimental technique in recent years have greatly assisted us in understanding reaction kinetics in coordination chemistry. Of special mention is mechanistic study of ligand substitution reactions, which has been strengthening with the assistance of rapid reaction methods that expand the range of observable rates. Of particular significance here are cobalt(III) and platinum(II) complexes, which are model systems for studying ligand substitution processes.

The theory behind these replacements is more and more supported by molecular orbital (MO) theory, which allows electronic interactions to be understood in a stereospecific way throughout the course of reaction. Not only does such integration define the role of bonding electrons as a determining factor in reactivity but also makes the ligand replacement reactions separable on the basis of their typical features and intrinsic mechanisms (Gray et al., 1966). Additionally, the cross-talk between ligand field theory and molecular orbital theory continues to determine stability for coordination compounds. It identifies how variation in field strength of the ligands—based on electronic structure and geometry—also varies the d-orbital energy levels directly, therefore influencing reactivity in the complexes. Stability can be engineered with the addition of strong field ligands, which lower the d-orbital energy and result in drastic electronic redistribution. The latter makes reaction prediction more difficult but also introduces avenues to rationally design new coordination compounds with targeted properties for a specific application, i.e., catalysis and material science. As the discipline evolves further, integration of the structural data into MO theory will yield improved tools for predicting and controlling reactivity and stability of coordination compounds efficiently (Gray et al., 1966).

4.6. CASE STUDIES OF LIGAND FIELD AND MOLECULAR ORBITAL THEORY

Within the framework of Ligand Field Theory (LFT) and Molecular Orbital Theory (MOT), case studies are used to describe the subtle electronic interactions between ligands and transition metal ions. Case studies merge theoretical prediction with experimental observation to probe the electronic nature of organometallic complexes. For instance, the sensitivity of the radiative decay rate to minor chemical perturbations highlights the sensitivity of electronic structure to molecular behavior. With the application of exact diagonalization models and the Configuration Interaction Singles (CIS) method, researchers are able to conclude that, for wide parameter ranges, the CIS is a reliable technique for mimicking low-energy excitations and photophysical features. This demonstrates the limited effect of electronic correlations, a finding that can influence theory foundations employed to describe such compounds, particularly in applications like optoelectronic devices (Jacko et al., 2011). Ruthenium(II) polypyridine complexes are also illustrated by case studies to reveal the advantages of merging Ligand Field

Theory with computational approaches like Density Functional Theory (DFT). By comparing a dataset of 98 complexes, researchers demonstrate how gas-phase DFT calculations correlate well with experiment. This correlation is key to maintaining theoretical models that are at loss with the complexity of molecular orbitals, i.e., the combination of antibonding metal orbitals with ligand orbitals. A Partial Density of States (PDOS) calculation introduces a nuanced understanding, indeed allowing for orbital energies to be derived for metal and ligand contributions. Not only does it identify a spectrochemical series of ligands but also offers a method for estimating Time-Dependent DFT (TD-DFT) absorption energies, showing how case studies expand our knowledge about ligand-metal interactions in MOT (Amolo et al., 2017).

Furthermore, results from PDOS analysis have implications stretching far beyond the realm of theoretical formulation. By connecting energies from PDOS-LFT to experimental luminescent properties, case studies unveil new avenues for application in materials science and photonics. The temperature dependence of luminescence lifetimes provides evidence for the applicability of these theoretical models in predicting real behavior. Case studies connecting elementary theories with experimental findings therefore serve as an essential bridge, allowing scientists to construct models explaining individual characteristics of complexes. The knowledge gained by way of these investigations not only adds to the comprehension of electronic correlations and bonding interactions but also enhances the potential for the development of novel materials with tailored photophysical properties, implying the salience of hybrid theoretical and experimental paradigms in the field (Amolo et al., 2017; Jacko et al., 2011) (Figure 4.11).

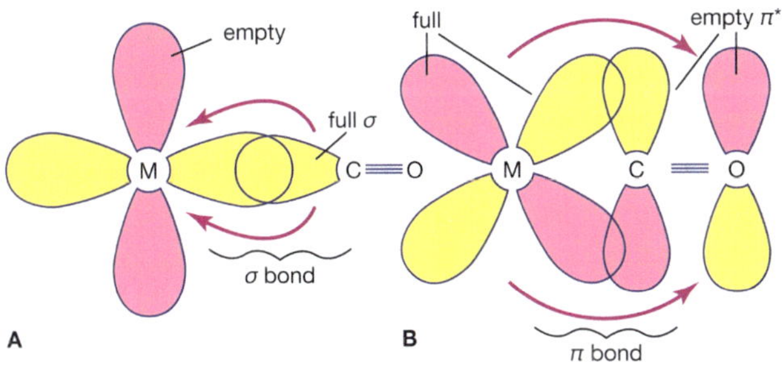

Figure 4.11. *The framework of ligand field theory (LFT) and molecular orbital theory (MOT).*

Source: https://cdn.britannica.com/91/7391-050-845D2240/ligand-Bonding-atom-bond-CO-electrons-pair.jpg.

5.6.1. Analysis of a Specific Transition Metal Complex

The investigation of transition metal complexes through the tools of ligand field and molecular orbital theories provides depth in understanding their electronic structure and chemistry. As an example, to study a specific ruthenium(II) polypyridine complex, computational methods like density-functional theory (DFT) can be employed. These methods provide valuable information about the geometry and electronic transitions described by experimental absorption spectra. DFT's predictive power for these complexes is underlined by good correlations to X-ray crystallography data, which suggests a firm foundation for theoretical study. But the efficiency of such computer techniques is complicated by widespread orbital mixing, particularly in heavy transition metals, and this introduces the need for sophisticated models describing the metal orbital and ligand contributions to the complexes energy profile, highlighting the interplay between ligand field theory and modern computer techniques. Further investigation of charge distribution in these complexes has revealed significant characteristics that differentiate them from their light counterparts. Employing a set of transition metal complex models, studies have documented a persistent tendency of charge depletion at the metal center in the process of substituting a more electronegative ligand. This phenomenon where metal atoms customarily exhibit charge deficiency, specifically from minority spin states, not only accounts for the type of bonding in the complex but also provides insight into its magnetic properties. These findings further reinforce the necessity for the unification of charge transfer knowledge with classical ligand field theory techniques, further supporting the electronic structure-magnetic property correlation that forms the basis of the objectives of ligand field and molecular orbital theory as stipulated in contemporary chemical studies (Gani et al., 2016). Moreover, by means of partial density-of-states (PDOS) analysis in combination with DFT calculations, researchers can infer useful information on the energetic order of molecular orbitals. The method offers an overall evaluation of the ligand field effects on the electronic structure of the metal center. Specifically, PDOS-LFT explains the metal $t_{2}g$ and e^{*}_{e} orbital and ligand π^{*} orbital energies, hence establishing a spectrochemical series on the basis of the field strength order of ligands. With this resonance, a clearer image emerges regarding the optical and electronic characteristics of transition metal complexes, disclosing their absorption characteristics as well as providing windows for more accurate predictions regarding luminescence and reaction kinetics under varying conditions. This comprehensive conceptualization underscores the interdependence of ligand field theory and molecular orbital treatment in providing a link between theoretical prediction and experimental observation (Amolo et al., 2017) (Figure 4.12).

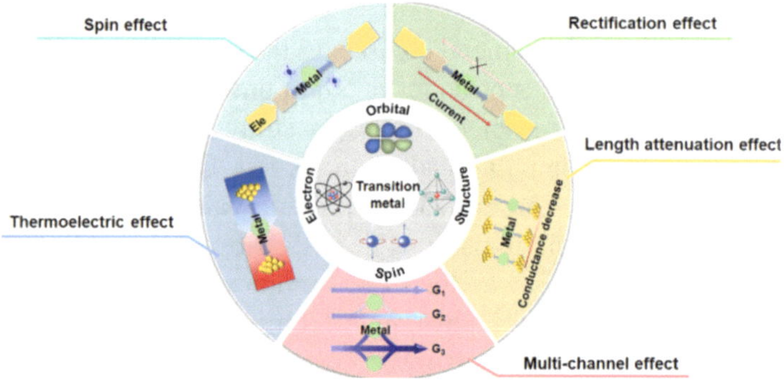

Figure 4.12. *The analysis of transition metal complexes.*

Source: https://media.springernature.com/lw685/springer-static/image/art%3A10.1007%2Fs12598-024-03127-4/MediaObjects/12598_2024_3127_Figa_HTML.png.

4.6.2. Comparison of Theoretical Predictions with Experimental Data

The comparison of theoretical predictions with experimental data is a cornerstone for the verification of ligand field and molecular orbital theories.

In the realm of coordination chemistry, particularly transition metal complexes, theoretical models provide information on the electronic structures and magnetic properties of these complexes. By employing computational tools such as Density Functional Theory (DFT), researchers are able to predict molecular shapes, electronic structures, and magnetic moments. When these theoretical predictions are contrasted with experimental techniques, such as spectroscopy or magnetic susceptibility measurements, discrepancies arise which reflect significant insight into the electronic environment around the complexes. Such similarity is crucial to enhance existing mathematical models and explore new theoretical models that will offer a better explanation of phenomena observed. Research on metal phthalocyanines (MPc) is a case in point and demonstrates how theoretical calculations compare well with experiment, thereby elucidating their electronic nature. Computational studies of various open-shell transition metals within the MPc model demonstrate the structural sensitivity of such compounds. For instance, spin density functional theory (SDFT) sheds light on electronic molecular gap and overall magnetic moment to be regulated through different metal replacements.

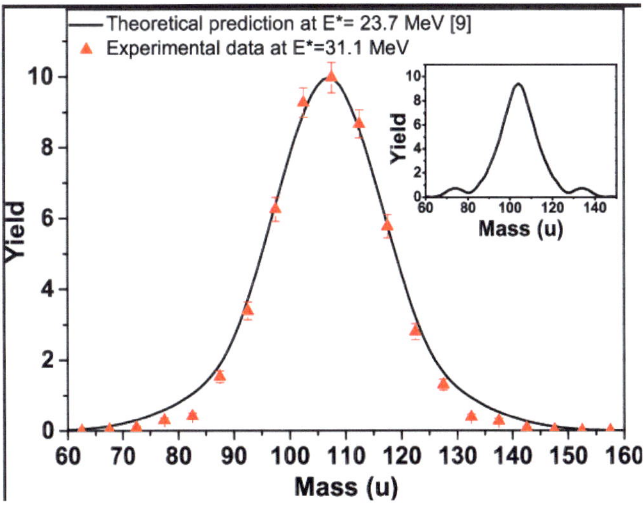

Figure 4.13. *The comparison of theoretical predictions with experimental data.*

Source: https://www.researchgate.net/profile/Dipen-Paul-2/publication/346800064/figure/fig1/AS:11431281079782954@1660842298131/Comparison-of-theoretical-predictions-by-Moeller-et-al-9-20-with-the-experimental-data.png.

This theoretical insight has direct correspondence with experimental detection of vibrational modes and magnetic behavior, thereby reinforcing the importance of this comparative analysis. In addition, when theoretical predictions are substantiated by experimental data, it fosters belief in the computational means employed and the resulting interpretations of chemical trends. Second, the application of advanced methodologies like the DFT+U method also serves to underscore the role of strong theoretical frameworks in elucidating transition metal complexes. Through transition metal porphine molecule adsorption on the gold surface, researchers were able to observe extreme electronic and magnetic behavior changes, which were based on the choice of the DFT+U parameterization. The findings showed extreme contrasts in ground states as well as interaction energies compared to DFT alone. This capability of theoretical models to accurately predict the influence of metal environments towards a direction suggests the necessity for continuous development and adjustment of these models based on experimental findings. Thus, when theory and experiments converge in close proximity to each other, scientists can gain profound understanding of the complex dynamics of ligand fields and molecular orbitals, ultimately advancing the discipline of coordination chemistry and materials science (Figure 4.13).

4.6.3. Implications for Future Research in Coordination Chemistry

The developments achieved in coordination chemistry have far-reaching implications for future work, particularly regarding the knowledge of electronic structures of transition metal complexes. As demonstrated by recent research activity, the intricate interplay of ligand-metal center interactions can be directly modulated to simulate the reactivity of metalloenzymes in biology, such as thiol dioxygenases (TDOs) and their models. Research on mononuclear non-heme iron complexes has shown that careful selection of the sets of ligands, e.g., tris(4,5-diphenyl-1-methylimidazol-2-yl)phosphine (Ph_2TIP), can effectively replicate pivotal active site features of naturally occurring metalloenzymes. This does more than just show that catalytic systems can be made, it also suggests the ligand field theory and principles of molecular orbitals upon which these interactions work. Further work built upon this can potentially introduce revolutionary advancements in catalyst design, illustrating further the fundamental chemical processes. With scientists delving more into ligand field and molecular orbital theory, it is increasingly clear that current computational methods play a crucial role in deciphering the complex electronic characteristics of coordination complexes. Current accounts of nitrogenase metal cofactors, particularly the P-cluster and iron molybdenum cofactors, are indicative of the relevance of many-electron wavefunction simulations in elucidating their electronic structures. These advanced theory models allow for more refined exploration of low-energy electronic states and the interplay between spin configurations and orbital excitations. The implications of the findings could revolutionize our understanding of electron transfer mechanisms in biological systems and direct the design of new coordination complexes mimicking them. Follow-up studies with similar computational approaches can give more insight into metal-ligand dynamics under various conditions. In addition, the scope of possible applications of future coordination chemistry is also in environmental science and sustainability. Development of earth-abundant metal complexes with reactivity similar to more expensive noble metals is a significant requirement, particularly with respect to renewable energy technology. Studies of the reactivity of these complexes, complemented by observations from ligand field and molecular orbital theories, might provide new solutions for energy conversion and storage devices. For instance, an investigation of the reactivity patterns of freshly synthesized ligands like cysteine bidentate derivatives could provide new catalytic pathways that are not only more efficient but also more environmentally friendly. This marriage of coordination chemistry with sustainability goals is an exciting frontier for future research, with the potential to make materials through synthesis that not only advance scientific understanding but also address important world problems.

SUMMARY

In Summary, the study of ligand field theory and molecular orbital theory provides valuable information on the electronic structures of metal complexes. The union of these theories not only further extends our understanding of coordination chemistry but also calls to the forefront the complications inherent in redox-active ligand-containing systems. The historical context of metal complex studies, particularly dithiolate-ligand-containing complexes, sets the stage for contemporary investigations that demonstrate classical oxidation-state characterizations are not sufficient. As observed from past outcomes, such systems also exhibit unique features, such as significant ligand–metal mixing within frontier orbitals, in contrast to widely accepted perceptions of electron transfer processes in such complexes (Eisenberg et al., 2011). Such a shift towards a more diverse comprehension necessitates sophisticated spectroscopic and computational approaches to further investigate the fundamental processes in coordination chemistry. In addition, the ramifications of such theories reach beyond an academic interest; they have very high prospects for hands-on applications within materials science and nanotechnology. For example, studies on molecular magnets, such as those containing Mn ions, show that unforeseen exchange interactions between oxidation states can take place. Ab initio calculations of the Mn_4 complex yield results that indicate the strongest exchange couplings are not only among identical ions but also among pairs of differently charged ions, which is opposite to the conventional assumptions (Anisimov et al., 2014). This new understanding of magnetic properties illustrates how molecular design can tap complex electronic interactions to facilitate technologies, with future possibilities for innovative materials with prescribed magnetic properties. Lastly, the joining of ligand field and molecular orbital theory is the basis of the multi-layered nature of metal complex chemistry. As scientists continue to build on these concepts through experimental evidence and theoretical modeling, there will certainly be ever-more advanced interactions that influence the stability, reactivity, and properties of these compounds. The first research to figure out the intricacies of electronic structures has paved the way for further questions that will make our understanding on complex systems more sophisticated. A complete grasp of these theories, thus, not just contributes to intellectual books but also births further investigations that might transform the discipline of inorganic chemistry and its application.

Understanding the difference and interrelationship between LFT and MO Theory is crucial in grasping the electronic character of coordination compounds. LFT is employed to depict d-orbital splitting of transition metal complexes by an external electric field generated by the ligands in their environment. This alters the energy of the d-orbitals, influencing the absorption spectrum and magnetic behavior of the complex. Conversely, MO Theory gives a more sophisticated picture through metal-ligand orbital interaction considerations, thereby allowing for a full investigation of bonding, structure, and reactivity. The integration of these perspectives gives a clearer understanding while supplementing predictive capability toward electronic transitions as well as coordination geometries. A key advancement in this respect is the development of Partial Density of States (PDOS) analysis, which has a close association with concepts based on LFT. It is reported that the use of gas phase density-functional theory (DFT) can explain the complex behaviors of ruthenium(II) polypyridine complexes, with excellent correlations with experimental results. The mirror of a spectrochemical series by PDOS-LFT energies demonstrates its ability to approximate the energies of metal t$_{2g}$ and e*$_{g}$ orbitals responsible for complex behavior, with the calculation of approximations of absorption energies in TD-DFT calculations. This approach emphasizes the need for an overall theoretical model that combines both the molecular descriptions arising from MO Theory and the empirical data delivered by spectrochemical measurement, highlighting the significance of comprehensive methods in current coordination chemistry. Also, the recent developments like the Effective Fragment Molecular Orbital (EFMO) approach have greatly changed our methodology to probe complex systems, offering new windows to reaction mechanism insight. The incorporation of the frozen domain method allows site optimization that considers polarization effects of the entire molecular system. This has been crucial in successful tracing of reaction paths, e.g., chorismate mutase. With the ability to trace out reaction coordinates and approximate energy barriers, the realism of employing molecular orbital methods in analyzing catalytic processes is reinforced. The power of these approaches not only testifies to the theoretical frameworks of LFT and MO Theory but also demonstrates their utility in solving difficult chemical problems, providing a unified view of ligand interaction and metal-ligand dynamics within coordination compounds.

REFERENCES

1. Abragam, A., Luchinat, C., Ravera, E., Parigi, G., Harriman, J. E., Jackson, J. D., Kaupp, M., et al. (2019). Pseudo contact shifts and paramagnetic susceptibility in semiempirical and quantum chemistry theories. http://arxiv.org/abs/1804.09055 (accessed on 13th September 2025).
2. Adamo, C., Arsenault, L. F., Balabin, I. A., Baldi, A., Bartók, A. P., Behler, J., et al. (2016). Machine learning, quantum mechanics, and chemical compound space. http://arxiv.org/abs/1510.07512 (accessed on 13th September 2025).
3. Alexandrova, A. N., Fuller, J., Kang, K., Reath, A. H., et al. (2019). Installation of internal electric fields by non-redox active cations in transition metal complexes. https://core.ac.uk/download/323078524.pdf (accessed on 13th September 2025).
4. Amolo, G., Casida, M. E., Kituyi, L., Magero, J., et al. (2017). Partial density of states ligand field theory (PDOS-LFT): Recovering an LFT-like picture and application to photo properties of ruthenium(II) polypyridine complexes. http://arxiv.org/abs/1707.03665 (accessed on 13th September 2025).
5. Anisimov, V. I., Anokhin, A. O., Khomskii, D. I., Korotin, M. A., et al. (2014). The nature of the ferromagnetic ground state in the Mn_4 molecular magnet. http://arxiv.org/abs/1312.6390 (accessed on 13th September 2025).
6. Arillo-Flores, O. I., Eckern, U., Fadlallah, M. M., Romero, A. H., et al. (2013). Magnetic, electronic and vibrational properties of metal and fluorinated metal phthalocyanines. https://core.ac.uk/download/212324854.pdf (accessed on 13th September 2025).
7. Azarbayejani, G. H., & Kikuchi, C. (2025). An MO theory of the cubic field splitting of $3d^5$ ^6S ions in II–IV compounds of Td symmetry. https://core.ac.uk/download/pdf/85248644.pdf (accessed on 13th September 2025).
8. Batista, V. S., Chandross, M. E., Leung, K., Medforth, C. J., et al. (2006). Density functional theory and DFT+U study of transition metal porphins adsorbed on Au(111) surfaces and effects of applied electric fields. http://arxiv.org/abs/cond-mat/0610742 (accessed on 13th September 2025).
9. Bertini, I., Gray, H. B., Lippard, S. J., Valentine, J. S., et al. (1994). *Bioinorganic Chemistry.* https://core.ac.uk/download/4890319.pdf (accessed on 13th September 2025).
10. Bogdanov, N. A., van den Brink, J., Büchner, B., Hozoi, L., et al. (2015).

Orbital reconstruction in nonpolar tetravalent transition-metal oxide layers. https://core.ac.uk/download/525036928.pdf (accessed on 13th September 2025).

11. Brunold, T. C., Fiedler, A. T., Fischer, A. A., Lindeman, S. V., et al. (2016). Synthesis, X-ray structures, electronic properties, and O_2/NO reactivities of thiol dioxygenase active-site models. https://core.ac.uk/download/213076498.pdf (accessed on 13th September 2025).
12. Chan, G. K. L., Guo, S., Li, Z., Sun, Q., et al. (2019). Electronic landscape of the P-cluster of nitrogenase as revealed through many-electron quantum wavefunctions. https://core.ac.uk/download/211385964.pdf (accessed on 13th September 2025).
13. Daniel, C., Dorn, J., Fumanal, M., González, L., et al. (2018). Quantitative wave function analysis for excited states of transition metal complexes. https://core.ac.uk/download/288360000.pdf (accessed on 13th September 2025).
14. Eisenberg, R., & Gray, H. B. (2011). Noninnocence in metal complexes: A dithiolene dawn. https://core.ac.uk/download/4892476.pdf (accessed on 13th September 2025).
15. Fedorov, D. G., Jensen, J. H., & Steinmann, C. (2013). Mapping enzymatic catalysis using the effective fragment molecular orbital method: Towards all *ab initio* biochemistry. https://core.ac.uk/download/pdf/50689172.pdf (accessed on 13th September 2025).
16. Fuhrer, M. S., Medhekar, N. V., & Yin, Y. (2019). Selective control of surface spin current in topological materials based on pyrite-type OsX_2 (X = Se, Te) crystals. http://arxiv.org/abs/1811.11330 (accessed on 13th September 2025).
17. Gani, T. Z. H., & Kulik, H. J. (2016). Where does the density localize? Convergent behavior for global hybrids, range separation, and DFT+U. https://core.ac.uk/download/141472821.pdf (accessed on 13th September 2025).
18. Golden, J. C., Ho, V., & Lubchenko, V. (2017). The chemical bond as an emergent phenomenon. http://arxiv.org/abs/1705.08429 (accessed on 13th September 2025).
19. Gray, H. B., & Langford, C. H. (1968). *Ligand Substitution Dynamics*. https://core.ac.uk/download/4886900.pdf (accessed on 13th September 2025).
20. Harden, I. (2023). *Computational Modeling of Complex Chemical Transformations in Homogeneous Catalysis*. https://core.ac.uk/download/618184695.pdf (accessed on 13th September 2025).

21. Held, K., Kresse, G., Liu, P., Tomczak, J. M., et al. (2017). Merging GW with DMFT and non-local correlations beyond. http://arxiv.org/abs/1703.08446 (accessed on 13th September 2025).
22. Hietschold, M., Shukrynau, P., & Smykalla, L. (2012). Investigation of ultrathin layers of bis(phthalocyaninato)lutetium (III) on graphite. http://arxiv.org/abs/1312.6509 (accessed on 13th September 2025).
23. Jacko, A. C., Powell, B. J., Baccouche, M., Bersuker, I. B., Bomben, P. G., Brocks, G., Butschke, B., et al. (2011). Electronic correlations in organometallic complexes. http://arxiv.org/abs/1102.2301 (accessed on 13th September 2025).
24. Jiménez Calzado, C., & Maynau, D. (2011). Evaluation of magnetic terms in Cu_4O_4 cubane-like systems from selected configuration interaction calculations: A case study of polynuclear transition-metal systems. https://core.ac.uk/download/322845459.pdf (accessed on 13th September 2025).
25. Ma, C. (2012). *Novel Algorithms and Tools for Ligand-Based Drug Design*. https://core.ac.uk/download/12212409.pdf (accessed on 13th September 2025).
26. Mitrushchenkov, A., Stoll, H., Hozoi, L., March, N. H., Ashcroft, N. W., Fulde, P., et al. (2007). *Ab initio* wavefunction based methods for excited states in solids: Correlation corrections to the band structure of ionic oxides. http://arxiv.org/abs/0704.1906 (accessed on 13th September 2025).
27. Natelson, D., & Scott, G. D. (2010). Kondo resonances in molecular devices. http://arxiv.org/abs/1003.1938 (accessed on 13th September 2025).
28. Pasquier, D., & Yazyev, O. V. (2018). Crystal field, ligand field, and interorbital effects in two-dimensional transition metal dichalcogenides across the periodic table. http://arxiv.org/abs/1810.01302 (accessed on 13th September 2025).
29. Reiher, M., & Weymuth, T. (2014). Inverse quantum chemistry: Concepts and strategies for rational compound design. http://arxiv.org/abs/1401.1512 (accessed on 13th September 2025).
30. Stickrath, A. B., Blankenship, R. E., Hoffman, B. M., Zhu, D., Lemke, H. T., Haldrup, K., Chen, L. X., et al. (2016). Ultrafast excited state relaxation of a metalloporphyrin revealed by femtosecond X-ray absorption spectroscopy. https://core.ac.uk/download/84003013.pdf (accessed on 13th September 2025).

CHAPTER 5

ISOMERISM IN COORDINATION COMPOUNDS

Contents

INTRODUCTION

The presence of isomerism in coordination complexes is a prominent area of study in coordination chemistry. Isomerism refers to the existence of compounds which have the same formula but have different structural or spatial geometry. This deviation leads to several alternate chemical properties and behavior of the isomers, hence why it's crucial for chemists to know about this topic. Different types of isomerism are realized for coordination complexes, most prominently structural isomerism and stereoisomerism. Complexes can be structurally different based on the ligand arrangement around the central metal cation, whereas stereoisomerism consists of distinct spatial arrangements of the same chemical entity. The complexity of coordination compounds through isomerism is imperative for use in most fields, including catalysis, biochemistry, and materials science, and emphasizes the need for further exploratory knowledge in this branch of study. Within the realm of coordination compounds, bonding or linkage dispositions are important in shedding light on the topic of isomerism. For instance, ligand coordination can lead to multiple isomeric species as in the aforementioned discussed linkage isomerism where the same ligand coordinates to the metal center in various ways. A notable example is found in the copper(II)-bearing nitrite reductases, where a tris(2-pyridyl)copper(II) complex exhibits the ability to show multiple modes of binding to nitrite anions. It is not only proof of the complexity of coordination chemistry but also that of isomeric form prediction and characterization. The study of such isomers can reveal information concerning the intrinsic properties of the compounds, for instance, their reactivity and stability, which is key in creating biomimetic catalysts in industrial processes, as discussed in the results of recent studies (Roger et al., 2017). The isomeric potential goes far beyond elementary chemistry to practical applications since it comes to play in fields such as pharmacology and materials science. The unique properties of isomers can dramatically alter the efficacy and performance of medicines. In such applications, getting to know the precise structural and functional properties of coordination isomers becomes extremely important for drug discovery and optimization. Moreover, techniques employed in synthesizing specific isomers constitute a crucial component in designing techniques in synthetic chemistry. According to definitions in university curriculum guides, thorough knowledge of coordination chemistry and isomerism result in innovative development to current challenges in chemistry and allied fields (Ashuyko et al., 2013). Combining theoretical knowledge and practical application, research on isomerism in coordination compounds has much potential for future scientific studies.

Definition of Coordination Compounds

Coordination compounds or coordination complexes are a type of compound that contains a metal atom or ion at the center with one or more ligands, which are ions or molecules. Anionic or neutral ligands can be used, and these ligands bind to the metal center by forming coordinate covalent bonds, in which both electrons in the bond originate from the ligand. The geometrical nature of coordination in these compounds results in a wide range of varied chemical properties and behaviors, enabling coordination compounds to play essential roles in catalysis, biology, and materials science. Distinguished by their unique geometrical structures, coordination compounds may vary in numerous forms of spatial arrangement, which creates a rich chemistry of isomerism that is necessary to account for their reactivity and function within chemical and biological systems. The coordination compounds' complexity enables it to exist in several classes of isomers, and it can be broadly categorized as structural and stereoisomers. Structural isomers feature differences in the bonding among the atoms, whereby a modification of the attachment of the ligands or coordination capacity leads to various compounds. Linkage isomerism appearing in a copper complex with nitrite ligands exemplifies how variations in ligand attachment could have significant impacts on a compound's properties. Reports indicate that such complexes can have more than one mode of binding, an indication being the κ2-ONO and κ1-NO_2 forms of the tris(2-pyridyl) copper complex $[Cu_1(NO_2)_2]$ (Roger et al., 2017). Such structural nuances are of essence since they have direct implications for the reactivity and selectivity of coordination compounds during various chemical reactions. Stereoisomerism also contributes to the diversity of coordination compounds because it introduces geometric and optical isomers, products with different spatial orientation of ligands in relation to the central metal atom. Geometric isomers, for instance, can have their configurations different based on the relative positioning of the ligands, hence variations in physical and chemical properties. By contrast, optical isomers, or enantiomers, are non-superimposable mirror images, which pose special challenges and opportunities in areas such as pharmacology, where chirality can be important for biological activity. The versatility of coordination compounds, such as the potential to form different isomeric forms, renders them subjects of ongoing scientific research and development, particularly in applications like biomimetic catalyst synthesis, which can utilize novel structural and electronic characteristics from natural systems (Hamaekers et al., 2017).

Importance of Isomerism in Chemistry

Isomerism plays a very significant role in the chemical discipline, particularly in coordination compounds since the positioning of the atoms can result in very disparate chemical behavior and properties. The differences in structure due to isomerism, like geometric and optical isomers, not only play a pivotal role in determining reactivity but also the efficiency of these compounds in various applications. For instance, the special spatial arrangements in coordination compounds may lead to selective binding specificity with biomolecules and therefore affect their biological activity. This applicability also applies to industrial processes, e.g., catalysis and materials science, where selectivity and efficiency of catalysts are often related to their isomeric forms, proving the significance of an utter understanding of isomerism in the design and application of coordination compounds. In addition, isomerism is a topic that provides useful information about molecular interactions of special significance in the comprehension of complex chemical systems. For example, the intricate networks within compounds such as $\{Zn[S_2CN(CH_2CH_2OH)_2]_2\}2(pyrazine)$ reveal the function of exact geometries and hydrogen bonding interactions. The different architectures, with their packing efficiencies and thermal stabilities, illustrate how isomerism can be correlated with the stability of compounds, as in the reported attributes of one and two in Arman et al. (2017).

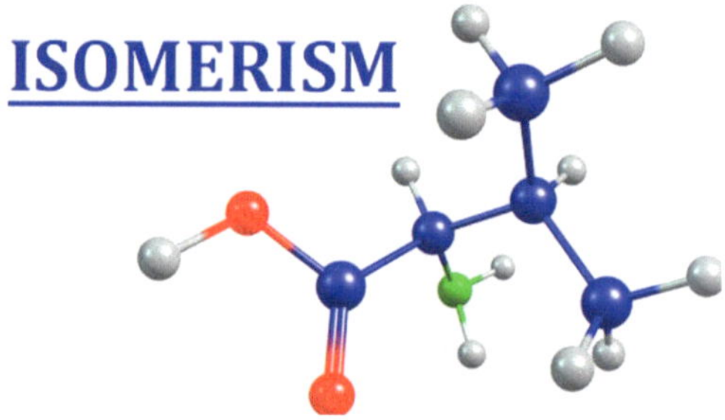

Figure 5.1. *Importance of isomerism in chemistry.*

Source: https://img.jagranjosh.com/imported/images/E/Articles/Isomerism_media.jpg.

Chemists can predict the stability and reactivity of coordination compounds by examining such structures and affect subsequent synthetic methods. This appreciation also further adds to the value of isomerism towards the advancement of coordination chemistry and its role in drug design, where the introduction of a single isomer can have dramatically different pharmacological effects. Additionally, the role of isomerism extends beyond mere structural variance and into the realm of synthetic chemistry, where the deliberate control of isomeric forms can lead to new methods of chemical transformation. For instance, the isomerization of lacunary polyoxoanions to single oxyanions, e.g., the β to γ isomerization of $[\beta 2\text{-}SiW_{11}O_{39}]8-$ to $[\gamma\text{-}SiW_{10}O_{36}]8-$, highlights the significance of isomerism in enabling reactions with vast structural changes (Rodr Aíguez-Fortea et al., 2016). The potential to utilize low-energy transition states for the transformation highlights the central interlink between isomerism and reaction mechanisms. As chemists continue to utilize isomerism in coordination compounds, they not only create new reactivity patterns but also underscore the importance of isomeric design in advancing the theoretical and practical dimensions of chemistry, ultimately enhancing scientific wisdom as well as applicability in numerous fields (Figure 5.1).

5.1. TYPES OF ISOMERISM

Isomerism is a key determinant of the structural variety and functional character of coordination compounds. The most significant classifications are structural isomerism and stereoisomerism, which both have a predictable very significant role in determining the behavior of these complexes. Structural isomerism occurs when compounds possess the same molecular formula but differ in the connectivity among their atoms. This type can be sub-classified into coordination isomerism, linkage isomerism, and functional group isomerism. For instance, in coordination isomerism, there can be multiple ions that are contained within the same coordination sphere and, therefore, possess different chemical properties. Linkage isomerism is a special one where a single ligand is capable of binding through different atoms, a good example being complexes such as $[Cu_1(NO_2)_2]$, which shows the multiple binding modes for nitrite discussed in Roger et al. (2017). Such variations show the intricacies to be encountered among coordination compounds. Stereoisomerism, the other major category, happens when molecules are not the same arrangement of atoms in space but the same connectivity. Geometric and optical are two broad types. Geometric isomerism typically exists in square planar or octahedral coordination complexes because of hindered rotation about a bond. For example, the ligand positioning in the form of cis and trans forms can greatly influence the reactivity and biological activity of the compound. Optical isomerism occurs when a compound can exist as non-superimposable mirror images, or enantiomers, and this plays a vital role in biological systems. This concept is important in understanding how the different isomers are capable of acting on biological molecules, as one enantiomer may be therapeutic and the other inert or toxic, and another reason why isomerism is essential in coordination compounds. Isomerism has even been investigated in the domain of computational chemistry, where theoretical models such as the one explained generator of virtual molecules have helped us better understand the behavior of coordination complexes. This model not only follows simple chemical rules like the octet rule but also efficiently creates valid and varied molecular structures. It also benefits from the capability of quickly recognizing duplicates, thereby speeding up the identification of chemical space. Virtual molecule generation and analysis enable new coordination compounds to be discovered, thereby substantiating the relevance of isomerism to materials science and synthetic chemistry. The findings on the prevalence of dipolar structures in the resulting molecules refer to undiscovered facets of isomerism, suggesting potential avenues to constructing new catalysts or medicines as discussed in Hamaekers et al. (2017). Isomerism thus remains an essential area of research, linking theoretical understandings to practical applications across diverse fields (Figure 5.2).

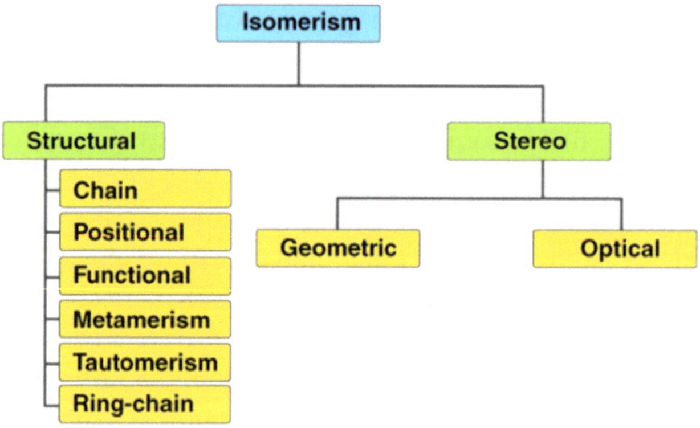

Figure 5.2. *Types of isomerism.*

Source: https://cdn1.byjus.com/wp-content/uploads/2016/01/Isomerism1-700x421.png.

5.1.1. Structural Isomerism

Structural isomerism plays a significant role in the identification and role of coordination compounds. Structural isomerism is present when molecules that contain the same molecular formula contain distinct structural forms of their atoms. Isomerism in coordination compounds occurs in several ways, including differences in ligand arrangement around the central metal atom or variation in the coordination bonds formed. These variations are responsible for adjusting the chemical property, reactivity, and physical parameters of the compounds. For example, the coordination sphere in metal complexes generates distinct isomeric species with distinct spectroscopic signatures. These variations are indicative of the utility of structural isomerism in the synthesis of complex molecules and their subsequent use in catalysis and material science. One of the intriguing features of structural isomerism in coordination compounds is the dynamic interrelationship among different isomers, particularly under varying conditions. Ligands may reorganize to change isomers, a process often facilitated by temperature or solvent changes. For instance, transition metal complexes with π-allylic ligands involve different isomeric forms—endo and exo—interconverting, showing dynamic behavior on external stimulation. Also, the fluxional character of these isomers enables distinctive interactions and reactivity patterns with other chemical species. Analytical methods, like multinuclear NMR spectroscopy, can clarify such structural changes, allowing us to access the mechanisms of isomerism. This underscores the broader significance of molecular flexibility in chemical design and reactivity, specifically in catalysis and organic synthesis (Cardin et al., 2017). Besides, studies on isomerism have uncovered the structural variability of coordination compounds based on various ligand structures and coordination forms. For example, studies on semi-rigid benzotriazole-based ligands demonstrated the preparation of greater than one coordination compound possessing different isomeric characters such as dimers or higher-dimensional coordination polymers (Abdul-Sada et al., 2018). These complexes illustrate how the geometric arrangement of space in the ligands is significant in dictating the properties and catalytic activity of the resulting complexes. Isomeric effects can even lead to the predominance of some isomers in solution regardless of crystallographic factors. Therefore, the comprehension of structural isomerism not only contributes to coordination chemistry's theoretical foundation but also aids practical use over a wide range from drug development to materials science, in which performance and effectiveness can be due to particular isomeric forms (Figure 5.3).

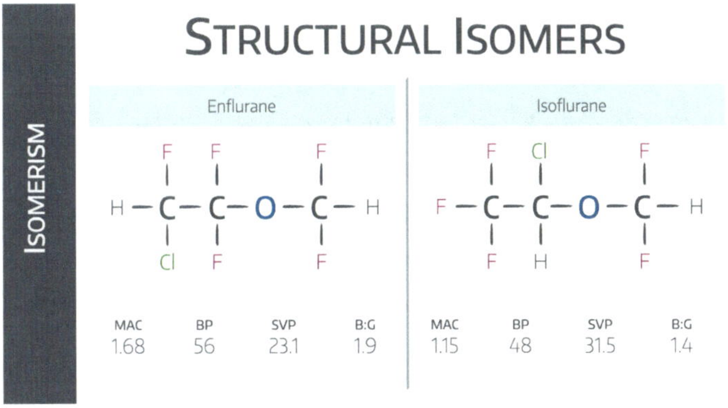

Figure 5.3. *Structural isomerism.*

Source: https://www.anaestheasier.com/content/images/2023/04/Isomerism-Structural.001--1-.jpeg.

5.1.2. Stereoisomerism

Knowledge of stereoisomerism is significant in the comprehension of coordination compounds because it discloses the spatial arrangements of ligands about a central metal atom. This isomerism occurs when molecules possess the same connection but vary in three-dimensional atom orientation. There are two primary forms of stereoisomerism, geometric (cis-trans) and optical isomerism. Geometric isomers can significantly contribute to the physical properties and reactivity of coordination compounds, such as their solubility and color. This geometric difference assumes an even higher level of importance when one considers metal complexes within biochemical systems, since the complex geometric arrangement of ligands within these compounds may determine the biological activity and interaction of the compound with other biomolecules.

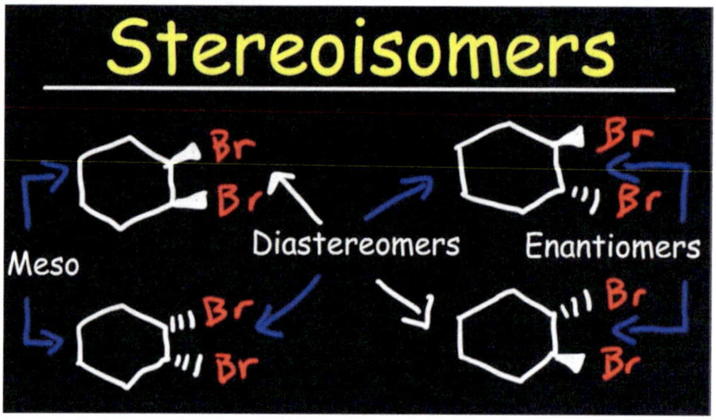

Figure 5.4. *The stereoisomerism.*

Source: https://i.ytimg.com/vi/gi55pE68dh4/maxresdefault.jpg.

Categorizing and knowing these types of arrangements allows chemists to more precisely predict and design the behavior of such coordination compounds within different chemical contexts. Increased to the stereoisomerism complexity is the addition of akamptisomerism, a new category found to possibly occur in coordination complexes. Research described in recent publications demonstrates that conventional systems of classifying compounds, including those provided by IUPAC, fail to account for this phenomenon, which complicates the understanding of the isomerism of these substances. Through analysis and synthesis of coordination compounds that fall outside the conventional definitions of stereoisomers, researchers become capable of defining lacunas within existing regulations. Building a Polytope Formalism model does help to integrate the various approaches previously employed in order to define stereoisomerism, and allow for akamptisomerism to be encompassed in a more general approach to understanding. This innovation not only enhances the characterization of compounds that already exist, but also presents the possibility of creating new compounds having typical stereoisomeric behavior. The educational relevance of the discovery and application of stereoisomerism in coordination compounds is vast, particularly with the inclusion of augmented reality (AR) technologies in educational curricula. As explicated in recent scholarship, the application of AR interventions offers students with hands-on learning spaces in which to engage with the abstract depictions of three-dimensional molecular shapes as tangible learning experiences. Such active engagement not only enhances learners' comprehension of stereoisomerism principles but also enhances intrinsic motivation and comprehension of the subject. During guided activities, such as those in coordination chemistry, learners are in a position to conceptualize stereochemical relationships, enhancing spatial perception and cognitive load management during learning. With improved instructional techniques, the application of AR tools helps ensure that pedagogical strategies in teaching chemistry are reformed to ensure that advanced concepts like stereoisomerism are easier and more engaging to learn for students in different learning institutions (Canfield et al., 2022; Elford et al., 2022) (Figure 5.4).

5.1.3. Geometric Isomerism

Geometrical isomerism, a significant isomerism phenomenon of coordination compounds, arises due to the varying spatial geometries of ligands in a metal center. The isomerism occurs more frequently in octahedral and square planar complexes, where positions of ligands can lead to isomers with differing chemical and physical characteristics. For example, in square planar complexes, cis disposition of two identical ligands results in diverse reactivity and stability compared to their trans analogs. The geometry orientation does not just affect coordination number but also very deeply impacts properties such as color, solubility, and biological activity. It is crucial when investigating the functionality and application of coordination compounds across catalysis to medicinal chemistry to consider such differences. The geometric isomerism observed in coordination compounds can be exemplified by transition metal complexes, the spatial arrangement of the donor atoms of which can lead to unique isomers. In octahedral complexes, geometric isomers can result in the form of cis or trans isomers, depending upon whether the similar ligands are adjacent or opposite. These geometries result in varied electronic and geometric surroundings that can vary stability and reactivity of the compounds. Steric and electronic interactions interact with one another, as exemplified in the $\{Zn[S_2CN(CH_2CH_2OH)_2]2\}2(pyrazine)$ compounds where five-coordinated geometry highlighted the significance of ligand orientation and the way it controls molecular stability and interaction. These analyses not just give a clearer picture of the phenomenon of geometric isomerism but also pave the way for the synthesis of coordination compounds with desired properties and functionalities (Arman et al., 2017). Furthermore, research on geometric isomerism is useful in gaining an overall understanding of the various forms of isomerism in coordination compounds. Coordination chemistry complexities seem even more highlighted when the subject involves the control by geometric shapes on optical isomerism, where geometric isomers of a compound could hold chiral properties. Investigating both geometric and optical isomerism not only enhances our base of chemistries, like students learning about coordination compounds, but also provides lines of new complex applications. With the central importance of isomerism to pharmaceutical and catalyst synthesis, the more thorough exploration of such molecular structures widens the chemist's arsenal (Ibodulloeva et al., 2023). This synoptic view takes in the manner fundamental notions like geometric isomerism illuminate that is critical to advancing coordination chemistry and its significance in science and business (Figure 5.5).

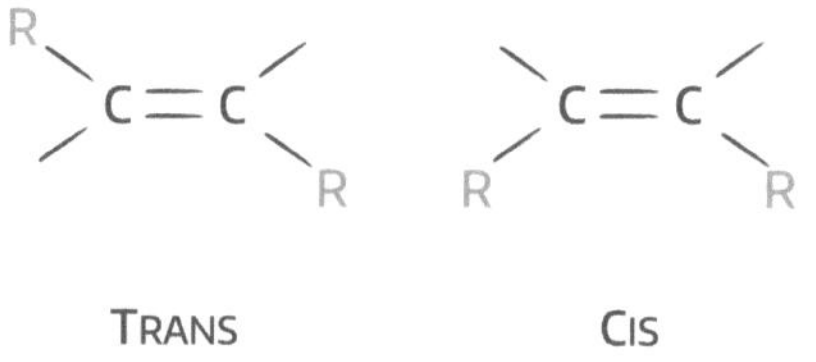

Figure 5.5. *Geometric isomerism.*

Source: https://www.anaestheasier.com/content/images/2023/04/Isomerism-Geo.001.png.

5.2. STRUCTURAL ISOMERISM IN COORDINATION COMPOUNDS

Structural isomerism in coordination compounds is fundamental to the knowledge of their chemical functionality and properties. Structural isomers are found where two compounds have the same formula but a different atom arrangement of the constituent atoms. Structural isomerism is particularly pertinent in coordination chemistry due to the intrinsic complexity of metal-ligand interactions. Isomers are grouped into numerous types, e.g., linkage isomers where the ligand attachment site on the center metal ion varies, and coordination isomers where ligands are transposed between two metal centers. The presence of such structural diversity not only renders the analysis and characterization of the compounds difficult but also causes them to possess a varied reactivity and stability, indicative of the intricate interplay between structure and function in coordination chemistry. Linkage isomerism is one such prime example of structural isomerism in coordination compounds, as amply proven by the properties of certain metal-nitrite complexes. When one talks of linkage isomers, one can use as an example the tris(2-pyridyl) copper complex that can bind nitrite in different ways—either as an asymmetric bidentate through the κ2-ONO mode or as a monodentate in various modes, including the κ1-ONO and κ1-NO_2 forms. The ability of ligands like nitrite to bind in different modes of binding makes structural isomerism crucial and its correlation with the reactivity and functionality of metal complexes. As seen in the copper complex, the presence of all three linkage isomers provides a deeper understanding of catalytic processes, particularly those engaged in the reaction such as nitrite reduction, thus revealing the intricate dynamics of metal-ligand interactions in biological and synthetic settings (Roger et al., 2017). Another fascinating phenomenon of structural isomerism is in the way of π-allylic ligands in coordination complexes, where conformational mobility allows for the generation of various isomers. π-Allylic ligated transition metal complexes exhibit dynamic features which can allow isomerization of exo to endo and vice versa, a feature that renders them even more effective in catalysis. Their fluxionality necessitates sophisticated analytical techniques, such as multinuclear NMR spectroscopy, to deconvolute their structural transformation. Knowledge of the isomerization mechanisms thus becomes central since they not only decide the stability and reactivity of the complexes but also tell us about how they can be utilized in organic synthesis. Such multi-dimensional analysis of π-allylic compounds encapsulates the rich mosaic of structural isomerism in coordination chemistry, creating new possibilities for research and application in both basic and applied sciences (Cardin et al., 2017).

5.2.1. Coordination Isomers

Coordination isomers are an interesting type of isomerism of coordination compounds with ligands illustrating a change in coordination modes around a central metal ion. The coordination isomers' structural heterogeneity can be the result of differing ligand orders, resulting in distinct geometries and perhaps divergent physical and chemical behavior. An example is the reaction of $[Co_2(H_2O)(TFMBz)4(py)4]$ (1) with 4,4-bipyridine (bpy) in diverse solvents to yield four coordination polymers with the same stoichiometry but highly disparate structures. This illustrates how minor differences in ligand interactions and coordination can create isomers that differ significantly in significant properties such as porosity and magnetic behavior, thereby highlighting the significance of the coordination environment in establishing the properties of resulting complexes (Ayllon et al., 2022). The peculiar behavior of coordination isomers highlights their value for applications ranging from catalysis to drug design. By exploiting the structural nuances inherent in such isomers, chemists can manipulate the properties of complex molecules to specific functional requirements. In drug design, for instance, the manipulation of coordination symmetry and ligand constitution permits specific therapeutic action. The underlying rationale for coordination isomerism thus also serves to connect foundational chemistry with its practical significance in areas such as medicinal chemistry, where the creation of compounds for targeted biological interactions is crucial to therapeutic action (Bruijnincx et al., 2009). Moreover, research on coordination isomers expands foundational understanding of coordination chemistry and metal-organic framework synthesis. As advanced research progresses into the intricate architectures formed by ligands and metal ions, novel isomers are being discovered with unanticipated structural and functional diversities. The application of sophisticated technologies, such as three-dimensional electron diffraction, further enhances the resolution of the structures, allowing researchers to realize their chemical behaviors. This ongoing exploration not only enhances knowledge at the theoretical level but also increases creativity in material science by highlighting the adaptive nature of coordination compounds. The synthesis conditions vs. isomer formation ultimately result in the dynamic nature of coordination chemistry, producing advanced systems with tailored functionalities (Ayllon et al., 2022) (Figure 5.6).

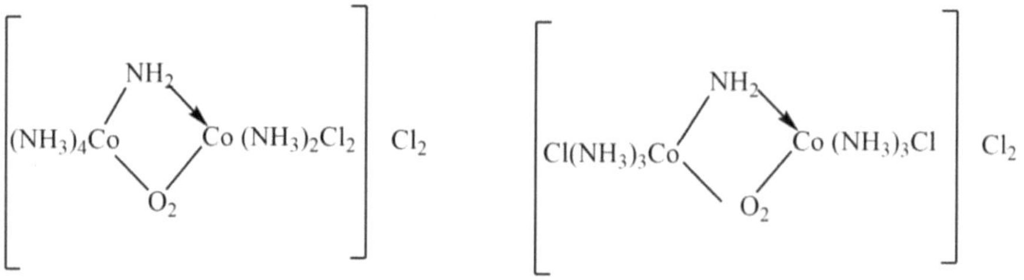

Figure 5.6. *Coordination isomers.*

Source: https://scienceinfo.com/wp-content/uploads/2022/11/Coordination-position-isomers.jpg.

5.2.2. Linkage Isomers

Linkage isomerism is a fascinating phenomenon of isomerism in coordination complexes, whereby ligands can coordinate to the metal center through different atoms. This is most often caused by bidentate ligands, since they possess two donor atoms capable of binding to the metal in different fashions. For example, in complexes with the nitrite ion (NO_2^-), the ligand can bind via the nitrogen or one of the oxygen atoms and yield various linkage isomers. Such structural variations have significant consequences in that they can affect the physical and chemical characteristics of the coordination compound, including stability, reactivity, and electronic properties. This intricacy emphasizes the multifunctional nature of coordination complexes and potential application in catalysis and materials science. Linkage isomer research often depends on the specific coordination sphere of the metal center. Recent research on palladium(II)-nitrite systems has demonstrated this concept through induction of complete conversion to a metastable endo-nitrito-(-NO) isomer under photonic excitation. These researches reveal that the main isomers arrangement can shift under specific conditions, when dynamic character of linkage isomerism becomes apparent. Such transient behavior, not only indicating metal-ligand interactions complexity, also opens new and innovative fields in photochemical processes. The easy availability of such isomeric species under mild conditions promises future enhancement in the preparation of smart materials responsive to light stimulation with adaptability in real-time applications (Hatcher et al., 2016). The structural flexibility provided through linkage isomerism can also have a significant impact on the efficacy of coordination complexes in catalytic reactions. For instance, ligand flexibility documented in complexes with complex architectures may introduce the variations in their coordination capabilities and, as a result, affect their catalytic activities. A perfect example is observed from studies on benzotriazole-derived ligands, in which coordination geometries variation introduced a variety of catalytic activities. Interestingly, ligands known to exhibit linkage isomerism exhibited enhanced activity in multi-component reactions, proving that even minor changes in ligand binding can play a vital role in reactivity and selectivity. This new insight into the structure-function relationship further highlights the importance of linkage isomers in the design of novel coordination compounds with specific catalytic applications (Abdul-Sada et al., 2018).

5.2.3. Ionization Isomers

Ionization isomers are of extreme importance in the analysis of coordination compounds as a series of isomers whose structure of the ions in solution varies. This occurs when the ligand and counterions have the capacity to replace each other within the coordination sphere to generate different ionic species with different properties. For instance, coordination compounds with the same central metal atom can have different reactive behaviors depending on whether specific ligands are included in the coordination sphere or as free ions in the solution. Such changes create challenges to classical assumptions of chemical identity, resulting in additional research into their utility in the stability and chemical reactivity of chemicals. As researchers learn more about this branch, a better understanding of ionization isomers is achieved, eventually leading us to a better understanding of the broader category of isomerism among coordination compounds. The investigation of ionization isomers contributes towards the broader discourse of isomerism in coordination compounds because the subcategory illustrates the intricate interplay between structure and function at the molecular level. Each isomer may have differing solubility, thermodynamic stability, and biological activity, depending on the geometric structure of the ligands and ions. Analysis of such isomers commonly employs advanced techniques such as electrospray mass spectrometry and computer calculations, providing significant information regarding the mechanisms by which such species equilibrate.

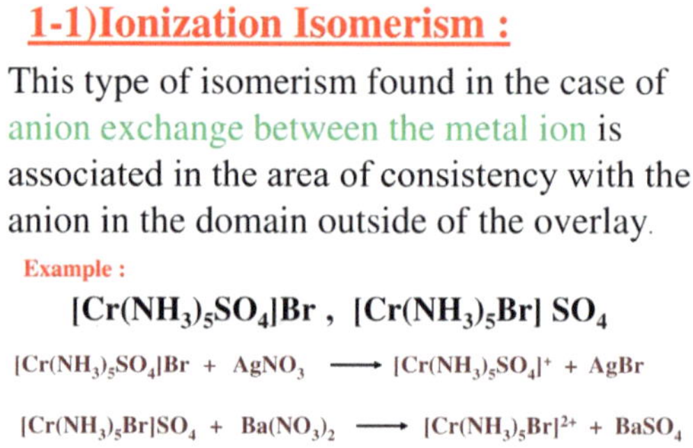

Figure 5.7. *Ionization isomers.*

Source: https://image1.slideserve.com/2228292/1-position-structural-isomerism-l.jpg.

Surprisingly, recent research has highlighted the role played by countercations in the stability and isomerization of ionization isomers, which emphasizes the intricacy entailed in these chemical compounds. This research trend addresses the connection between ionization isomerism and other forms of isomerism, thus emphasizing the significance of understanding ionization isomers as part of the entire coordination chemistry framework (Ibodulloeva et al., 2023). The investigation of ionization isomers not only adds to the knowledge of coordination compounds but also has uses in actual fields such as catalysis and materials science. The unique characteristics that accompany ionization isomers can lead to variations in reactivity, and such ionization isomers are of significant use in catalyzing some reactions. For example, certain ionization isomers can exhibit higher catalytic efficiency due to the presence of certain ions that facilitate reaction mechanisms. Also, the specific characteristics of the isomers can be applied to guide the synthesis of new materials so that it can be tailored to specifications for end use. As indicated in more recent research in polyoxotungstate chemistry, the pathways of transformation of these complexes can be broken down to determine critical information on structural control and reactivity (Rodríguez-Fortea et al., 2016). Therefore, the ionization isomers form part of the greater study of isomerism in coordination compounds, which reinforces a sense of understanding about their chemical diversity and utility (Figure 5.7).

5.3. STEREOISOMERISM IN COORDINATION COMPOUNDS

Stereoisomerism in coordination compounds is one of the key features of isomerism that focuses on the spatial geometry of the ligands around a central metal atom. This phenomenon is significant, as it not only affects the chemical reactivity of the compounds but also their physical properties, like color, magnetism, and biological activity. For example, the coordination compound cis-diaminedichloroplatinum(II) or cisplatin exhibits stereoisomerism, in which the cis form is therapeutically active in cancer treatment but the trans form does not exhibit the same degree of effectiveness. The understanding of stereoisomerism is essential to chemistry students since it is the bridge to the understanding of more complex concepts in coordination chemistry and other areas of study (Blake et al., 2018). This detailed understanding of molecular pattern configuration can prove to be invaluable in its application towards material science, pharmacology, and catalytic reactions. Stereoreplacement in coordination compounds can mainly be categorized under geometric and optical isomers. Geometric isomers arise from different spatial arrangement of ligands to the metal center, often described as cis and trans configurations. For instance, in octahedral compounds, ligands can have either fac (facial) or mer (meridional) arrangements that yield different properties and functions. Optical isomers arise when a compound presents non-superimposable mirror images, a characteristic commonly encountered with chiral complexes. These stereoisomeric differences reveal an even more intimate relationship between structure and function. Furthermore, recent literature has introduced definitions such as akamptisomerization, a novel type of conformational isomerism that subjects traditional nomenclature to a test and enlarges the understanding of stereochemical disparities in coordination compounds (Blake et al., 2018). The implications of stereoisomerism extend beyond theoretical realm, impacting areas such as pharmacology and materials science. In pharmacy chemistry, the dissimilar biological activities of enantiomers raise basic questions in drug development and regulation. For example, the efficacy and safety of chiral medications depend essentially on their stereochemical structure. Stereoisomers may lead to dissimilar physical properties in materials science, which affects catalyst and electronic material design. In return, a comprehensive understanding of stereoisomerism allows chemists to create and utilize the best possible coordination compounds in numerous applications. As the science evolves further with research, it will be important for future chemists to understand classical as well as novel stereoisomeric forms (Nelson et al., 1965).

5.3.1. Optical Isomerism

Optical isomerism, which is also frequently found in coordination compounds, arises when two or more compounds possess non-superimposable mirror forms, known as enantiomers. This isomerism finds its special significance in describing the spatial arrangement of ligands around a central metal ion in coordination compounds. The compound's rotatory power for plane-polarized light distinguishes these enantiomers and thereby proves them to have differential optical behavior. For all practical purposes, particularly those related to drugs and biochemical reactions, differential optical activity of such isomers can lead to vastly different biological behaviors. As coordination chemistry progresses, an understanding of optical isomerism becomes increasingly important in the design of compounds that are able to interact with biological systems, affecting the efficacy of drugs to toxicity. The molecular complexity of coordination compounds will, in most situations, generate the phenomenon of optical isomerism, particularly for the bidentate or polydentate ones. For instance, for semi-rigid benzotriazole-based ligand complexes, the structural design and symmetry can lead to enantiomerism. It has been proved from research that modifications of the flexibility of ligands, as exhibited by compounds synthesized with L1-L3 ligands, can significantly influence their ability to form optical isomers. The presence of different coordination environments leads to multiple structures. These differences, therefore, explain the ensuing optical activity of the compounds, which shows an elegant correlation between ligand structure and optical isomerism. This kind of complexity is critical for the synthesis of complex materials with desired optical properties for use in optoelectronic applications, as can be seen in studies of the conduction mechanism of fullerene derivatives (Macovez et al., 2016). Moreover, the importance of optical isomerism extends beyond basic chemistry to applied science, where the intriguing optical properties of the compounds are employed in the development of technological advances. For example, optical isomer interaction can influence the performance of photovoltaic cells and photodetectors. Coordination compounds exhibiting optical activity are also researched in various applications ranging from catalysis to materials science. The catalytic activity of some optical isomers in reactions, as with coordination compounds of silver cations, serves to emphasize their functional importance. To demonstrate, one such compound that was valued for its catalytic activity also exhibited this isomerism effect, serving to emphasize the direct correlation between structural characteristics and catalytic efficacy (Abdul-Sada et al., 2018). By further discovering these properties, researchers can maximize the application of optical isomers in a variety of high-tech technologies, which could lead to dramatic advancements in industry and science (Figure 5.8).

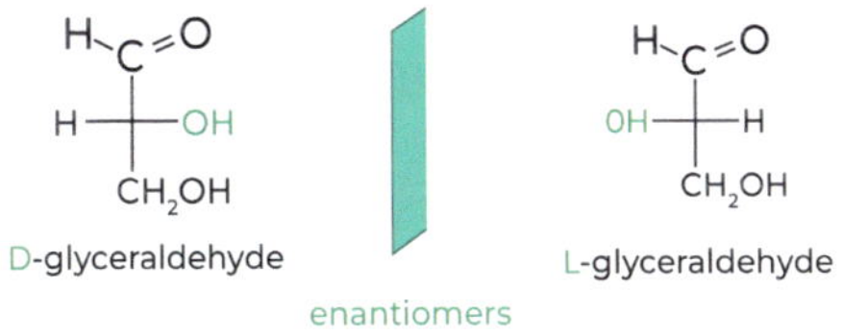

Figure 5.8. *Optical isomerism.*

Source: https://media.geeksforgeeks.org/wp-content/uploads/20231226122502/Optical-Isomerism-01.png.

5.3.2. Geometric Isomerism

Geometric isomerism, a fundamental aspect of coordination compounds, results from the direction of ligands about the central metal atom. This isomerism occurs when ligands are capable of taking up diverse positions relative to one another without altering the connectivity of the compound. In octahedral complexes, for instance, ligands are capable of occupying positions either adjacent (cis) or opposite (trans) relative to one another. Such spatial diversity gives rise to different physical and chemical properties between the isomers. Geometric isomerism's significance extends beyond mere aesthetics in molecular design since its presence can significantly influence stability, reactivity, and biological activity of coordination compounds, causing a richness of complexity in the systems. Detection of geometric isomers will normally involve the use of spectroscopic methods to identify the different structural features between isomers. Techniques such as IR and NMR spectroscopy, and X-ray crystallography enable chemists to investigate the distinctive vibrational and electronic properties associated with each isomer. These analytical techniques form the key to understanding the minute but significant differences in behavior between trans and cis forms. For instance, geometric isomerism in certain transition metal complexes has a great impact on redox potentials and, by extension, the reactivity of these compounds in industrial and biological processes. As Aljohani et al. (2016) observe, second coordination sphere interactions also alter the properties of such isomers, demonstrating the interdependent relationship between geometry and reactivity in coordination chemistry. Besides, biological and industrial applications of the effects of geometric isomerism are broadened to where a specific ligand configuration can dictate functionality. For instance, certain geometric isomers are selectively recognized by receptors or enzymes, illustrating the intricacies of molecular recognition processes. Preparation of specific isomers, therefore, is important in the synthesis of catalysts, pharmaceuticals, and materials science. As coordination compounds continue to progress, the means of selective production and manipulation of geometric isomers become increasingly relevant. By understanding more about the principles and applications of geometric isomerism, chemists can best use such phenomena to create new finds within fields of study such as biomimetic systems and catalytic reactions. This exploration is necessary, especially as there are now studies that cite the possibility of second coordination sphere interactions as noted in Кривоносова et al. (2016).

5.3.3. Chiral Complexes

Chiral complexes represent a fascinating group of coordination compounds due to their non-superimposable mirror images, just like right and left hands. This phenomenon of chirality is important in the realm of science, from the pharmaceutical industry to materials science. The significance of chiral complexes arises from their unique properties that arise from their asymmetric nature. Octahedral and tetrahedral coordination structures offer a platform on which metal centers may display diverse spatial arrangements with chiral ligands. The relative positioning of ligands with respect to the center metal determines the global reactivity, stability, and selectivity of such compounds, making them pivotal in the fields of catalysis and drug discovery, where enantiomeric purity is a critical criterion. Recent advances in the synthesis and analysis of chiral complexes have been responsible for unearthing their behavior and applications. For instance, 2-substituted oxetan-3-ones illustrate how chirality can be induced using specific metal coordination modes. Through lithiation and selective electrophile reactions, researchers were able to synthesize these compounds successfully in high enantiomeric excess, showing the potential of chiral complexes in the creation of sophisticated molecular structures. This process explains the degree to which metalation processes can offer more chiral control on conversions, with subsequent synthesis pathways that would have been very difficult in organic chemistry. Such processes point to the accessibility of synthesized chiral complexes, hence synthetic methodology innovations. Interactions of chiral complexes with metal ions are also research areas for chiral complexes, and they explain advanced coordination behavior with great implications on separation science. Most notably, the quadridentate N-heterocyclic ligand $CyMe_4$-hemi-BTBP is a prime example of the chiral complexity of metal-ligand interaction. Studies have shown that the ligand spontaneously produces the resolved 1:1 complexes of distinct compositions with selected lanthanide and transition metal ions, yielding a paramagnetic mixture of isomeric helical 2:2 chiral arrangements. Additional studies have revealed that said arrangements can display an affinity for one directional isomer, attesting to the subtle nature of chiral complexes in coordination chemistry. Knowledge of these interactions is the key to future uses in selective ion extraction and separation processes, highlighting the endless possibilities of chiral complexes to further advance both fundamental and applied chemistry (Albrecht et al., 2012; Beasley et al., 2013) (Figure 5.9).

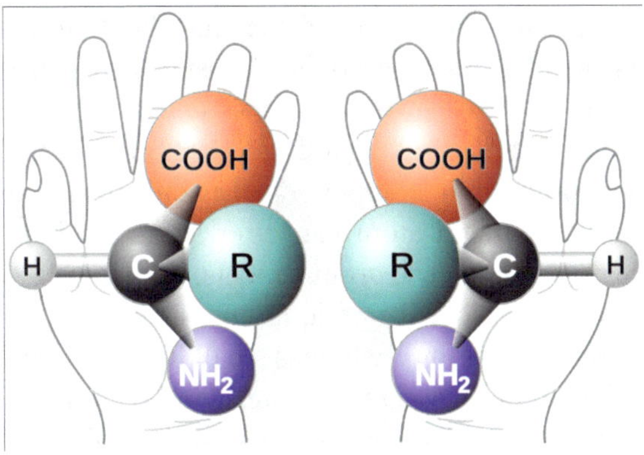

Figure 5.9. *Chiral complexes.*

Source: https://upload.wikimedia.org/wikipedia/commons/thumb/e/e8/Chirality_with_hands.svg/1200px-Chirality_with_hands.svg.png.

5.4. FACTORS INFLUENCING ISOMERISM

Isomerism in coordination compounds is decided by several key factors, most of which are the nature of ligands and their coordination environment. Ligands, which can be monodentate or polydentate, decide what geometric and optical isomers are possible. For instance, while monodentate ligands will only produce a limited number of isomeric forms since their single bond site on the metal center is restricted, polydentate ligands can bind through multiple sites and produce a more diverse range of structural arrangements. This is particularly relevant in the case of π-allylic ligands, in which structural flexibility will produce varied isomeric forms, for example, varying endo and exo configurations. The specific chelation and coordination sphere structures are critical to isomerism, demonstrating that the character of the ligand is a fundamental characteristic to regulate the possible isomers of a compound (Cardin et al., 2017). The oxidation number of the central metal ion is yet another vital factor influencing isomerism in coordination complexes. Different oxidation numbers may change the electronic structure, thereby modifying the coordination sphere geometry and bonding character by the ligands. For example, a higher oxidation state of a transition metal may favor tetrahedral coordination due to higher positive charge, while octahedral coordination can be favored in lower oxidation states. Such versatility can also lead to different isomeric types since different geometric structures (e.g., facial and meridional in octahedral complexes) can be adopted based on the metal's oxidation number.

Besides, the dynamic behavior of the coordination compounds, such as fluxionality, is dictated by these oxidation states to permit isomerization processes to exchange or stabilize various forms of isomers (Nelson et al., 1965). Finally, steric factors and ligand sterics also contribute considerably to isomerism in coordination compounds. Ligand size, shape, and spatial arrangement around the metal atom can offer steric hindrance that, in turn, influences the preferred geometric arrangements. Bulky or large ligands might restrict the number of possible isomeric structures since they cannot occupy some of the metal center positions, whereas small ligands might be more elastic and therefore produce more isomers. The steric influences can also strengthen or destabilize the coordinating ligand interactions, and this leads to variations in reactivity and stability between the isomers. This sterics interaction indicates the intricacy of coordination chemistry, demonstrating how physical properties of ligands can lead to isomeric forms in such compounds (Cardin et al., 2017).

5.4.1. Ligand Types and Their Properties

Ligand categorization is essential in understanding behavior and properties of coordination compounds, particularly in isomerism. Ligands may be classified based on denticity, i.e., the degree of bonding sites present for coordination with a metal center. Monodentate ligands, such as ammonia or water, coordinate through one atom, while bidentate ligands like ethylenediamine coordinate through two atoms. Polydentate ligands, particularly chelating agents, coordinate through more than one bond to a metal, enhancing stability and influencing isomeric forms. The type of ligand can have a significant impact on the electronic character of the formed complex, which in turn can affect its geometry and possible reactivity. Flexibility in polydentate ligands, for instance, could produce several coordination modes, thus emphasizing the role that ligand type has in determining isomerism in coordination compounds. The ligand character is more than just a simple classification, revealing their functions in coordination chemistry and isomer expression. One such great example is in the case of π-allylic ligands with their typical dynamic character and structural reactivity.

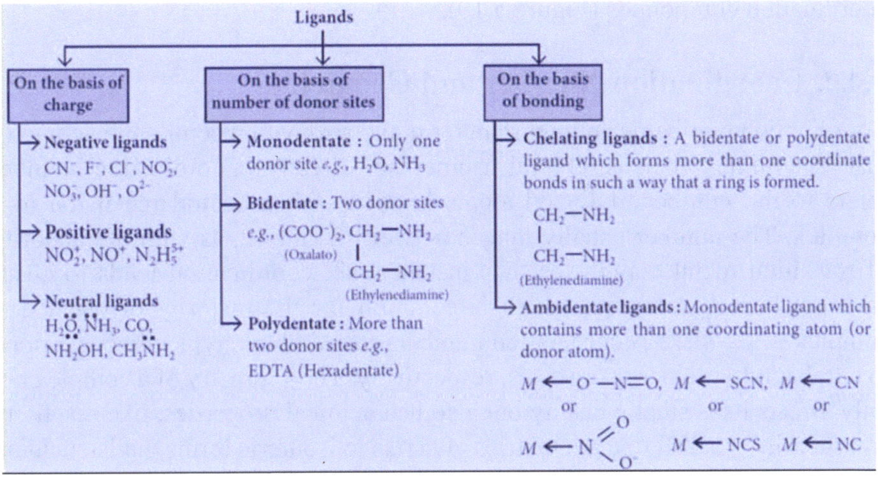

Figure 5.10. *The classification of ligands.*

Source: https://cn.edurev.in/ApplicationImages/Temp/8743297_a0801cbd-bed1-48e6-a54a-b71780c18162_lg.jpg?w=400&dpr=2.6.

The ligands come in various configurations, such as exo or endo isomers, leading to complex reorganization. Their coordination involves special bonding and fluxionality, which could contribute to the complex's overall stability. The isomerization ability of these ligands is at the core of catalysis since the reactivity of these ligands can be managed by adjusting the oxidation state of the central metal atom and thus guiding favorable reaction channels. Understanding the principles governing such ligand interactions is central to designing and creating effective catalysts and new coordination compounds, as evidenced in studies involving dynamics of π-allylic ligands (Cardin et al., 2017). Further, the nuances of coordinating isomerism, particularly ligand-binding modes, contribute significantly towards the versatility of coordination compounds. Alternate linkage modes can completely transform the nature and properties of a complex. For instance, when considering the nitrite ligand, it is found in monodentate or bidentate coordination, influencing connectivity as well as the overall stability of the complex. This is realized in the studies on copper-containing nitrite reductases where ligand behavior illustrates how varying configurations of binding result in linkage isomers in one complex. These findings emphasize the role of ligand properties in isomerism expression, revealing the potential for controlling selective reaction and guiding synthetic strategy in coordination chemistry (Roger et al., 2017). By this, the issue of ligand type and isomerism must be handled with care while exploring coordination compounds (Figure 5.10).

5.4.2. Coordination Number and Geometry

Coordination number is an important term in coordination compound geometry that determines their structural isomerism directly. Coordination number refers to the number of ligand atoms bonded with a central metal ion in a complex. The number usually ranges between 2 and 12, as with the majority of transition metal complexes. For instance, six coordination tends to occur with octahedral geometry, which is common for the majority of octahedral complexes. However, coordination numbers four and two typically correspond to tetrahedral or linear geometries, respectively. The geometry of a complex not only affects its aesthetic beauty but also its chemical properties, like reactivity and stability. As different geometries give rise to isomeric forms, understanding how coordination numbers determine such structures is essential while learning isomerism of coordination compounds.

Geometric isomerism among coordination compounds tends to be present when ligands are able to take on alternate positions in the space surrounding the central metal. Particularly in octahedral compounds, various settings can become evident—such as facial and meridional isomers when there are various types of ligands involved. Tetrahedral complexes also exhibit some level of geometric isomerism, with variety less pronounced because the number of ligand arrangements accessible is less. The nitrite reductases contained in copper provide a good example of this rule in action because it involves the inclusion of different modes of binding, ranging from bidentate to monodentate interactions. In the studied tris(2-pyridyl) copper complex, linkage isomers (κ1-ONO and κ1-NO_2) arise due to varying geometries depending on the metal center coordination number (Roger et al., 2017). This shows not just how coordination numbers affect geometry but also how the flexibility of ligand coordination to coordinate can lead to isomeric derivatives. The metal ion-ligand interactions are not rigid but may dynamically switch to lead to isomerization. Such interactions are particularly common in complexes containing π-allylic ligands, where fluxionality and agostic interactions become paramount. Different geometries of π-allylic ligands, for example, may generate exo and endo isomerism, profoundly affecting the chemical nature of the resultant complex. In addition, the oxidation state of the metal center can dramatically alter the π-allylic ligand's bonding characteristics as well as the reactivity pattern (Cardin et al., 2017). As these properties vary in relation to coordination numbers and geometries, the resulting collection of potential isomers illustrates the intricate nature of the relationship between chemical functionality and structural character. This interaction is vital in understanding the overall importance of isomerism in coordination complexes and highlights the richness of their chemistry.

5.4.3. Metal Center Characteristics

Metal center properties in coordination compounds play a central role in determining their isomeric behavior along with their overall properties. Metal ions, typically transition metal ions, possess more than one oxidation state which contributes to the richness of their coordination chemistry. This variation allows for the creation of different coordination shapes, such as octahedral, tetrahedral, and square planar geometries, and all of these are able to yield different isomers. The intrinsic properties of the metal center, e.g., size, and electronic structure also contribute significantly to the determination of the strength of the ligand field and the stability of complexes formed. Structural studies of various metal complexes, particularly silver ion complexes, demonstrate how the characteristic coordination numbers and geometries lead to distinctive isomers, emphasizing the significance of considering metal center characteristics in coordination chemistry (Abdul-Sada et al., 2018). In addition, the steric and electronic character of ligands bound to metal centers also enhances isomerism within such compounds. Ligands are classified as bidentate or polydentate, which affects the coordination number and thus the potential for isomer development.

For instance, ligands like benzotriazole are semi-rigid with varying coordination modes with metal ions that lead to richness in structure for the resulting compounds. The intrinsic plasticity of ligands, e.g., in the variation of benzotriazole units, introduces a dynamic metal center-coordinating ligand interaction that is capable of delivering exceptional isomeric diversity. Such structural malleability is imperative for the actualization of coordination processes of coordination compounds and how they further interact in chemical reactions (Macovez et al., 2016). Finally, the presence of more than one coordination site at one metal center has a tendency to cause the phenomenon of geometric isomerism and other types of isomerisms such as optical isomerism. Such features are particularly highlighted by chiral ligand or asymmetric framework-containing systems. Because coordination compounds can adopt many coordination modes and geometries, understanding their isomerism is essential for materials science and catalysis applications. The study of isomerism not only enhances our understanding of metal-ligand interactions but also allows for the development of more effective catalysts and functional materials. Therefore, knowledge of the nature of metal centers is invaluable for the advancement of coordination chemistry as well as for the full realization of various coordination compounds (Abdul-Sada et al., 2018) (Figure 5.11).

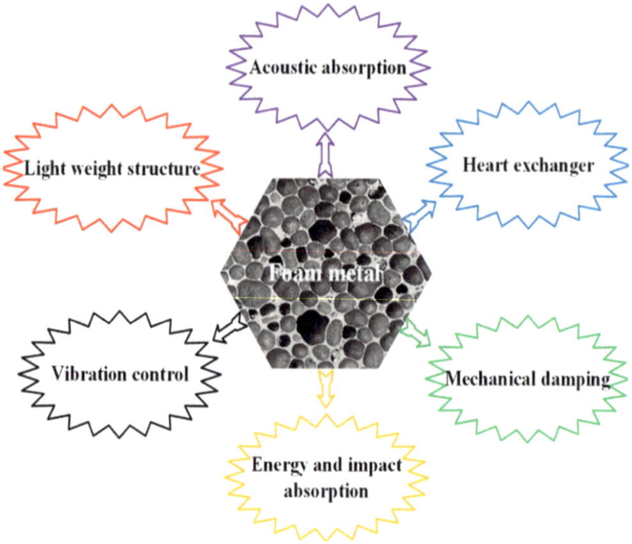

Figure 5.11. *The characteristics of metal centers in coordination compounds.*

Source: https://www.researchgate.net/publication/371637843/figure/fig1/AS:11431281173795562@1689040404845/The-main-characteristics-of-foam-metal.png.

5.5. METHODS OF ISOMER IDENTIFICATION

An in-depth understanding of isomerism in coordination compounds calls for effective techniques for isomer identification. Nuclear magnetic resonance (NMR) spectroscopy is one of the primary techniques adopted, which yields valuable information regarding the molecular dynamics and isomer structure. By the method of chemical shifts and nuclear splitting patterns, particularly of hydrogen and carbon nuclei, chemists can distinguish among the various isomeric forms on the basis of their typical environments. Multinuclear NMR spectroscopy has proved particularly illuminating, and most of all, for the fluxional chemistry of compounds with π-allylic ligands. Such methods unveil rearrangements and determine the occurrence of exo and endo isomers, showing how varying bonding environments influence molecular properties and thus increase our understanding of coordination compounds (Cardin et al., 2017). In addition to NMR spectroscopy, various chromatographic methods are necessary for isomer separation and identification. Techniques such as gas chromatography (GC) and high-performance liquid chromatography (HPLC) facilitate efficient separation of isomers because they allow differential interactivity between stationary and mobile phases. They are especially critical in coordinating complex analysis where geometric and optical isomerism may lead to compounds with distinct retention times and separation patterns. Furthermore, mass spectrometry (MS) is an adjunct technique that provides data on molecular weight and fragmentation patterns to assist in the determination of structural characteristics. Combined, these techniques are a valuable analytical tool that assists in differentiation of isomeric coordination compound complexes, thereby enabling improved insight into their diverse chemical behavior. Application of computational chemistry and molecular modeling represents the third emerging trend in the determination of isomers. These approaches enable chemists to predict and see the structural characteristics of isomers, making it possible for virtual screening and energy calculations consistent with experiments. The addition of density functional theory (DFT) calculations also provides information on the stability of isomeric forms, showing interconversion routes, particularly in compounds that exhibit dynamic behavior. Such interdisciplinarity enriches overall knowledge in coordination chemistry, not only showing the existence of isomers but also their potential reactivity and application in various areas, e.g., catalysis and material science. This new method focuses on synergy between predictive calculations and experimental observation as crucial for building our knowledge on isomerism in coordination compounds (Macovez et al., 2016).

5.5.1. Spectroscopic Techniques

Spectroscopic methods have a central role to play in the clarification of coordination compound structural complexities, especially in isomerism. The importance of these methods is reflected in their capacity for insight into the geometric and electronic structures of metal complexes. Techniques such as nuclear magnetic resonance (NMR), ultraviolet-visible (UV-Vis) spectroscopy, and X-ray crystallography allow chemists to examine not just the ligand connectivity to the metal but also the ligand spatial relationship. With the help of these techniques, researchers can distinguish between geometric isomers, such as cis and trans isomers, and assess dynamic behavior of complexes, which have a tendency to isomerize quickly in solution. Finally, the data acquired via spectroscopic experiments provide a person with a better understanding of coordination chemistry principles and the reasons underlying why various isomers exhibit some specific properties. Apart from NMR and UV-Vis spectroscopy, electron paramagnetic resonance (EPR) spectroscopy turns into a significant instrument, particularly for the analysis of coordination compounds with an unpaired electron. EPR enables one to examine the paramagnetic centers and gain valuable information on electronic environment as well as ligand disposition geometry.

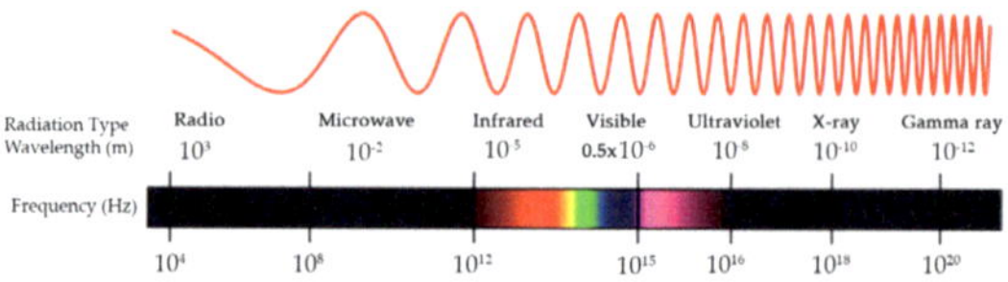

Spectral Region	Technique	Transition	Units
Gamma Ray	Mixed Gamma	Nuclear	MeV/keV
X-ray	Absorption/Fluorescence	Core Electrons	keV/eV
UV-VIS	Absorption/Fluorescence	Valence Electrons	nm/
Infrared	Absorption	Vibrations	cm^{-1}
Microwave	Absorption	Rotations	MHz/GHz/THz/cm^{-1}

Figure 5.12. *Spectroscopic techniques.*

Source: https://www.oxinst.com/learning/uploads/inline-images/figure1-spectrum-20230425130826.jpg.

As an illustration, it is not possible to overstate the role played by EPR spectroscopy in examining tetranuclear complexes like those found in Photosystem II. Electronic structure studies of the complexes, as attested from recent works, reflect spin state transitions and show the correspondence with isomerization and oxidation state. EPR provides a unique perspective on the behavior of metal-ligand interaction under various conditions, contributing to the understanding of the mechanistic processes accompanying isomerism in coordination compounds (Agapie et al., 2020). Furthermore, new advances in spectroscopic techniques, particularly through the use of synchrotron radiation, have enabled higher resolution and accuracy of structural determinations in coordination chemistry. The use of X-ray absorption spectroscopy (XAS) is, for example, a technique that enables it to infer detailed information concerning oxidation states and local symmetry of metal centers in coordination complexes. When combined with other spectroscopic methodologies, XAS facilitates the construction of a detailed picture of how isomerism interacts with reactivity. This can be easily observed with studies of π-allylic ligands, where dynamic behavior and fluxionality of isomers can be charted definitively. Through such combined approaches, scientists are not only characterizing isomers but are also finding new reactivity patterns that can be created by the fine differences in ligand orientations and metal modes of coordination, thereby increasing our worldwide knowledge of isomerism in coordination compounds (Cardin et al., 2017) (Figure 5.12).

5.5.2. Chromatographic Methods

Chromatographic techniques are at the core of isomer analysis and isomer separation in coordination compounds. These techniques are based on the ability to differentiate complex mixtures by exploiting differences in physical and chemical properties of the components. In particular, liquid chromatography as high-performance liquid chromatography (HPLC) has emerged as a critical tool applied to the separation of several isomeric forms of coordination complexes. The effectiveness of HPLC is especially useful when determining the sensitive balance between incompatible isomers so that their electronic and structural features can be determined. Furthermore, the method facilitates the determination of ligand coordination modes that are crucial to the description of metal complex behavior in solutions. The interpretation of chromatographic data is vital for the identification of structural isomers, such as geometric and optical isomers, that arise from different arrangements of ligands around the central metal atom. Scientists can derive useful information regarding the stability and reactivity of coordination compounds through chromatographic techniques. For instance, solvation effects in research have shown that solvent polarity plays an important role in defining isomer distribution of a particular complex, as seen in cases of luotonin A with various metal ion complexes (Abraham et al., 2013). How such factors cooperate to influence each other can make a great impact in shaping synthesizing approaches and application potential for such substances, for instance, their roles in biological systems and catalysis. Additionally, chromatographic methods are used in the study of ligand exchange reaction kinetics and thermodynamics in coordination chemistry. Resolution of the transient species formed within such processes enables researchers to quantifiably measure reaction rates and equilibria. For example, relative stabilities of 1:1 and 1:2 metal-ligand complexes are explained through extensive chromatographic analysis, which gives details regarding their binding capacities and effects caused by competitive ligands (Bridges et al., 1966). So, not only do chromatographic methods enhance our understanding of isomerism in coordination compounds but also contribute to the designing of certain ligands for certain metal ions, which eventually help in material science and drug discovery.

5.5.3. X-Ray Crystallography

Molecular structure elucidation by techniques such as X-ray crystallography has helped significantly in our capacity to understand isomerism in coordination compounds. By applying this method, researchers can determine the 3-D shape of atoms in a crystal material and determine how ligands coordinate around a metal cation. The ability to visualize the various geometries assumed by coordination complexes clarifies the principle of stereoisomerism, such as cis-trans and optical isomerism. An example is how X-ray crystallography has been important in unearthing how specific complexes exhibit diverse modes of ligand binding, which can result in different isomeric forms, thereby highlighting the complex interrelation of molecular structure and isomeric properties. In enzyme catalysis, X-ray crystallography has provided insightful information about metal–ligand interactions and their role in biological systems. An example of this is studies on copper-contained nitrite reductases that have employed this technique in describing active site structures that facilitate the conversion of nitrite to nitric oxide.

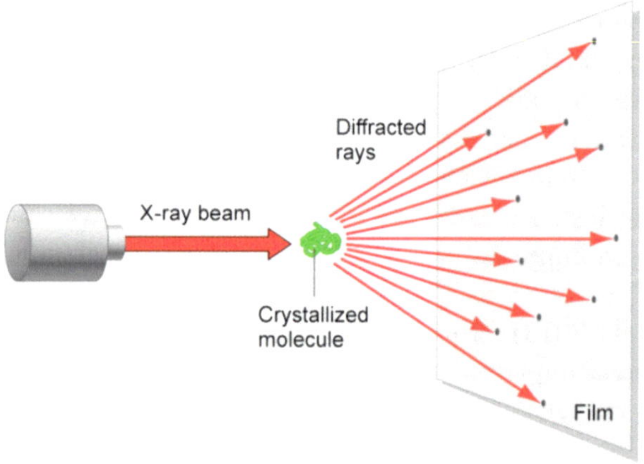

Figure 5.13. *X-ray crystallography.*

Source: https://260h.pbworks.com/f/X-ray+Crystallography-image-2.png.

Notably, Roger et al. (2017) provides the example of how some copper complexes can exhibit more than one binding mode towards nitrite, indicative of linkage isomerism within a single molecule. Such structural details not only enlighten us about the mechanistic processes of biologically occurring enzymes but also aid in designing biomimetic catalysts to mimic such activities, thereby solidifying coordination chemistry towards beneficial outcomes in the world of catalysis. In addition, the use of X-ray crystallography extends from inorganic chemistry into biochemistry with knowledge of amyloid formation and correlation with disease states. An example can be gleaned from research into immunoglobulin light chains, wherein X-ray crystallography has been central in the determination of the dimeric and monomeric structures of these proteins. Such structure elucidation is important, as it highlights dynamic equilibrium between stable and amyloid-forming configurations. The finding of ligands to stabilize light chain dimer state, such as in Brumshtein et al. (2015), shows how vital accurate structural information is in combating diseases related to protein misfolding. Generally, X-ray crystallography is an extremely useful method to investigate the complex interplay between structure and isomerism of coordination compounds and biomolecular systems (Figure 5.13).

5.6. APPLICATIONS OF ISOMERISM IN COORDINATION COMPOUNDS

Isomerism in coordination chemistry has a specific influence on the uses and properties of compounds involving metals. Of notable significance is linkage isomerism, which occurs when a ligand can bind to a metal center via more than one atom. It heavily influences the operation of many catalysts. This is adequately demonstrated in copper complexes where nitrite can coordinate in different modes of binding and exhibit distinct catalytic activities. For instance, a recent study on a copper nitrite reductase complex where they documented several binding modes for nitrite, not only pointing to the intrinsic richness in the compound but also emphasizing the significance of such isomeric forms in promoting catalytic process optimal to the highest order. Pondering the implications of such isomerism is important for the optimization of improved biomimetic catalysts for reduction of nitrite in environmental systems, thereby addressing significant ecological problems (Roger et al., 2017). Isomerism also arises with π-allylic ligands in transition metal complexes that are of critical importance in applications such as synthesis and catalysis. Structural flexibility based on exo and endo isomerism and dynamic nature of these ligands offers new prospects for designing novel catalysts. Understanding the interconversion mechanism between isomers is important in organizing the reactivity of such complexes during design. Understanding this allows chemists to predict and manage consequences during processes such as allylic alkylation and other conversions that are central in organic synthesis and industrial processes. In addition, their interaction with other ligands, such as phosphines and α-diimines, with π-allylic systems is portrayed to underscore their potential in catalysis through demonstrating the importance of isomerism in enhancing catalytic selectivity and efficiency (Cardin et al., 2017). The diversity and complexity provided by isomerism can lead to electrocatalysis being pushed to new horizons, an area of utmost importance in energy conversion and molecule synthesis. By analyzing the role played by isomers in enabling particular bonding interactions and courses of reaction, researchers can engineer improved catalysts more appropriate for electrochemical applications. As an example, the investigation of transition metal π-allylic ligands has provided convincing fluxional chemistry by which structural modifications can be employed to direct desirable reaction courses. The discovery and characterization of such new isomeric species result in more effective catalytic systems and play a big role in green chemistry and the development of alternative energy technologies.

Generally, the knowledge about the application of isomerism in coordination compounds triggers innovation in many fields, emphasizing the fundamental role of such compounds to solving some of the current scientific challenges (Cardin et al., 2017).

5.6.1. Role in Catalysis

The role of isomerism in contributing to the coordination compounds is key to controlling their catalytic activity. Isomerism can potentially significantly affect the electronic and steric environment on the metal center, which controls the way the complexes react with substrates of relevance. Various isomers of the same coordination compound may differ in their reactivity profile, and their differences must be understood when it comes to catalyst design. For instance, geometric isomers are able to alter the three-dimensional coordination of ligands, which influences reagent accessibility to the metal center. This equilibrium may alter catalytic reaction rates, emphasizing the way in which isomerism supports catalytic efficiency and selectivity in numerous chemical transformations (Ashuyko et al., 2013). Flexible ligands have recently been discovered to perform pivotal functions in catalytic applications, particularly in coordination compounds. The semi-rigidity of ligands, as simulated by benzotriazole-based ligand coordination compounds, offers room for variability in coordination geometries and modes. Such flexibility has a tendency to generate heterogeneous isomers that can perform diverse catalytic activities. For example, it has been established that some isomers in a series of coordination compounds have better catalytic activity than their counterparts. One such case is the complexes $[Ag(L_1)(CF_3CO_2)]$ and $[Ag_2(L_3T_2)(CF_3SO_3)_2]$, whose isomeric nature affects their catalytic activity towards processes like hydration of alkynes and A3 coupling reactions (Abdul-Sada et al., 2018). Isomerism dependence of catalysis emphasizes the significance of selective synthesis and characterization of coordination compounds. Understanding how isomers engage differently in catalytic reaction can translate into better, more effective catalysts, engineered for specific reactions. For example, the better efficiency of certain isomers in the multi-component A3 coupling reaction indicates the potential of optimized coordination compounds in synthetic chemistry. By isomeric diversity, the chemist can unveil new avenues to clean and high-yielding reactions, testifying to the ever-relevant role of isomerism in advancing catalyst technology in coordination chemistry (Abdul-Sada et al., 2018).

5.6.2. Importance in Pharmaceuticals

In pharmaceuticals, the role of isomerism in coordination compounds cannot be overstated. Isomers, which are molecules with the same molecular formula but different structural make-up, are important to drug efficacy and safety. For instance, the therapeutic effect of a drug can be drastically changed based on whether it is an isomer or not. An isomer may be the sought-after pharmacological entity, or another one may be inert or even toxic. This is to highlight the importance of broad characterization and understanding of isomeric species in the drug pipeline so that only the very best compounds proceed to clinical trials. For this purpose, the application of coordination compounds is particularly relevant, as their complex shape and diverse bonding environments can produce a rich isomeric fabric that must be thoroughly explored in pharmaceutical application (Кривоносова et al., 2016).

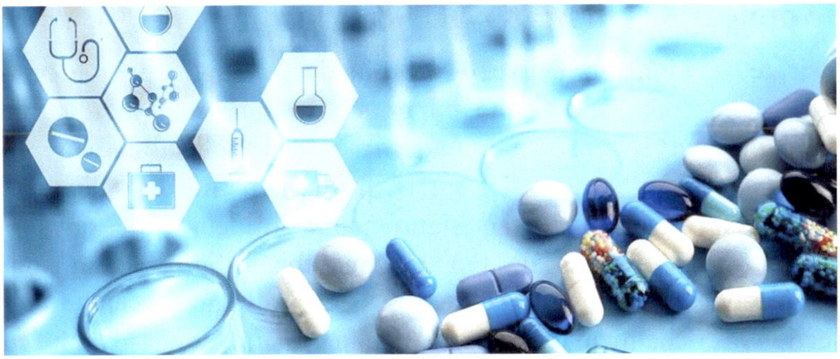

Figure 5.14. *Importance in pharmaceuticals.*

Source: https://trityenviro.com/assets/blogs/importance-of-pharmaceutical-industry-effluent-treatment-plants.jpg.

The intimate connection between pharmaceutical activity and isomerism is amply illustrated in nitrogen heterocyclic synthesis, which is a component of most modern drugs. They are typically scaffolds to hold several pharmacophores and show a range of biological activities that are determined largely by geometry. Straightforward functionalization of nitrogen heterocycles via novel catalytic pathways has been responsible for expanding the inventory of pharmaceuticals. The syntheses of the compounds are under control, allowing for regioselective modification that introduces a variety of functional groups to optimize their pharmacokinetic and pharmacodynamic profiles. Development in catalytic method development, as shown through work on Rh-catalyzed reactions, is an example of how an understanding of isomerism can contribute to the development of drug candidates by selective modification of their molecular structures (Alberico et al., 2008). Further, the significance of isomerism comes also from the fact that the way drugs interact biologically influences ADME profiles. Enantiomeric purity comes into play exactly because the various enantiomers can behave distinctively against biological targets, which influences therapeutic effects. This places a focus on drug preparation and drug regulatory approval procedures, where the exact analysis of isomeric forms must be undertaken. Pharmaceutical industries spend significant amounts on analytical methods for efficiently isolating and characterizing isomers and determining their stability and safety profiles. The intricacies of isomerism dictate a multidisciplinary effort that includes chemistry, biology, and engineering, highlighting the inalienable significance that isomerism in coordination compounds plays in the progression of pharmaceutical science and guaranteeing the production of safe, effective therapeutic agents (Кривоносова et al., 2016) (Figure 5.14).

5.6.3. Applications in Materials Science

The intricate relationship between coordination compound isomerism and materials science underscores the importance of information regarding the structural complexity of these compounds in practical applications. Coordination complexes have many structural isomers, from geometric to optical isomers, and linkage isomers, which can significantly influence the physical and chemical properties of materials. For example, the ability of some isomers to undergo selective interactions makes it possible to upgrade catalytic activity and selectivity for chemical reactions, which in turn makes possible the creation of more effective catalysts for different industrial processes. Inspired by the possibility of isomerism to material design, researchers have focused more on coordination compound design for specific functions and hence enhanced the function and performance of materials in areas such as electronics and photonics (Ashuyko et al., 2013). As the advancing developments in synthetic techniques, isomeric adjustments in coordination compounds have led to exciting applications in materials science, particularly in creating advanced materials with desired properties. Magnetic properties in coordination compounds, for instance, can be accessed using isomerism to enhance the functionality of magnetic materials. The coordination geometries and ligand architectures can be tuned to yield materials with novel magnetic properties, which can be tailored for applications in spintronics and data storage. Furthermore, due to the advent of computational models that generate virtual molecules based on rules for regular chemical behavior, researchers can effectively predict and examine novel isomeric forms. This technique encourages innovation due to the capacity to quickly select good materials for materials science application, including highly magnetic and electronic materials (Hamaekers et al., 2017). The intersection of isomerism and green materials science becomes even clearer as researchers work to develop ecologically friendly materials that are similarly optimized according to performance metrics. The efficient design of the coordination compound through isomeric control can generate biomimetic material or green catalysts that guarantee minimal wasteful side products and utilization of energy. Isomerism presents a pathway towards designing compounds to catalyze reactions under mild conditions effectively, ultimately tackling the sustainability goals of many chemical processes. By highlighting the structural diversity offered by isomers, scientists are able to enhance overall material efficiency greater, and thereby advance the green chemistry ideas in material science further. This placing emphasis on isomerism utilization highlights the importance of interdisciplinarity in addressing global issues via innovative materials (Ashuyko et al., 2013).

SUMMARY

It is evident that isomeric species significantly influence not only structural diversity but also reactivity and activities of such complexes. Existence of linkage isomerism, as outlined in studies performed on metal-ligand interactions, demonstrates the importance of coordination geometry and ligand location in determining chemical behavior. For example, the copper complex illustrated in Roger et al. (2017) has different binding modes that result in different coordination geometries, which have a direct impact on its efficiency as a nitrite reduction catalyst. This isomer complexity forms a foundation for understanding molecular function, paving the way for future advances in synthetic chemistry and biomimetic catalysis. Furthermore, the presence of isomerism provides new avenues for developing new coordination compounds that meet specific technological and biological needs. The formation of different chemical entities, as shown in Hamaekers et al. (2017), illustrates the potential of finding new coordination complexes that adhere to traditional principles of chemistry but further extend the boundaries of known chemistry. These advances could be the key to isolating materials with better properties or completely new applications in the fields of electronics, pharmacy, and environmental chemistry. Thus, the study of isomerism not only provides depth to our knowledge of coordination compounds but also pushes the horizons of research in chemical science. Finally, the consequences of isomerism are more than theoretical significance; they are significant for day-to-day applications in numerous industries. An understanding of isomeric forms' delicacy can be utilized to refine catalyst, pharmaceutical, and high-performance material design by enabling chemists to create compounds for specific applications. The application of such is proven in the necessity of efficient catalysts as reported in Roger et al. (2017) and demonstrates the necessity of isomeric diversity in inducing the necessary reactivity. In addition, the methods outlined in Hamaekers et al. (2017) enable groundbreaking achievements in chemical space, pushing coordination compound chemistry to new horizons. Therefore, encouraging research on isomerism is not only important for academic purposes but also for practical solutions to real-life issues.

Research on isomerism in coordination compounds is important to understand how they act and what chemical properties they possess. Coordination complexes can also show geometric, optical, and linkage isomerism. Geometric isomerism exists when the ligands are able to adopt other relative positions with regard to one another around the central metal atom, e.g., in octahedral complexes. For instance, if a complex contains two identical ligands, the structure may be either cis or trans, which significantly alters the physical and chemical properties of the compounds.

Optical isomerism, though, occurs in non-superimposable mirror images with varying interactions, particularly in biological systems in which chiral structures have important functions. The diversity of the isomer forms suggests the intricacy of coordination compounds and the need for specific study of their arrangements to be able to predict functionality. Linkage isomerism is a very interesting area of coordination chemistry because it deals with the coordination of ligands to the metal atom by different donor atoms. A nice example exists in certain copper complexes where nitrite ion can coordinate in more than one mode, resulting in intriguing structural variety. The complex $[Cu_1(NO_2)_2]$, for instance, is an interesting example of coexistence of both monodentate and bidentate modes of coordination. This bimodal binding interaction exerts a significant influence on the stability and reactivity of the molecules, suggesting that subtle structural changes can result in substantial differences in chemical activity. Understanding how these isomeric forms interact within a compound can provide critical information for the development of novel materials and catalysts that mimic life processes, advancing the field of coordination chemistry's applicability across the board. Along with that, the structural effects of isomerism are also achieved in the stability and packing ability of coordination compounds. As an example, examination of the binuclear molecule $\{Zn[S_2CN(CH_2CH_2OH)2]_2\}2$(pyrazine) brings to the fore the contribution of hydrogen bonding towards the three-dimensional orientation of molecules in a crystal lattice. The prevalence of typical hydroxyl-O–H···O(hydroxyl) interactions is accountable for the stability of the structure to a great extent. The variation in arrangements of varying shapes, such as crystal 1 and the dioxane solvate 2, show the way minor variation in arrangement can build higher stability in packing density. These findings become important when considering the material properties of coordination compounds, where the capacity to recognize the nuances of isomerism can lend itself to material design and optimization for specific chemical use.

REFERENCES

1. Abdul-Sada, A., Kostakis, G. E., Loukopoulos, E., Lykakis, I. N., et al. (2018). Structural diversity and catalytic properties in a family of Ag(I)-benzotriazole based coordination compounds. https://core.ac.uk/download/159994917.pdf (accessed on 13th September 2025).
2. Abraham, A., Babu, C., Cagir, A., Carmona, C., Cheng, T., et al. (2013). Effect of hydrogen bonding and complexation with metal ions on the fluorescence of luotonin A. https://core.ac.uk/download/11857388.pdf (accessed on 13th September 2025).
3. Agapie, T., Britt, R. D., Chatterjee, R., Lee, D., et al. (2020). S = 3 ground state for a tetranuclear $Mn(IV)_4O_4$ complex mimicking the S_3 state of the oxygen evolving complex. https://core.ac.uk/download/286730722.pdf (accessed on 13th September 2025).
4. Alberico, D., Alonso, F., Alvarez, E., Bellina, F., Bercaw, J. E., Bergman, R. G., Campeau, L. C., et al. (2008). Direct functionalization of nitrogen heterocycles via Rh-catalyzed C−H bond activation. https://core.ac.uk/download/77929542.pdf (accessed on 13th September 2025).
5. Albrecht, M., Alexander, J., Amoroso, A., Andreiadis, E. S., et al. (2012). Complexation of lanthanides, actinides and transition metal cations with a 6-(1,2,4-triazin-3-yl)-2,2':6,'2"-terpyridine ligand: Implications for actinide(III)/lanthanide(III) partitioning. https://core.ac.uk/download/9990624.pdf (accessed on 13th September 2025).
6. Aljohani, F. (2016). *Secondary Interactions Which Modify Redox and Other Properties of Organometallic Systems.* https://core.ac.uk/download/41994084.pdf (accessed on 13th September 2025).
7. Arman, H. D., Jotani, M. M., Poplaukhin, P., Tiekink, E. R. T., et al. (2017). Supramolecular association in (μ_2-pyrazine)-tetrakis(N,N-bis(2-hydroxyethyl)dithiocarbamato)dizinc(II) and its di-dioxane solvate. https://core.ac.uk/download/148366784.pdf (accessed on 13th September 2025).
8. Ashuyko, V. A. (2013). *Chemistry of Complex Compounds*. https://core.ac.uk/download/143997703.pdf (accessed on 13th September 2025).
9. Ayllon, J. A., Domingo, C., Gemmi, M., Lanza, A., et al. (2022). Supramolecular isomerism in cobalt(II) coordination polymers built from 3,5-bis(trifluoromethyl)benzoate and 4,4'-bipyridine. https://core.ac.uk/download/532374790.pdf (accessed on 13th September 2025).
10. Beasley, B., Clarkson, G. J., Geden, J. V., Shipman, M., et al. (2013). Asymmetric synthesis of 2-substituted oxetan-3-ones via metalated

SAMP/RAMP hydrazones. https://core.ac.uk/download/19210704.pdf (accessed on 13th September 2025).

11. Blake, I. M., Cai, Z. L., Canfield, P. J., Crossley, M. J., et al. (2018). A new fundamental type of conformational isomerism. https://opus.lib.uts.edu.au/bitstream/10453/125589/1/main_text_edit_1517588129_return4%20with%20figs.pdf (accessed on 13th September 2025).
12. Bridges, K. L. (1966). *Investigation of the Cis-Trans Isomers of Tetrachlorodiaquorhodate (III) anion.* https://core.ac.uk/download/651435598.pdf (accessed on 13th September 2025).
13. Bruijnincx, P. C. A., & Sadler, P. J. (2009). Controlling platinum, ruthenium, and osmium reactivity for anticancer drug design. https://core.ac.uk/download/44701.pdf (accessed on 13th September 2025).
14. Brumshtein, B., Cascio, D., Eisenberg, D. S., Esswein, A. J., et al. (2015). Inhibition by small-molecule ligands of formation of amyloid fibrils of an immunoglobulin light chain variable domain. https://core.ac.uk/download/323075783.pdf (accessed on 13th September 2025).
15. Canfield, P. J. (2022). The polytope formalism: Isomerism and associated unimolecular isomerization. https://core.ac.uk/download/543819659.pdf (accessed on 13th September 2025).
16. Cardin, D., Hartl, F., & Ryan, D. E. (2017). η3-Allyl carbonyl complexes of group 6 metals: Structural aspects, isomerism, dynamic behaviour and reactivity. https://core.ac.uk/download/74207026.pdf (accessed on 13th September 2025).
17. Elford, D. (2022). Augmented reality in chemistry higher education. https://core.ac.uk/download/552965807.pdf (accessed on 13th September 2025).
18. Hamaekers, J., Israels, R., & Maaß, A. (2017). The octet rule in chemical space: Generating virtual molecules. http://arxiv.org/abs/1709.06746 (accessed on 13th September 2025).
19. Hatcher, L. E. (2016). Raising the (metastable) bar: 100% photo-switching in $[Pd(Bu_4dien)(\eta 1\text{-}O_2)]^+$ approaches ambient temperature. https://core.ac.uk/download/288393885.pdf (accessed on 13th September 2025).
20. Ibodulloeva, M. I., & Narzullayeva, V. D. (2023). Study of the isomerism of coordination compounds. https://core.ac.uk/download/586795684.pdf (accessed on 13th September 2025).
21. Macovez, R., Manesh, Z., Mitsari, E., et al. (2016). Solid state physicochemical properties and applications of organic and metallo-organic fullerene derivatives. https://core.ac.uk/download/81571877.pdf (accessed on 13th September 2025).

22. Nelson, C. (1965). Coordination compounds and complex ions. https://core.ac.uk/download/232825983.pdf (accessed on 13th September 2025).
23. Rodríguez-Fortea, A., Contant, R., Iijima, F., Oshio, H., Cameron, J. M., Poblet, J. M., Murillo, J. C., et al. (2016). Investigating the transformations of polyoxoanions using mass spectrometry and molecular dynamics. https://core.ac.uk/download/46165611.pdf (accessed on 13th September 2025).
24. Roger, I., Senn, H. M., Sproules, S., et al. (2017). An investigation into the unusual linkage isomerization and nitrite reduction activity of a novel tris(2-pyridyl) copper complex. https://core.ac.uk/download/84340544.pdf (accessed on 13th September 2025).

CHAPTER 6

STABILITY OF COMPLEXES IN COORDINATION COMPOUNDS

Contents

INTRODUCTION

Coordination compounds are the subject of study in inorganic chemistry as far as the stability of metal complexes is concerned, and this is an area of immense interest. Coordination compounds consist of a central metal ion surrounded by more than one ligand, which can influence the properties and stability of the resulting product complexes. Stability, in this context, means the thermodynamic preference for the formation and stability of such complexes in solution. The intricate behavior of metal and ligands underscores the importance of understanding variables that affect stability, such as ligand denticity, metal oxidation states, and solvent effects. Study of these properties enables us to better comprehend the formation of complexes and illuminate many applications ranging from catalysis to biomimetic processes. Thus, research of the stability of coordination compounds is not only of academic value but also a foundation for enhancing practical application in chemistry. Stability is frequently evaluated in coordination compound chemistry through rigorous experimental studies and computational simulation. Researchers have developed good methods, including solution studies and thermally stimulated analyses, to explore complex behavior under varied conditions. For instance, the systematic analysis of imidazole-substituted ligand complexes yields a novel paradigm for predicting stability from induced behavior under heat stress. The integration of Thermally Induced Evolved Gas Analysis by Mass Spectrometry (TI-EGA-MS) has proven to be imperative toward ascertaining the solid-state features and thermal stability of such complexes. Such rigorous approaches not only increase our understanding of the coordination complexes' thermodynamics but also allow for the development of predictive models that can guide future research directions toward the synthesis and exploration of new compounds, thus multiplying their usefulness in modern chemistry. Moreover, complex stability is often determined by the distinctive structural properties of metal-ligand interactions. For example, studies involving the coordination of organophosphorus compounds with alkali metal ions have highlighted the subtle interaction of coordination geometry, ligand site, and solvent effect on the stability of the complexes. By the application of calorimetric evaluation in concert with molecular mechanics computations, scientists can recognize how stoichiometric variations influence thermodynamic stability. These findings indicate that the complex structural characteristics, together with solvent motion, are significant determinants of the global stability of coordination compounds. The employment of additive schemes in this work highlights the complexity of metal-ligand interaction and validates the observation that a multidisciplinary effort is necessary to an enhanced understanding of stability in coordination chemistry. Lastly, these results offer avenues towards rational design and utilization of coordination complexes in a wide array of chemical applications.

Definition of Coordination Compounds

Coordination compounds consist of a large number of chemical substances based on a central metal atom and peripheral molecules or ions referred to as ligands. The complexes may contain transition metals, actinides, and lanthanides and possess characteristic properties based on their capacity for coordination with various ligands. The ligands are classified into unidentate, bidentate, or polydentate based on the donor atoms they possess, which in turn can be engaged with the metal center. This interaction leads to the coordination complex having a specific geometry, typically octahedral, tetrahedral, or square planar. It is worth understanding the construction and definition of coordination compounds because it lays the preparatory foundation for probing the stability of these complexes, a subject of firm interest in coordination chemistry. Stability of coordination complexes is significantly influenced by the character of ligands and metal center involved. Electronegativity of the donating atom, charge, and sterics of the ligand all contribute significantly towards deciding the overall final stability of the coordination compounds. Additionally, the chelate effect, or increased stability of the chelating ligands compared to their non-chelating analogs, is also a crucial concept here. Compounds synthesized with ligands that are bidentate or polydentate are typically more stable as they are capable of creating several coordination bonds with the metal center. It has been proven through experiments that ligand substitution reactions can further ascertain the inherent stability trends in various coordination complexes. Thermal and chemical stability experiments of such complexes, for example, those involving Thermally Induced Evolved Gas Analysis by Mass Spectrometry (TI-EGA-MS), aid us in comprehending their behavior under different conditions and provide us with information about coordination chemistry (Buiarelli et al., 2019). Advances in computation over the recent past also allow us to explore coordination compounds by applying machine learning and high-throughput computations. Such methods enable researchers to screen vast material spaces with ease and hence identify patterns and correlations among different structural and compositional attributes of coordination complexes. Graph convolutional neural networks enable in-depth knowledge of coordination environment stability and similarity, crucial for the prediction of behavior of diverse complexes. For instance, visualization of coordination property emergence in various materials solidifies theoretical foundations of complex stability. Through the application of these modern techniques, researchers can prepare better models for predicting the stability of coordination compounds and thereby make effective utilization in materials science and medicinal chemistry (Grossman et al., 2018).

Importance of Stability in Coordination Complexes

The stability of coordination complexes plays a critical role in their applications in many fields, most significantly in medicinal chemistry. For instance, metal drug compounds such as platinum complexes are of clinical importance as they can form stable complexes with biological macromolecules. Complex stability is critical as it determines the effectiveness and selectivity of drug molecules, certain coordination complexes being cytotoxic by targeting DNA. The process of coordination facilitates extremely precise control over the behavior of drugs in biological systems, an important feature for anticancer treatments. This is seen in the design of platinum compounds, which not only show remarkable stability but also can be designed to be selectively reactive, promoting more effective treatments. Hence, understanding coordination complex stability mechanisms is essential for the evolution of pharmaceutical applications and the enhancement of treatment processes in cancer therapy. Also, stability in coordination complexes is a major aspect in catalysis, with the coordination compounds acting as catalysts to enhance different chemical conversions.

$$\text{(i) } M + L \rightleftharpoons ML, \qquad K_1 = \frac{[ML]}{[M][L]}$$

$$\text{(ii) } ML + L \rightleftharpoons ML_2, \qquad K_2 = \frac{[ML_2]}{[ML][L]}$$

$$\text{(iii) } ML_2 + L \rightleftharpoons ML_3, \qquad K_3 = \frac{[ML_3]}{[ML_2][L]}$$

....

$$\text{(iv) } ML_{n-2} + L \rightleftharpoons ML_{n-1}, \qquad K_{n-1} = \frac{[ML_{n-1}]}{[ML_{n-2}][L]}$$

$$\text{(v) } ML_{n-1} + L \rightleftharpoons ML_n, \qquad K_n = \frac{[ML_n]}{[ML_{n-1}][L]}$$

Figure 6.1. *The stability of coordination complexes.*

Source: https://dcx0p3on5z8dw.cloudfront.net/Aakash/s3fs-public/inline-images/2_48.PNG?VgaCok5No02rI1LPUQzotYORv3Pr1c68.

As an example, some ternary systems of inorganic platinum salts and oxides exhibit active catalytic activity for methane oxidation, a reaction with substantial energy and environmental impact. Their catalytic activity is defined by the stability of the complexes in various conditions and, therefore, methane selectivity to methanol conversion. This is particularly important with regard to the necessity of more water-resistant systems, as highlighted in recent research. Stable and robust coordination complexes in catalytic reactions not just improve the conditions of the reaction but provide access to efficient and sustainable industrial chemistry. Hence, stability in coordination complexes forms the core of synthetic methodology development and environmental friendliness in chemical practice. Moreover, the relationship between stability and structural characteristics of coordination complexes cannot be overstated. Ligand conformation, steric influence, and electronic properties all play a role in the total stability of a complex, determining its behavior and reactivity in many applications. The introduction of bidentate or polydentate ligands will generally increase stability through the chelate effect, reducing the dissociation tendency. This kind of stability is very important, particularly within biological systems, as coordination compounds must withstand a variety of physiological conditions. Besides that, ongoing research into new metal complexes, for example, organometallic ruthenium complexes, is revealing new pathways to the development of highly selective and stable agents for medical application. The ongoing research into these new structures not only emphasizes the importance of stability in coordination complexes but also illustrates its central theme in both developing our scientific understanding and applying that understanding to useful purposes in a range of areas (Figure 6.1).

Overview of Factors Influencing Stability

Coordination complex stability is regulated primarily by the metal ion in use. The transition metals that possess variable oxidation states and coordination numbers exhibit varied behavior that continues to regulate stability of the complex. For instance, the geometry and distribution of electrons in the complex are regulated by the kind of metal used, thereby influencing reactivity and interactions with ligands. Increased metal-ligand interaction can lead to increased stability. More significantly, d-orbitals offer the potential for stable complexes through effective orbital overlap. The study of ligand type shows that chelating ligands—those that can bind more than one bond to one metal center—will greatly enhance stability due to the chelate effect, which reduces the entropic costs of coordination of the ligand. The electronic nature of ligands also plays a significant role in determining the stability of coordination complexes. Those ligands which are effective at donating electron density to metal ions through σ-donation or π-backbonding are likely to stabilize the resulting complexes. For instance, bidentate and tridentate ligands are more stable compared to their monodentate counterparts because they possess multiple points of attachment, thereby minimizing the chance of dissociation. Detailed examples, such as the stabilization of copper complexes in drug use, illustrate how different ligands influence the stability of metals in physiological conditions. Stability in such metal-ligand interactions has been shown to lead to high biological activity, which shows that the structural design of the ligand is of significant importance in maximizing the stability of the complex in biological systems, as covered in recent studies (Enyedy et al., 2018). Environmental parameters like solvent and temperature also influence the stability of coordination complexes. The choice of solvent can influence metal-ligand interactions significantly due to variations in polarity, solvation energies, and competitive intermolecular interactions with the solvent molecules. Polar solvents stabilize ionic complexes through the generation of solvation shells that influence dissociation kinetics. Additionally, changes in temperature can influence complex stability by altering the energy landscape, which may cause shifts in equilibrium positions between different coordination states. Higher temperatures will drive molecular motion, which can enhance the breakdown of less stable complexes. Experimental techniques, such as thermogravimetric analysis, have also given insight into these processes so that a greater appreciation of how metal ions interact with ligands under a wide range of conditions can be achieved (Nováčková et al., 2009). Thus, both intrinsic chemical properties and extrinsic environmental factors collectively explain coordination compounds' complicated stability profiles (Figure 6.2).

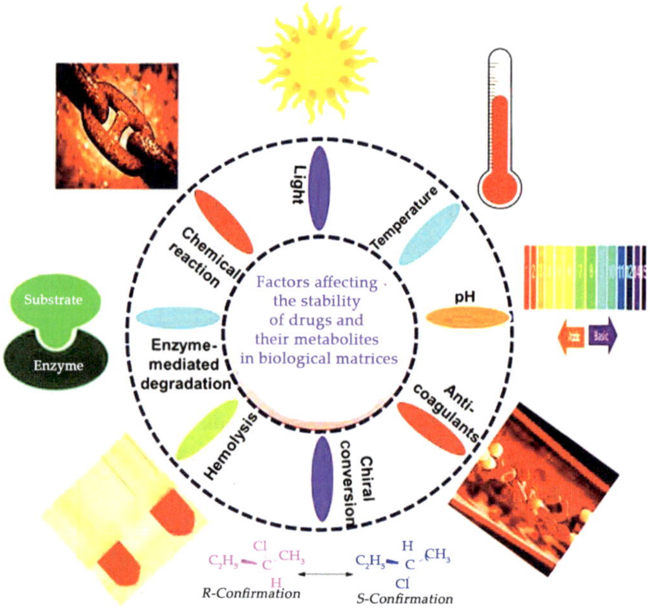

Figure 6.2. *The factors influencing stability.*

Source: https://ars.els-cdn.com/content/image/3-s2.0-B978012814425100022X-f17-01-9780128144251.jpg.

6.1. THE NATURE OF COORDINATION BONDS

The character of coordination bonds is at the core of understanding the stability of coordination compounds' complexes. Coordination bonds are formed by the interaction of a metal ion and one or more molecules or ions, known as ligands, that possess lone pairs of electrons. The interaction forms a coordinate covalent bond where the ligand contributes both the electrons in the bond. The character and strength of the bonds can potentially have an important role in deciding the stability of the resulting complex. For instance, highly charged metal ions have the tendency to establish stronger coordination bonds with the ligands due to greater electrostatic attraction. Besides, factors such as ligand denticity, steric hindrance, and electronic effects also strongly contribute to determining the overall stability of these complexes, and hence the study of coordination bonds is crucial to predict the coordination compounds' behavior under various chemical conditions. The composition of the ligand is also capable of having a significant influence on the properties of coordination compounds. Monodentate ligands, which bind via a single donor atom, confer special properties other than polydentate ligands, which can bind via several donor sites. The latter will form more stable chelate complexes due to the chelate effect as the formation of a ring within the complex makes the entity less willing to dissociate. Experiments have revealed that polydentate and bidentate ligands increase stability by minimizing the entropy cost of complex formation, as they bind to the metal ion more tightly. More intricate coordination environments can be achieved using ligands designed for specific purposes, such as organometallic drugs to combat disease. These advancements highlight the richness and significance of ligand structure in regulating coordination complex stability and reactivity in medicinal chemistry (Bruijnincx et al., 2009; Habtemariam et al., 2012). In addition, variations in coordination geometry—e.g., octahedral, tetrahedral, or square planar structure—quite importantly impact coordination compound stability. The metal ions oxidation state and charge and size of ligands are governed by geometry. For example, d-block metals often exhibit octahedral geometries due to the spatial arrangement of electron cloud to achieve maximum overlap with orbitals of ligands. The geometrical preference not only influences stability but also reactivity of coordination complexes like ligand substitution and other kinetical behaviour. Understanding of such geometric structures can lead to the optimization of the structure of complexes with desired therapeutic properties, giving rise to new forms of treatment. The interrelationship between coordination bonds, ligands, and geometry underlies a comprehensive picture of the stability of complexes in coordination compounds (Bruijnincx et al., 2009; Habtemariam et al., 2012).

6.1.1. Description of Coordinate Covalent Bonds

The coordinate covalent bond principle is vital in explaining stability of complexes in coordination compounds. As compared to the regular covalent bonds, in which two atoms share an equal number of electrons to create a bond, coordinate covalent bonds occur when an atom donates both electrons to bond with another atom. This unique bonding process is typically observed in metal coordination complexes, where a metal center becomes coordinated with ligands, which are electron-pair donors.

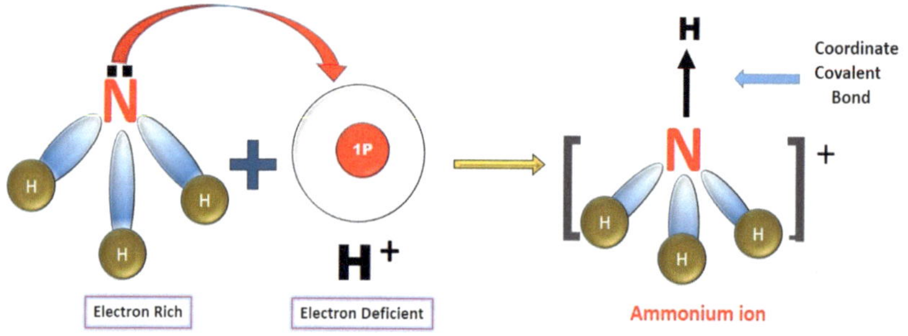

Figure 6.3. *The concept of coordinate covalent bonds.*

Source: https://i.ytimg.com/vi/2EJ-u729QWk/maxresdefault.jpg.

The ability of the ligands to bond in such a way enhances the stability of the overall complex by allowing more interaction between the metal and the ligands, so more desirable electronic conditions may be formed around the metal center. Coordinate covalent bonding stability is essential, as it generally dictates reactivity and properties of the complex in varying chemical environments. The function of coordinate covalent bonds is not limited to the structural and dynamic aspects of coordination compounds. Coordinate covalent bonds assist in a network of electronic interactions that can stabilize complex structures, such as octahedral or tetrahedral geometries common in coordination chemistry. Their development usually mimics the inherent electronic structure of the metal ion and its ligands. In metals with d-orbitals, for example, the ability to stabilize specific oxidation states through the formation of coordinate bonds is important in catalysis and materials properties. The study shows that the overall symmetry and stability of complexes result from the interplay of ligand field effects and coordinate covalent bonding, in which the importance of the interactions in coordination chemistry is stressed (Golden et al., 2017). Coordinate covalent bonds are also not limited to metal-ligand interactions. These bonds also share the wider universe of bonding with multi-center interactions and how they impact the stability of complex structures. A thorough understanding reveals coordinate covalent bonding can be regarded as a special instance of multi-center bonding where symmetry and electronic localization play vital roles (Albright T et al., 2010). Such coordination compound stability can thus better be evaluated by analyzing how coordinate covalent bonds promote electronic delocalization and can satisfy geometrical configurations to adjust for changing electronegativities and densities. The perspective not only illuminates us about the inherent coordination complex stability but also directs us toward new ways of synthesizing compounds with particular characteristics, underlining the use value of such elementary chemical principles (Figure 6.3).

6.1.2. Role of Ligands in Bond Formation

The role of ligands in bonding in coordination compounds is the key to stability and activity of the complexes. Ligands, by which one can understand ions or molecules giving electron pairs on a central metal atom, play an immense role in the geometric and electronic coordination sphere of the metal center. This interaction not only creates bonds but also stabilizes the resulting coordination complex. The character and strength of the bond formed between the ligand and the metal ion are dependent on factors like the character of the ligand, denticity, and oxidation state of the metal. By way of guided ligand interactions, such as σ donation or π back-donation, the overall stability of the complex may be enhanced, giving rise to special properties that can be further exploited in various applications, such as catalysis and drug discovery. Design and selection of ligands have become extremely important in medicinal chemistry, particularly as researchers seek to create molecules of selective therapeutic activity. The intricacies of metal center coordination by ligands provide insight into how to optimize drug leads against specific biological targets. Ligands can modulate the reactivity and overall stability of metal-based drugs, thereby improving efficacy and lowering side effects. With modulation of structural features of ligands, including steric and electronic properties, chemists can optimize the pharmacokinetic properties of coordination compounds. The molecular manipulation through such means not only creates a comprehension of ligand's behavior but also fills the gap of understanding between biology and chemistry, resulting in compounds that are able to target precise biochemical pathways being concerned with disease pathology (Bruijnincx et al., 2009). The role of organometallic chemistry in the formation of targeted therapeutic agents shows the innovative potential of the ligand for the formation of the bond. Specifically, ligands can be designed to facilitate triggered activation of metal complexes, modulating their pharmacological activities. For instance, differences between varying ligand environments can alter redox properties of metal sites or influence the kinetically of ligand substitution reactions, and further tune the interaction of these complexes with biological molecules. This approach has opened up new avenues to the understanding of drug action mechanisms, particularly through outer-sphere recognition processes and ligand-induced stability aspects. Research into such complexes evidences the ongoing advancement of bioorganometallic chemistry, demonstrating the pivotal role ligands play in not only coordination compound stability but also expanding the boundaries of therapeutic applications (Habtemariam et al., 2012).

6.1.3. Comparison with Ionic and Covalent Bonds

Coordination compound stability is basically provided by the type of bonds that stabilize these complexes. Comparing ionic and covalent bonds, it is of great significance to indicate that ionic bonds, characterized by the electrostatic attraction between oppositely charged ions, bestow distinctive stability properties upon complexes. Coordination complexes are likely to form through ionic interactions, particularly when metal ions that are likely to have a positive charge coordinate with ligands that are negatively charged or neutral. The interaction leads to strong electrostatic forces of attraction, and this contributes to the overall stability of the resulting complex. However, the effectiveness of ionic interactions can vary significantly with the environment, such as solvent polarity or with the influence of other competing ions that could intervene with these forces and influence coordination compound stability. Covalent bonds, on the other hand, are more reliant on the sharing of electrons among atoms to yield more localized and directional forces that can further optimize complex stability even better than ionic bonds in certain cases. In coordination complexes, there is covalent tendency whenever ligands donate electron pairs to the metal centers and develop coordinate covalent bonds. This type of bonding results in a more rigid and more stable structure, which can withstand external changes like temperature and pressure changes. The nature of the ligand—size, electron-donating power, and steric hindrance—also influences the stability of the complex to a great extent. For instance, chelating ligands that possess more than one point of attachment can form more stable complexes with metal ions compared to monodentate ligands due to the phenomenon known as chelate effect. The correlation between ionic and covalent bonding in coordination compounds is central to analyzing their stability. As seen from the analysis of transition metal sulfides as well as their surface chemistry, competition between ionic and covalent characteristics may not only define the electronic structure but also the functional properties of such complexes. Difficulty in functionalization of such compounds, as reported in studies, lies in the complexity of their bonding. Deeper understanding of such interactions has the potential to unlock new paths for beneficial applications, suggesting that powerful functionalization strategies are essential for the stabilization of colloidal dispersions of metal complexes for applications such as bioimaging and catalysis (Dengo et al., 2017). Therefore, the application of an intimate knowledge of ionic and covalent bonds to coordination compounds is crucial to the growth of materials science and coordination chemistry (Nováčková et al., 2009) (Figure 6.4).

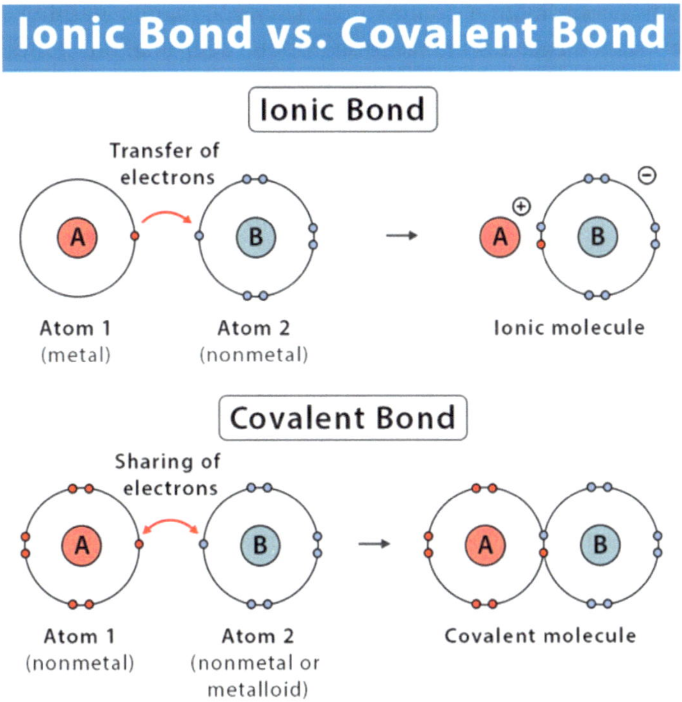

Figure 6.4. *Comparison with ionic and covalent bonds.*

Source: https://www.chemistrylearner.com/wp-content/uploads/2020/09/Ionic-vs-covalent-bonds.jpg.

6.2. FACTORS AFFECTING STABILITY

Electronic properties of the metal center mostly control the stability of coordination complexes by dictating how readily it will achieve coordination with ligands. Transition metals, for instance, are known to exist in variable oxidation states, which facilitate them to coordinate an immense range of ligands. Their type is also crucial, as their denticity and electronic properties are able to stabilize or destabilize the complexes that are formed. Those ligands that are electron-donating, for example, nitrogen or oxygen donor atoms, typically enhance stability through more secure metal-ligand bonds. This is what occurs when trying to grasp concepts like hard and soft acid-base theory, which put forward the suggestion that metals prefer to coordinate with ligands of identical electronic nature. Thus, the interplay between the electronic configuration of the metal and the nature of the ligands gets combined to yield the overall stability of the coordination compounds. Thermodynamic stability is another factor to be remembered while learning about coordination complexes. The stability constants of these kinds of complexes, derived from metal-ion and ligand interactions, provide measurements of solution stability. Large stability constants have been found to be coupled with positive enthalpic contributions which imply the stronger bonds are being established upon complexation. A pertinent example may be found in humic substance research, in which the addition of transition metal ions has been shown to alter thermal stability and binding behavior, indicating a multidirectional interaction of variables controlling stability. For instance, the use of humic acid together with sodium and copper ions is not only observed to exhibit saturation variations in the levels of binding sites but also to give insights into the stability of such novel complexes under heat stress, thus providing improved insight into coordination chemistry and reactivity in various environments (Nováčková et al., 2009). Steric hindrance and geometric positioning also significantly influence the stability of coordination compounds. The geometrical positioning of ligands around the metal center can be facilitative or hindering of stability based on the accessibility and location of the ligands. As established in other research, certain geometries, for instance, octahedral or tetrahedral geometries, are more appropriate for particular ligands based on their size and shape. For example, the presence of bulky ligands around a metal center introduces steric strain and thus decreases the overall stability of the complex. Moreover, ligand lability, or the facility of ligand exchange, can also be a key determining factor of stability. Lability usually indicates the kinetic factors involved in bond formation and dissociation, influencing the thermodynamic stability of the complexes obtained. Therefore, assessing stability within coordination compounds involves a profound knowledge of steric and geometric considerations, which can ultimately decide the effectiveness and reactivity of these metal-ligand processes (Burr et al., 2013).

6.2.1. Electronic Effects of Ligands

Electronic ligand effects are significant in accounting for stability and reactivity of coordination compounds. Ligands may also be grouped according to their electronic attributes, such as their ability to donate or withdraw electrons from the metal center. Electron-donating ligands, for instance, stabilize metal ions through σ-donation, leading to a net stability of the complex. Electron-withdrawing ligands destabilize the metal-ligand association to form weaker complexes. An understanding of these electronic effects is important in the design and synthesis of coordination compounds designed for specific applications, e.g., drug discovery and catalysis. With drug chemists looking for novel molecular architectures, control over electronic properties via choice of ligand results in phenomenal advances in therapeutic strategy, particularly disease targeting by enhancing complex stability and activity (Bruijnincx et al., 2009). Another important aspect of the electronic ligand effects is their influence on coordination compound geometry and steric arrangement.

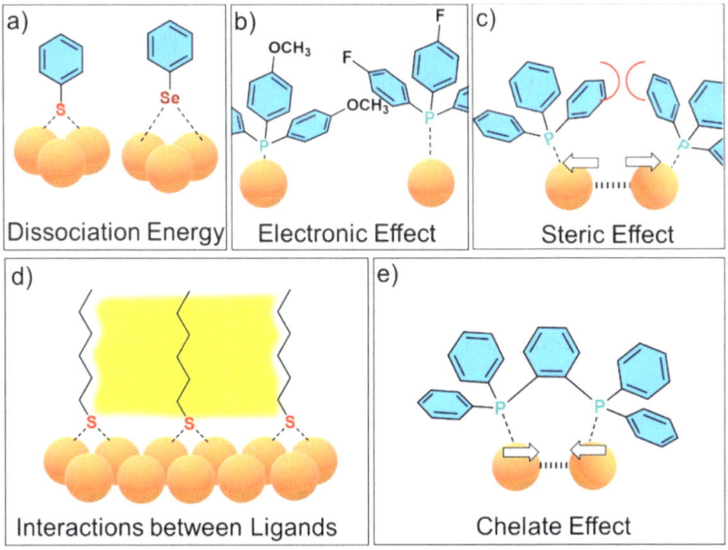

Figure 6.5. *The electronic effects of ligands.*

Source: https://www.researchgate.net/publication/372990419/figure/fig2/AS:11431281255236080@1719411533414/Ligand-effects-in-stability-of-AuNCs-a-dissociation-energy-between-gold-and-ligands-b.png.

Geometric issues are immediately linked to electronic structure of ligands, and therefore ligand field strength and consequently metal complex stability are affected. For example, bidentate ligands will form chelate effects, yielding higher stability compared to their monodentate counterparts. This stability is a result of the electronic overlap between the ligand and the metal, enabling higher bonding interactions. Ideal ligand design in such a case could result in complexes that not only afford additional stability but also tailor-made binding properties important in designing effective therapies (Habtemariam et al., 2012). With the use of these electronic geometric parameters, coordination compounds can be designed that exhibit unparalleled stability and reactivity trends. Finally, the electronic effects of the ligand also encompass the redox characteristics, which are of fundamental significance in biochemical reactions. There are ligands that can impose redox character on metal complexes so that they can participate in electron transfer reactions. This aspect becomes increasingly crucial in bioorganometallic chemistry, where electronic and steric effects are regulated to achieve favorable catalytic processes. Organometallics, for example, exhibit differentiated reactivity due to possessing pre-defined ligand environments to facilitate ligand exchange and redox processes. This is significant in drug design since it enables one to create targeted therapeutics that can exhibit selective biological interactions. Exploitation of these electronic effects not only enhances the stability of the resulting complexes but also provides scope for the development of novel therapeutic agents capable of addressing challenging clinical issues (Habtemariam et al., 2012) (Figure 6.5).

6.2.2. Steric Hindrance and Its Impact

Steric hindrance is one of the key contributions to coordination compound stability, both in the geometric arrangement and reactivity of the complexes. The size of the substituents and their spatial arrangement around the central metal ion can play an important role in determining the ability of the ligand to coordinate in the correct manner. The larger the ligands, the greater the steric overcrowding will be with greater strain in the metal-ligand bond. This strain has a tendency to destabilize the complex and result in altered coordination geometries. It is for this reason that having an understanding of how steric hindrance affects stability allows chemists to predict reactivity patterns of such complexes and allow compounds' design with traits desired for catalytic or synthesis purposes. The effect of steric hindrance extends a long way to impacting the dynamics of ligand exchange processes in coordination compounds. The presence of sterically bulky ligands may prevent the access of incoming ligands, thus slowing down substitution reactions. For instance, in the presence of bulkier sterically hindered ligands, the coordination center becomes less accessible, and this stabilizes the complex given against associative or dissociative processes. The fluxionality process, through which some complexes are able to quickly exchange between geometrical isomers, is also steric. In this context, less bulky ligands should facilitate faster exchange reactions and give rise to more dynamic behavior in the complex, eventually affecting its solution stability (Cardin et al., 2017). Moreover, the cooperation between electronic and steric effects should not be overlooked. Bulker ligands have electronically competent donor atoms but are often highly constrained by the bulk of their donor atoms from entering stabilizing interactions with the metal center. This weakened interaction can render the compound less stable in general, particularly in mixed-ligand complexes when there are sterically constricting and smaller ligands. This has substantial ramifications in bioinorganic chemistry as well, where sterically crowded metal complexes have been proven to display hindered interactions with biological macromolecules. For instance, studies on gold(III) complexes reveal that steric interactions could dictate ligand binding to proteins, illustrating how both electronic and steric characteristics must be taken into consideration in tandem in order to understand the activity of coordination complexes comprehensively (Barone et al., 2017).

6.2.3. Metal Ion Characteristics and Their Influence

The inherent characteristics of metal ions play paramount roles in the determination of the stability of coordination complexes. Factors such as ionic size, charge, and electron configuration influence the binding affinity of metal ions to ligands and consequently the overall stability of the coordinated complex. Transition metals, for instance, have a tendency of showing variable oxidation states that allow them to stabilize varying ligands. The ability of these metal ions to adopt a number of different coordination geometries enhances their capacity to stabilize a number of different structures. This complex stability and metal ion property relationship is also evidenced by systematic studies that link thermally induced properties with the behavior of these complexes under applied stress, confirming the relevance of the selection of metal ions in coordination chemistry (Buiarelli et al., 2019). Sterics and the nature of bonding also determine the coordination of the metal ions with ligands surrounding them. Soft metal ions such as Ag(I) and Hg(II) coordinate with soft ligands and, consequently, form stable complexes through favorable interactions with covalent bonding. Hard metal ions such as Li(I) and Na(I) bond strongly with hard ligands to form more stable complexes. In addition, steric effects, which determine spatial relationships and spatial requirement of ligands, also play an important role in coordination compound formation and stability. Bulky ligands, for instance, may hamper coordination, producing less stable complexes or even reactivity under specific conditions. This fine interplay between bonding nature and steric effect points to the complexity entailed in forming stable coordination complexes (Nováčková et al., 2009). Lastly, thermodynamic and kinetic stability must be considered in the examination of the impact of metal ion properties on complex stability. Stability constants can quantify the thermodynamic stability of a complex, showing the tendency towards complex formation at equilibrium. Kinetic stability, however, refers to how stable a complex is with respect to dissociation over time. The metal ions such as Co(III) or Pt(II) are highly kinetically stable with long-lived complexes due to the slow rate of ligand substitution. Such stability factors are known and can be utilized for beneficial applications in numerous fields, such as catalysis and drug development, wherein the stability of metal complexes is the predominating factor. Therefore, the varied effect of metal ion properties ultimately establishes the landscape of coordination chemistry, revealing the significant implications for theoretical as well as applied advancement in the field (Buiarelli et al., 2019) (Figure 6.6).

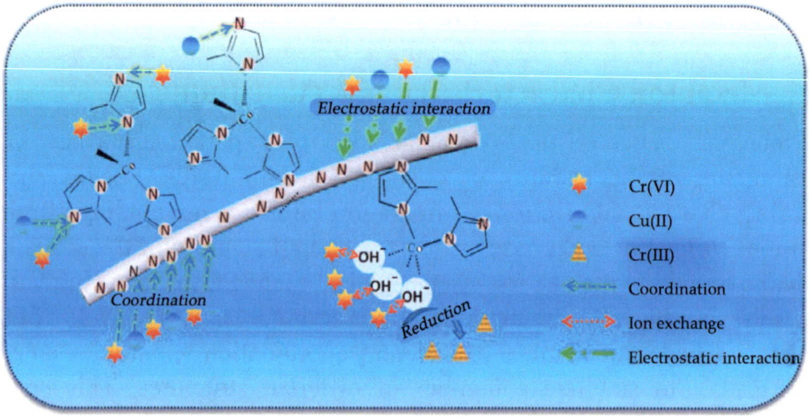

Figure 6.6. *Metal ion characteristics and their influence.*

Source: https://ars.els-cdn.com/content/image/1-s2.0-S0959652623008673-gr9.jpg.

6.3. THERMODYNAMIC CONSIDERATIONS

It is imperative to realize the thermodynamic considerations underlying the stability of coordination complexes for unveiling their reactivity and usefulness in a variety of fields, including medicinal chemistry. The intricate relationship between enthalpy and entropy takes a pivotal position in identifying the stability of such complexes. Contributions from enthalpy arise due to interactions between the ligands that surround the central metal ion and the metal ion itself, as well as the overall geometry of the complex. However, contributions related to entropy rely on the number of degrees of freedom associated with the freedom of motion of the ligands and the solvation shell. Also, coordination compound stabilization is normally a delicate balance among such energetic inputs where attractive interactions must counteract the entropic price of structural immobilization. As interest in targeted drug design grows, the information gained from such thermodynamic parameters gains leverage to optimize medicinal compounds in therapeutic processes and align with the interdisciplinary nature discerned in Bruijnincx et al. (2009). An important aspect regarding coordination complex stability is ligand substitution reactions, which are necessarily under thermodynamic control. These reactions, where there is dynamic exchange of metal-bound ligands, are responsible for determining the stability profile of coordination compounds to a large extent. New advances in rapid-reaction methodologies have enabled ligand substitution processes to be investigated in depth and what makes ligand exchange thermodynamically favorable to be understood. Such studies highlight the fact that even though kinetic processes play a role, the intrinsic thermodynamic topography ultimately decides the favorable pathways and the equilibrium states of these complexes. For example, strong field ligand complexes tend to be more stable due to increased lattice energy as well as a more favorable Gibbs free energy of formation, establishing their significance in forming rugged materials and drugs as suggested in Gray et al. (1966). In addition to these early thermodynamic issues, one cannot overstate the impact of molecular orbital (MO) theory on coordination complex bonding character understanding. The MO model provides a beautiful explanation for the process of stabilization occurring, particularly regarding electronic interaction between metal and ligand orbitals. By studying the distribution and bonding interaction energy, scientists can clarify how different ligand types and structures affect the stability of complexes. The combination of MO theory and thermodynamic calculation not only increases the predictive capability of stability evaluation but also leads to the design of ligands with desired binding characteristics. With the revolution in the face of coordination chemistry, the interaction between thermodynamic principles and molecular orbital treatment continues to be an invaluable aid in understanding how to build successful complexes, trending towards a general approach to coordination compounds.

6.3.1. Gibbs Free Energy and Stability

Gibbs free energy principle lies at the nucleus of the understanding of stability in complexes in coordination compounds. Gibbs free energy, which is expressed as $G = H - TS$, where H is enthalpy, T is temperature, and S is entropy, provides a measure of reaction spontaneity and the equilibrium position. In coordination chemistry, a reduction in Gibbs free energy corresponds to an enhancement in stability of a complex. A stable coordination complex generally has negative Gibbs free energy changes, and hence the synthesis of such a compound is thermodynamically favorable. Ligand identity and metal center both play crucial roles in determining the enthalpic and entropic contributions that govern the overall free energy change, thereby directly influencing the stability of the complexes formed. The choice of ligands has a major impact on the Gibbs free energy for coordination complex stability. Strong σ-donor ligands, for instance, raise the extent to which metal-ligand contacts can lower the Gibbs free energy, making more stable complexes. Research has shown that the addition of strongly coordinating ligands to metal ions can significantly change their water reactivity and stability.

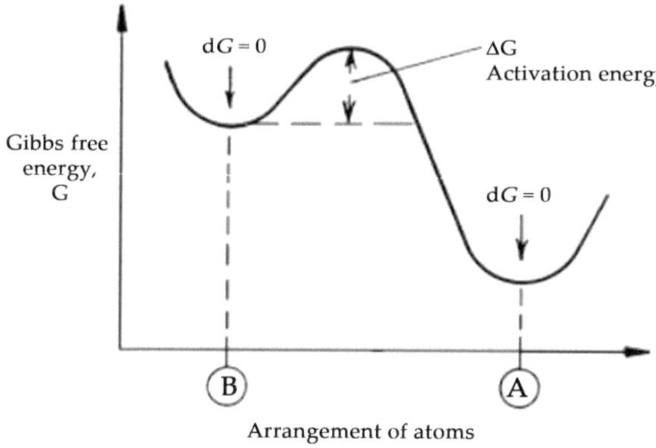

Figure 6.7. *The Gibbs free energy and stability.*

Source: https://ars.els-cdn.com/content/image/3-s2.0-B0080431526012146-gr1.gif.

For example, in the study of uranyl complexes, coordination of dimethylsulfoxide (dmso) instead of acetone (aco) resulted in a bigger decrease of Gibbs free energy due to the fact that it provides more ligand donation, hence enhanced bonding with other reactants like dioxygen and water. This finding underscores the fact that ligand choice not only defines reactivity but also has a significant impact on the stabilization of coordination compounds as is shown from experimental results and theoretical calculations in ligand assessments (Bryantsev et al., 2009). Coordination compound stability is also further explained with research on metal hydrides, particularly those of actinium. Theoretical predictions of actinium hydrides are highly stable and superconducting, suggesting their future use and basic significance in coordination chemistry. The evolutionary algorithm used to predict the hydrides has showed that different structural phases possess variable Gibbs free energy profiles that correspond to stability under different pressure conditions. This implies that as one alters the molecular configuration of coordination complexes, the entropy change related to the changes in conformation can have a substantial impact on Gibbs free energy. Hence, it is essential to understand the subtle relationship of enthalpic and entropic change to design stable coordination complexes, particularly superconducting ones wherein phase stability is the deciding criterion (Kruglov et al., 2018) (Figure 6.7).

6.3.2. Enthalpy and Entropy Contributions

The enthalpy and entropy contributions play a crucial role in stability of metal-ligand complex in coordination compounds. Enthalpy, or the heat content of a system, plays an important role in how ligands interact with the metal center. Greater negative enthalpic contribution from stronger metal-ligand interactions such as covalent or ionic bonds tends to raise the overall stability of the complex (Jackson et al., 2013). The entropy factor, however, defines the randomness or disorder in the system. When a coordination compound is formed, there may be differences in the number of molecules or ions in the reactants and products that affect the systems entropy. Thus, a greater appreciation of these thermodynamic parameters is required in order to predict and explain the stability of various coordination complexes. Moreover, the interaction between enthalpic and entropic contributions is crucial in attaining stable coordination compounds. Even though a highly negative enthalpic contribution can stabilize the complex, increasing order with the complex formation process might mean the loss of entropy (Micoulaut et al., 2016). This interaction normally controls the thermodynamic favorability of complex formation. For instance, in systems with numerous ligands binding to one metal center, the subsequent loss of disorder can have a considerable effect on the entropy term, potentially counteracting the favorable enthalpic contributions. Identifying such dynamics not only enables evaluation of complex stability but also helps in designing new coordination compounds with pre-designed stability characteristics for catalysis, materials science, and bioinorganic chemistry. In addition, entropy-enthalpy optimization plays a key role in realizing the full potential of coordination compounds in practical applications. Selection of solvent, temperature, and pressure are parameters that can significantly impact both enthalpic and entropic contributions. For example, some coordination compounds can be stabilized by solvation effects when a solvent is present, which increases the enthalpic contributions by raising the ion-dipole interactions. At the same time, the solvation may also have an influence on the entropy since there would be additional molecules of solvents interacting with the complex (Jackson et al., 2013). Thus, comprehensive thermodynamic models that include both enthalpy and entropy, as well as other pertinent parameters, become increasingly valuable in predicting complicated behavior under various conditions, eventually allowing for better control of the design and optimization of coordination complexes for new uses in modern chemistry.

6.3.3. Stability Constants and Their Significance

The role of stability constants in the study of coordination compounds cannot be overemphasized because stability constants are numerical measures of metal-ligand complex stability. Stability constants, commonly symbolized by the letter K, represent the equilibrium between free ligand and the related metal ion within a solution. A larger stability constant indicates a stronger interaction between the metal cation and the ligand, and hence makes the complex stable under varying conditions. These constants provide us with insight into the reactivity as well as selectivity of coordination compounds, which becomes crucial for real life applications ranging from materials science to biochemistry. Because metal-ligand interactions dictate much of the behavior of these complexes, stability constants thus play a key role in predicting the functionality and efficacy of metal compounds, particularly in biological systems (Nováčková et al., 2009).

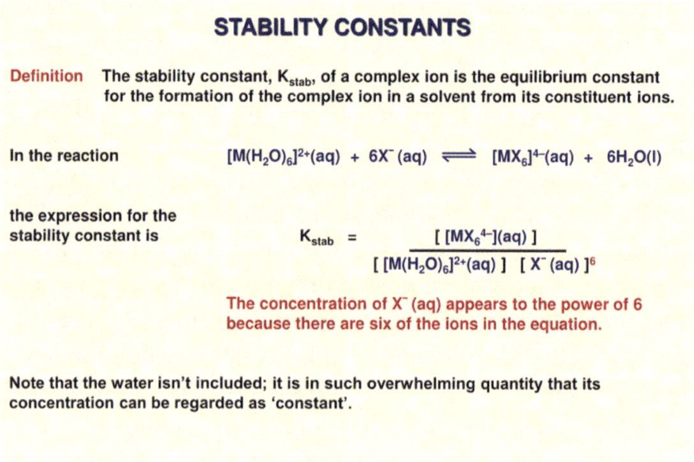

Figure 6.8. *The significance of stability constants in the study of coordination compounds.*

Source: https://image.slideserve.com/1450146/slide5-l.jpg.

Stability constant determination is typically a function of many various analytical methods, including spectrophotometry and potentiometry, whereby systematic inspection of equilibrium relationships is enabled. Through measurement of the free coordination and complexed species concentrations at equilibrium, scientists can obtain stability constants that offer critical information about the metal ion complexation behavior. Such information can be extremely revolutionary when it comes to examining the functions of metal ions in biological and environmental systems. Stability constants, for instance, can be used to evaluate the bioavailability of soil nutrients or the mobility of heavy metals in aquatic environments. The need for correct determination of stability constants is applicable to drug design, where the extent of interaction between a metal drug and its target can have significant influence on therapeutic responses (Gray et al., 1966). The significance of stability constants is also brought out by their application in determining reaction mechanisms in coordination compounds. The mechanistic study of ligand substitution reactions is a nascent area of study which is enabled by a complete understanding of stability constants. By measuring the relative stabilities of complexes, researchers can classify and predict reaction pathways for ligand substitution reactions. This is growing recognition of stability constants as not only static measures but as dynamic values that inform coordination chemistry principles at a more general level. With the emergence of new fast-reaction techniques, the ability to probe less frequent central metal atoms increases once more, emphasizing again the importance of these constants in establishing fundamental insight into advanced behavior (Gray et al., 1966; Nováčková et al., 2009) (Figure 6.8).

6.4. KINETIC STABILITY VS. THERMODYNAMIC STABILITY

In coordination compound research, it is essential to appreciate the distinction between kinetic stability and thermodynamic stability to predict the reactions of complexes. Kinetic stability refers to the rate at which a complex will undergo chemical reaction, e.g., by substitution or decomplexation. Kinetically stable compounds are not susceptible to these reactions taking place, although they may not be therapeutically stable. Conversely, thermodynamic stability measures the likeliness of the formation and existence of a complex under conditions of equilibrium. Thermodynamically stable is a complex that will, once formed, prefer to exist in that state for long periods of time unless perturbed. This duality illustrates that a complex can be kinetically stable but not thermodynamically stable, and the dynamics of coordination chemistry become very complicated and influence many applications, including catalysis and drug design. The complexity of thermodynamic and kinetic stability can further be explained from research that has been done on transition metal coordination compounds. Studies such as the one explored in Das et al. (2025) demonstrate the role of differences in stability at the core of understanding ion-association reactions and substitution reactions. For instance, the arsenic (III) and (V) complexes show remarkable variation in their stability profiles depending on the reactant environment, which impacts their kinetic and thermodynamic characteristics. The findings point out that a complex that can show high thermodynamic stability can also undergo rapid kinetic reactions, making it suitable for applications with rapid response demands, e.g., treatment application. Thus, kinetic and thermodynamic properties must be investigated together in order to sufficiently take into account the behavior of coordination complexes under various conditions. Also, investigation of the structural and functional properties of coordination compounds can yield data on kinetic and thermodynamic stabilities. Characterization and synthesis of novel coordination compounds are examples of studies in El-Helw et al. (2024), with results showing that the choice of ligands and metal ions play a significant role in both types of stability. Certain metal complexes were found to be highly kinetic inert and yet possess good thermodynamic parameters, thus finding more relevance in many applications such as drug delivery and environmental treatment. This interaction between stability types not only identifies the suitability of such complexes for specific functions but also prescribes new material and strategy design in coordination chemistry. Through appropriate analysis, researchers can create targeted strategies in order to manage the stability of complexes based on industrial and medicinal needs.

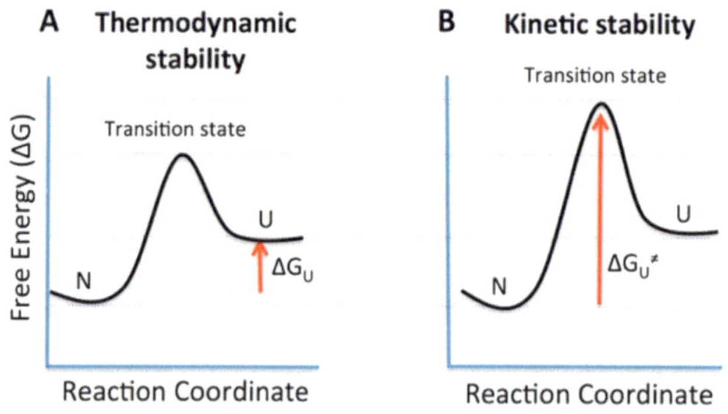

Figure 6.9. *Kinetic stability vs. thermodynamic stability.*

Source: https://www.creative-biostructure.com/upload/image/5-1-1-3-protein-stability-and-dynamics-1.jpg.

6.4.1. Definitions and Differences

To comprehend the stability of complexes in coordination compounds, one must first clearly define coordination compounds and their components. Coordination compounds consist of a central metal atom or ion, typically a transition metal, surrounded by ligands that are molecules or ions donating electron pairs to form coordinate covalent bonds. The nature of these ligands significantly influences the stability of the resulting complexes. Ligands may be quite varied in their nature, ranging from monodentate ligands that bind via one atom to polydentate ligands that bind through several different sites. This difference in ligand type is the basis for how to approach molecular interaction in terms of stability. The electronic structure of the metal, the chelation power of the ligands, and the geometry of the coordination compound are interrelated factors that govern the stability of various complexes as established in previous works on coordination chemistry (Nováčková et al., 2009). Coordination complexes vary in stability based on varying strengths in interactions among the central metal ion and ligands. In general, chelating ability by ligands significantly enhances stability by forming chelate rings, which considerably lowers the energy of the system. This is what is known as the chelate effect and is a result of the entropic reasons that are associated with releasing solvent molecules in the process of binding to the ligand. Moreover, different coordination geometries, such as octahedral or tetrahedral geometries, produce different electronic environments that can affect the nuances of stability. For example, transition metal complexes with strong-field ligands tend to have a stronger tendency toward the presence of stable low-spin states than complexes with weak-field ligands, which favor high-spin states. These theoretical backgrounds inform the synthesis and design of stable complexes, where rigid criteria are employed to select ligands (Acerete et al., 2014). The stability of coordination complexes is ultimately related to the definitions and distinctions of the pieces that make it up. With a knowledge of the fundamental concepts of coordination compounds, we can ascertain the influence of ligand and central metal atom properties on overall stability. The disparate behavior of different ligands under different circumstances also indicates the complexity of coordination chemistry. Scientists studying thermal stability in these complexes have also shed light on the specific interactions that underlie their durability, with ramifications extending across catalysis and materials science. This intricate relationship highlights the importance of fine appreciation of coordination chemistry definitions and distinctions in facilitating the field towards new breakthroughs and uses (Acerete et al., 2014; Nováčková et al., 2009).

6.4.2. Examples of Kinetically Stable Complexes

The existence of kinetic stability for coordination compounds can be proved through compounds like cobalt(III) ammine, which has been proven to resist ligand substitution reactions enormously. This stability derives from the slow rate at which ligands dissociate from the cobalt center, a property that is very regulated by the electronic and steric environment surrounding the metal ion. Studies have proven that the energy barrier to ligand exchange reaction for these complexes is very high and calls for specific conditions for effective reactivity. The formation of such kinetically stabilized complexes is not only significant in inorganic chemistry but also holds promise in drug development, where stability can equate to longer-term therapeutic action. Cobalt(III) ammine complexes' stability, for example, addresses the possibility of building drug candidates that are stable in biological conditions and release active therapeutic agents efficiently (Bruijnincx et al., 2009). Another striking example of kinetically stable complexes exists for platinum(II) derivatives, and that is cisplatin, a widely used cancer chemotherapeutic drug. The cisplatin geometry allows it to bind preferentially with DNA, forming extremely stable adducts with high efficacy in preventing DNA replication and transcription. The kinetic stability of cisplatin emanates primarily from the availability of its bidentate ligands, which form a very stable skeleton above the metal center. This stability is not merely theoretical; it also has practical implications on the drugs therapeutic activity. Understanding the mechanisms of the ligand substitution reactions of such complexes may lead to the formulation of more effective chemotherapeutic agents, and therefore, the general success rates of the treatments for cancer (Gray et al., 1966). Further studies on the reactivity profiles and stability of these complexes continue to offer insight into the synthesis of drugs in the future. Besides, the kinetic stability in coordination complexes significantly contributes to the field of coordination chemistry and its allied material science. To illustrate, supramolecular structures composed of kinetically stable transition metal complexes display elegant catalytic and sensing characteristics. These types of complexes are resistant to harsh environmental conditions without degrading, thereby maintaining their functional integrity for a very long time. The presence of stable ligands enhances kinetic stability, providing a basis for designing materials for specific applications, such as sensors that detect environmental changes or catalysts that induce selective reactions under mild conditions. Researchers can tailor stability and reactivity for application in industrial and environmental chemistry by synthesizing new ligand frameworks. Thus, the exploration of kinetically stable complexes enriches our theoretical foundation and inspires innovation in various scientific disciplines (Bruijnincx et al., 2009).

6.4.3. Implications for Chemical Reactions

The implications of chemical reactions involving coordination complexes go well beyond mere stability and encompass catalytic activity and versatility during synthesis. Catalytic applications of metal complexes are crucial to the advancement of organic synthesis, i.e., carbon-hydrogen (C–H) bond functionalization. These transforamtional processes exhibit pronounced economic and environmental advantages compared to traditional routes, as evidenced by the current trends in employing enzymes as biocatalysts for hydroxylation and hydroalkylation reactions. The current research activities on these enzymic processes signal the need for further research on metal-ligand interactions that enhance reaction yield and selectivity. Furthermore, since coordination compounds can be systematically improved, their use in preparative and industrial chemistry is a revolution towards more environmentally friendly procedures in chemical synthesis while ensuring high efficiency levels (Arnold et al., 2011). Moreover, knowledge of ligand substitution reactions is of utmost importance in the elucidation of mechanisms controlling coordination chemistry and the stability of these complexes. New advancements in fast-reaction techniques have enabled researchers to study the kinetics of such reactions, specifically reactions involving monodentate ligands and metals such as cobalt(III) and platinum(II).

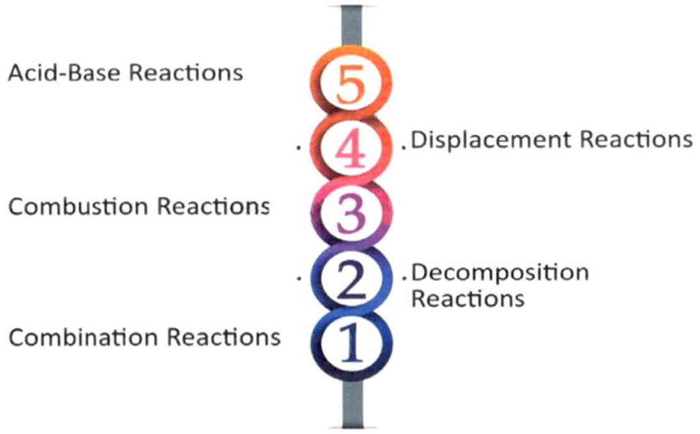

Figure 6.10. *The implications for chemical reactions involving coordination complexes.*

Source: https://fastercapital.co/i/Chemistry--Form4-Chemistry--Delving-into-Elements-and-Reactions--Types-of-Chemical-Reactions-and-Their-Significance.webp.

By creating a classification system for such ligand substitution reactions, researchers are able to make more precise predictions and gain a better understanding of reactivity trends specific to different coordination compounds. This foundational piece of research is significant because it connects present work with tested theory, most notably molecular-orbital theory. The application of this theory provides insight into electronic factors governing reaction dynamics, thus allowing further manipulation and prediction of coordination complex behavior under given conditions (Gray et al., 1966). Lastly, stability in coordination compounds is a significant area of study regarding the creation of new materials and chemical processes. As metal-ligand interactions regulate the stability and reactivity of these complexes, understanding how varying conditions in the environment, including solvent and temperature, influence their behavior can lead to new applications in drug delivery and catalysis.

Systematic examination of these interactions allows chemists to design more stable complexes for special purposes, paving the way towards increased potential for the exploitation of coordination compounds in industry. Also, as the progress of coordination chemistry goes on, the knowledge that will be gained will, naturally, lead to the innovation of new ways of making chemical reactions more efficient and sustainable, as the interrelated nature of reactivity and stability in coordination compounds (Arnold et al., 2011; Gray et al., 1966) (Figure 6.10).

6.5. APPLICATIONS OF STABLE COORDINATION COMPLEXES

Applications of stable coordination complexes vary to multiple fields, especially catalysis. Coordination complexes with their stable metal-ligand bonding interactions facilitate numerous chemical reactions as catalysts. One such instance is the covalent triazine framework (CTF) as possible materials for heterogeneous catalysis. They have their development to achieve high surface area, permanent porosity, and stability toward aggressive thermal and chemical environments. These features are complemented by functionalization methods that enable incorporation and stabilization of nanoparticles and immobilization of molecular complexes in CTFs (Krishnaraj et al., 2019). Thus, the flexibility and stability of such coordination compounds provide a firm platform for catalyzing various reactions, bringing them into prominence in industrial processes as well as research studies. In addition to catalysis, stable coordination complexes also play an important role in medicine, particularly in drug delivery and imaging processes. Several metal complexes have been investigated for medicinal use based on their ability to bind selectively to biological targets. Stability in complexes such as these is crucial because it ensures that the metal ions remain enclosed in the complex until they arrive at the target site of action. For example, organophosphorus podand mimics have been synthesized to investigate their complexing potential with ions such as Li+ and Na+ in varying solvents. These studies not only uncover the thermodynamic nature of the interaction between metal ions but also highlight the significance of the complex structure and the solvent effects on drug efficacy enhancement (Baulin et al., 1992). Thus, the stability of coordination complexes is not only good but imperative for the advancement of medical applications. Moreover, stable coordination complexes are the foundation for sensors and detection systems, particularly environmental monitoring and analytical chemistry. Coordination complexes' selective binding properties make them effective sensors for the identification of specific ions or molecules like heavy metals and toxic organic chemicals. The architecture of these sensor systems is often to incorporate stabilization of coordination complexes that are environmentally sensitive, such as pH or temperature, and thereby enhance selectivity and sensitivity. Through continued research, incorporation of coordination complexes into sensor technology has great potential for developing novel solutions to existing environmental challenges. By virtue of this stability of these complexes, scientists are able to develop and construct sophisticated detection methods that provide real-time monitoring of contaminants, noted for their versatility in actual applications to ensure public health and environmental wellness.

6.5.1. Role in Catalysis

The use of coordination compounds as catalysts is also governed by the stability of the complexes. Stability will both influence a catalyst's performance and longevity in chemical reactions. Transition metal complexes, which are ubiquitously utilized as catalysts, have high coordination environments that improve their ability to catalyze reactions without degrading themselves. Transition metal complexes form stable complexes with reactant molecules and thereby lower the activation energy necessary for reactions to take place. Increased stability is associated with increased control of reaction pathways, leading to increased selectivity and yield. Understanding of the stability of these coordination compounds is thus crucial, as it determines their potential application in catalytic processes and overall viability under reaction conditions. The functionalization of coordination compounds also enhances their catalytic activity, particularly in heterogeneous catalysis. For instance, covalent triazine frameworks (CTFs) serve as an enhanced scaffold for metal complex and nanoparticle immobilization, greatly promoting their catalytic activity. These supports have good thermal and chemical stability, allowing them to be used for longer periods during reactions with continuous sites for catalysis. Their permanent porosity and high surface area allow higher catalytic site density, promoting easier substrate molecule interaction. By functionalization of CTFs, researchers are able to develop catalysts for particular reactions, hence resulting in more effective and sustainable catalytic processes in various sectors, as mentioned in Krishnaraj et al. (2019). The dynamic relationship between electronic and steric effects in coordination compounds also adds additional complexity to their catalysis application. Organometallic complexes, particularly those in which there is a transition metal such as Ru II, have proved to possess fine catalytic qualities due to their ability to undergo facile ligand substitution and redox reactions. Both of these reactions are finely controlled through the ligand sphere surrounding the metal center so that reaction mechanisms are under control. This capability has resulted in new drug design approaches wherein organometallic complexes act not only as drugs but also as effective catalysts. Their capacity to identify outer-sphere biological targets is one mechanism by which coordination chemistry principles are applied beyond industrial catalysis into biochemistry and medicine, evidencing the scope and utility of these complexes in current science, as addressed in Habtemariam et al. (2012).

6.5.2. Use in Medicinal Chemistry

The integration of coordination compounds into medicinal chemistry has revolutionized drug discovery the way scientists envision now. Coordination complexes, that are a central metal ion surrounded by ligands, offer chemical properties that are unique and can be used for treating disease. Because of their versatility, coordination compounds allow medicinal chemists to design molecules that are specific to bind with biologic targets of disease. According to the immense bulk of literature, with the structural insights obtained from the developments in molecular biology and genomics, chemists can now develop tailor-made drugs that effectively disrupt the disease-associated biochemical pathways in most illnesses. This intersection of fields highlights the importance of an interfaculty effort in the quest for beneficial therapeutics, paving the way for new breakthroughs in drug therapies (Bruijnincx et al., 2009). Metal utilization in medicinal chemistry is well demonstrated by the use of metal-based drugs such as cisplatin, which has been extremely successful in cancer therapy. Cisplatins mode of action, whereby it forms covalent bonds with DNA, is an excellent example of the careful utilization of coordination compounds for therapeutic impact. However, the limiting aspect demonstrated through the drug's toxic side effect has prompted researchers to reassess its design and use. As emphasized in recent reviews, a trend also exists towards the development of new metal-based drugs that are endowed with therapeutic activity and reduced toxicity, encapsulating a strategy more and more focused on drug effect. Along with stress laid upon selectivity of delivery systems, this trend also encompasses maximization of interaction mechanisms of such complexes, thereby enhancing the stability and efficacy of new therapeutics (Ang et al., 2008). The study of coordination complexes in medicinal chemistry illustrates the paradigmatic significance of integrating metal ions into therapeutic treatments. Through the unmasking of the delicate relationship between the reactivity and structure of such compounds, chemists can facilitate the introduction of more effective and selective drug candidates. Further, advances in synthesis approaches and discovery of novel biological targets have increased the velocity at which new metal-based drugs are being advanced to clinical trials. This growing volume of evidence not only emphasizes the relevance of coordination chemistry to drug discovery but also directs the reader to the excitement in the future of designing next-generation therapeutics. While medicinal chemists delve deeper into the subtleties of drug designing, the stability of these complexes will remain an essential factor in determining their overall efficacy and safety for use in the clinic (Bruijnincx et al., 2009) (Figure 6.11).

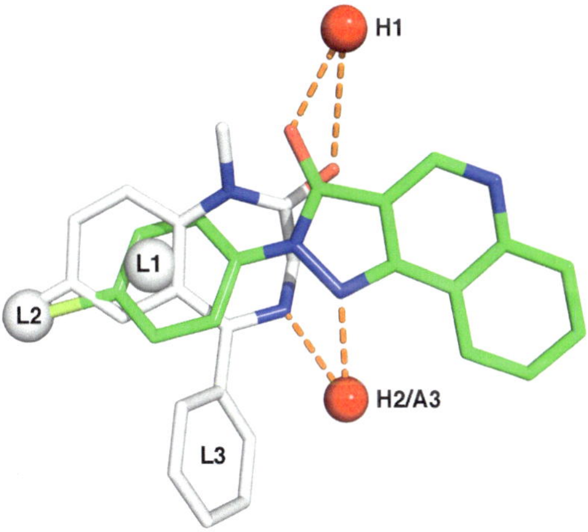

Figure 6.11. *Use in medicinal chemistry.*

Source: https://upload.wikimedia.org/wikipedia/commons/6/6f/Bzr_pm.png.

6.5.3. Importance in Materials Science

The significance of coordination compounds in materials science primarily has to do with their structural stability and functional versatility. Stability in such complexes not only determines their performance in a majority of chemical processes but also affects their usability in many areas. Stable coordination compounds, for example, are important in the development of advanced materials with design-specific properties, such as self-healing hydrogels. The application of specific ligands, e.g., the 2,6-bis(1,2,3-triazol-4-yl)-pyridine within polymer scaffolds, illustrates the way coordination chemistry can improve the mechanical and rheological material properties. These advances create opportunities for self-healing materials, which reveal an essential overlap between materials science and coordination chemistry (Hoogenboom et al., 2020). The relevance of this study spans many industries, from biomedical to coatings, and isolating the pivotal role of stability in the advancement of modern material solutions. In addition to structural stability, coordination complexes provide unique means to improving material performance by incorporating metal complexes. Aromatic bis-[1,3,5]triazinyl diazenes, for instance, become more thermally stable as different substituents are introduced on the triazinyl rings. This demonstrates how coordination of the metals to organic scaffolds can impart significant thermal property enhancement, which in turn impacts the materials utility in products such as fire-resistant polymers and coatings. This relative comparison of the materials and metal complexes demonstrates how certain electron-withdrawing groups enhance thermal stability, yet the coordination of the metals induces performance fluctuations. Studies of such relationships reveal their value in material innovation and safety, pointing to the importance of the stability of coordination complexes in driving the frontier of useful applications forward (Aelmans et al., 2014). Furthermore, the study of coordination compounds in materials science gives profound insight into how material properties can be tailored for functions. Through the understanding of the stability of complexes, scientists have succeeded in regulating molecular-level interactions, which has played a pivotal role in advancing various technologies. For example, the observation of differentiated efficacy of some zinc and copper complexes as flame retardants emphasizes the value of wise ligand and metal choices in creating new materials. These molecules are not spectators; they are part of composite development that plays a role in determining performance parameters critical to industrial applications. While materials scientists continue to expand the scope of coordination chemistry, stability inherent in such complexes will be at the center of developing safer, more efficient, and functionally diverse materials that address the needs of the present across different industries.

6.6. EXPERIMENTAL METHODS FOR STUDYING STABILITY

Knowledge of complex stability in coordination chemistry is vital if one has to project how such complexes will behave and what applications they will find in coordination chemistry. Experimental methods applied to explore stability typically consist of a variety of techniques that allow scientists to study the thermodynamic and kinetic properties of such systems. One of the methods includes the application of isothermal titration calorimetry (ITC), enabling the precise assessment of heat changes associated with complex formation. With this approach, the binding affinity as well as the stoichiometry of coordination compounds are disclosed, providing critical information regarding their stability. Moreover, spectroscopic methods such as UV-visible, infrared, and NMR spectroscopy may provide complementary information about electronic and structural properties of complexes. Through use of all these tools, researchers are able to build a general picture of stability with respect to different external parameters and intrinsic character of the coordination compounds themselves. Yet another powerful means of probing coordination complex stability is through structural characterization methods. X-ray crystallography serves as a gold standard for determining three-dimensional structures of atoms in a complex. It is accurate in identifying stability with stable correlations. Coupled with computational methods, such as density functional theory (DFT), scientists are able to foresee the impact of structural parameter modifications on stability. Further, techniques like dynamic light scattering (DLS) can identify aggregation behavior of complexes in solution and give information about stability under varying conditions. Understanding structural dynamics is vital since it often has a role in defining the kinetics of ligand exchange processes, which are an intrinsic component of the stability profiles of coordination compounds. Finally, molecular dynamics simulations have been a useful experimental tool in probing stability in coordination complexes. Such simulations allow one to probe energetic landscapes and determine stable and metastable states under various conditions, such as temperature and solvent composition. The use of effective medium theory (EMT) potential parameters allows scientists to measure how molecular interactions at the microscopic level affect stability. As suggested via investigation into metallic glasses, structural properties and their transitions are an important factor in the determination of the total stability of a compound. Via investigation into thermodynamic properties and diffusivity, scientists are able to make conclusions regarding coordination compounds' fragile or strong liquid nature and hence comprehend the intricate dynamics of their stability (Bailey et al., 2003; Dobrosavljević et al., 2018) (Figure 6.12).

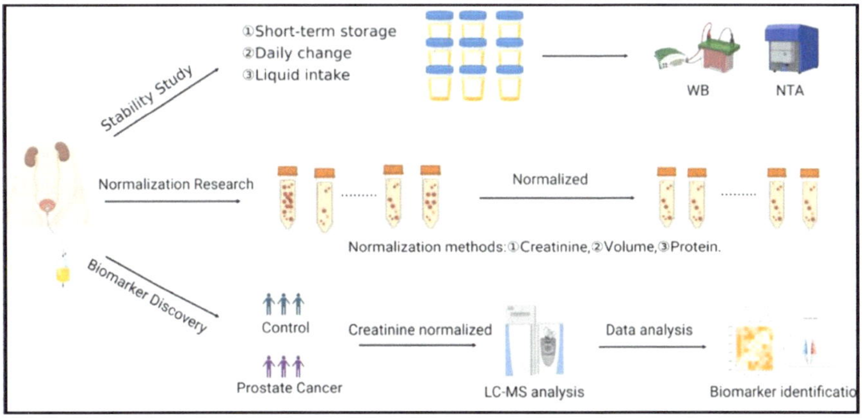

Figure 6.12. *Experimental methods for studying stability.*

Source: https://www.researchgate.net/publication/379886100/figure/fig1/AS:1143128123678141 0@1713375992800/Experimental-Design-and-Workflow-of-the-Study-First-the-stability-of-urine-EVs-during.jpg.

6.6.1. Spectroscopic Techniques

The field of spectroscopic techniques has a significant contribution to making the complexity of complexes understood in terms of stability. These techniques, for example, methods such as X-ray absorption spectroscopy, infrared spectroscopy, and nuclear magnetic resonance, enable scientists to study electronic and structural properties of complexes. Among these, X-ray absorption near edge structure (XANES) stands out for its ability to provide information on the oxidation state and coordination environment of metal centers within complexes.

For instance, the application of Ti K-edge XANES to analyze Ti-bearing silicate glasses makes it possible to have a fine understanding of variable coordination environments, in which composition is observed to influence stability. The data provided by such spectroscopic techniques yield valuable information about the geometric and electronic features responsible for dictating the properties of coordination compounds, all of which are important in catalytic-related, materials-related, and biochemical applications and thereby justify the relevance of knowing how stable they are. Apart from understanding coordination numbers, spectroscopic techniques enable research on interatomic interactions within coordination complexes. With multidimensional approaches, researchers can identify minor electronic transitions and vibrations modes associated with several ligands.

The merging of theoretical models with spectroscopic data enhances intricate phenomena in coordination chemistry. For example, advanced simulations can relate experimental parameters and theoretical calculations to offer a comprehensive understanding of stability factors. Notably, findings from spectroscopic studies can offer insight into predictive models used in the synthesis of new coordination compounds. This correlative collaboration between experimental and theoretical work underscores the pivotal use of spectroscopic methodologies in creating the field as well as clarifying the subtle dynamics that result in stability in these complexes. Besides, the diversity of spectroscopic techniques encompasses the detection of complexing and decomplexing kinetic processes, which directly affects coordination compound stability. Techniques such as time-resolved spectroscopy allow researchers to directly observe transient species, which plays a very significant role in learning about the kinetics and reaction mechanisms. For example, reactive intermediates determination may reveal paths to stable or unstable configurations in coordination compounds. The integration of real-time spectroscopic observations with strict theoretical models ultimately advances the understanding of stability as an intricate function induced by environmental factors and ligand movement. As studies continue to evolve, the integration of new spectroscopic techniques with emerging computational tools is certain to improve coordination chemistry knowledge, facilitating the fabrication of long-lasting materials for specific use.

6.6.2. Electrochemical Methods

Electrochemical methods have become key in explaining coordination complex stability since they provide information about redox properties controlling complex behavior. Such methods commonly enable the examination of electron transfer between metal centers and ligand units. More specifically, studies on redox-active ligands have enlightened electronic interactions behind stabilizing or destabilizing different complex geometries. Notably, the research highlighted in advance reports on dithiolate ligands demonstrated such complexities. Chemical preparation and spectroscopy of homoleptic bis and tris complexes evidenced that conventional oxidation-state nomenclature was not sufficient in describing their electronic structures' complexity and thus emphasized the contribution of electrochemical studies toward paving the path to an improved understanding of complex stability (Eisenberg et al., 2011). Application of electrochemical techniques is also seen to investigate intricate stability in high-performance battery materials, elucidating potential mechanisms for boosting energy density.

The new use of lithium bis(fluorosulfonyl)imide (UFSI) as an electrolyte indicates the capability of electrochemical techniques to deliver solutions to pressing issues in energy storage. Electrolyte composition optimization can result in significant breakthroughs in terms of interface stability when paired with lithium metal anodes and high-capacity cathodes. These studies not only improve our insight into electrochemical performance and complex compound formation, but they also underscore how electrochemical methods can be employed to achieve more efficient and safer energy-related materials (Borodin et al., 2020). This correspondence between stabilization and performance offers worthwhile information about coordination compound behavior at operational conditions. Moreover, electrochemical methods not only help in the definition of stability but also allow real-time monitoring of the dynamic process of coordination complex transformation under different conditions. Current response and potential variations provide valuable information regarding electron transfer kinetics and thermodynamic stability. These results help in the identification of optimal conditions for the preparation of stable coordination compounds. A detailed knowledge of how solvent interactions, or changes in concentration, affect the stability of the complex compounds can lead to more sophisticated coordination chemistry designs. As electrochemical techniques improve further, they will increasingly expose more nuanced aspects of the behavior of coordination complexes, further solidifying their role in theoretical as well as applied chemistry (Figure 6.13).

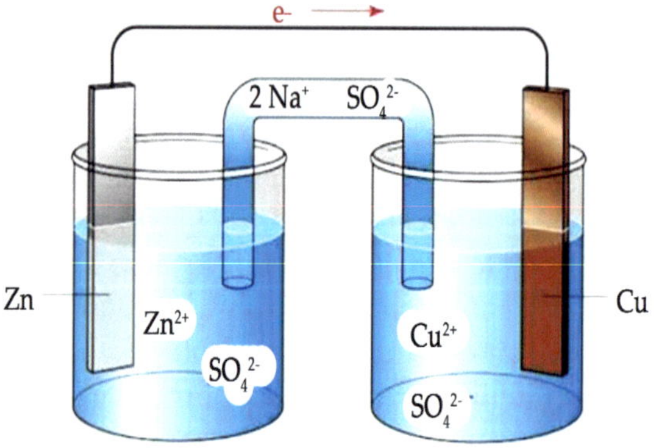

Figure 6.13. *Electrochemical methods.*

Source: https://cdn.slidesharecdn.com/ss_thumbnails/electrochemicalmethods-180131093438-thumbnail.jpg?width=640&height=640&fit=bounds.

6.6.3. Computational Chemistry Approaches

The stability of coordination complex compounds is greatly affected by the emerging technology in computational chemistry, which employs intricate algorithms and simulations to predict molecules. Computational processes provide information on the electronic structure, geometric structures, and energy topography of coordination complexes with visualizations of interactions between ligands and metal centers. By computer modeling of such interactions using sophisticated methodologies, scientists are able to ascertain stable conformations and compare their relative stabilities. Such predictive capacity is crucial, not only for familiarity with compounds that already exist, but also for designing new ones with desired properties. As the discipline evolves, the union of machine learning with computational approaches holds the potential to revolutionize material design, helping to examine structural and chemical spaces heretofore unknown. High throughput computational approaches streamline the discovery process by allowing researchers to screen vast libraries of candidate coordination compounds quickly. The preparation and characterization of new materials can be time-consuming and resource-intensive; however, computational protocols can identify which metal-ligand combinations are likely to yield stable complexes. New methodological advances, like graph convolutional neural networks, introduce new avenues of visualization and analysis of the extremely complex interrelations among different materials to make it easier to determine patterns that indicate stability and reactivity. As the literature presents to us, these methods not only enhance the understanding of coordination chemistry but also offer the way to automated material design, which drastically changes the methodologies of researchers in designing new coordination compounds (Grossman et al., 2018). Furthermore, computational chemistry packages enable a detailed study of the interactions governing the stability of coordination complexes, particularly when strong electron correlations are involved. Systematic studies of metal and ligand electronic properties reveal the dominant influence of orbital interactions on binding energies and hence stability. The balance between electron correlation and charge transfer processes accounts for the way both are accountable for overall structural integrity of complexes. It is what is required to forecast metastability and polymorphism in coordination chemistry so that one may have an advanced understanding of how stability varies with different conditions. The understanding gained from calculations is priceless, improving predictive capacities and enabling design of coordination compounds with individualized applications (Dobrosavljević et al., 2018).

SUMMARY

In Summary, complex stability in coordination compounds is a complicated subject matter that adopts both theoretical and pragmatic elements. As shown throughout this essay, a number of factors such as ligand character, metal ion identity, and environmental conditions all play a crucial role in the stability of the complexes. More understanding into these interactions not only complements our fundamental knowledge of coordination chemistry but also offers the potential for advances in many areas. For instance, the structure of organometallic complexes has been shown to provide a rich source for new catalysts, which may help improve the efficiency in chemical processes and drug development. The structures and functions in coordination complexes must be well understood to achieve the highest level of stability and effectiveness towards their respective applications, as indicated by the types of functionalities exhibited by complex systems in biological and industrial processes. In addition, progress in complementary fields, such as bioorganometallic chemistry, shows that the uses of stable coordination compounds extend far beyond the traditional chemical uses. The application of organometallic agents to therapeutic processes will likely bring targeted therapies with reduced side effects. As mentioned, these complexes may revolutionize drug delivery systems by utilizing outer-sphere interactions for selective binding and activation with minimal harmful interaction with non-target tissues. The stability requirements thus established in coordination compounds provide a vital framework for such innovation, highlighting the significance of knowledge about ligand substitution dynamics and thermodynamic stability. Therefore, the findings in this essay encourage future studies into how advanced stability may be leveraged towards medicinal targeting biomedical applications and catalysis. Finally, the study of covalent triazine frameworks (CTFs) and functionalization provides further dimensions to our understanding of complex stability in coordination compounds. These new structures not only depict the theory of stability but also demonstrate how these can be utilized in heterogeneous catalysis, where stable and high-performance materials are the priority. CTFs offer a medium for stabilizing nanoparticles and immobilizing molecular complexes to enhance their surface interactions and catalytic efficacy. Success in CTF design is reflected in the broader theme of this essay: that stability in coordination compounds is the key to their uses in various fields. Through step-by-step analysis of how structural design influences stability and function, we can further transfer these advances to breakthroughs in materials science, greatly impacting applications ranging from catalysis to biology.

Coordination compound stability is a multifaceted topic comprising many factors that influence their formation and dissociation. Critical determinants of complex stability are the metal ion nature, the ligands, and the net geometric structure of the complex. Transition metals, which display a unique ability for multiple oxidation states and coordination numbers, play a commanding role in determining stability in these compounds. Metal ions with high oxidation state, in particular, tend to form more stable complexes due to strong electrostatic attraction with coordinating ligands. This basic context sets the stage for the consideration of how different ligands may stabilize or destabilize metal-ligand interaction stability, an important consideration across applications ranging from catalysis to the development of drugs. Ligands as bidentates or chelating agents add complexity to coordination complex stability. Chelating ligands are more stable than their monodentate counterparts due to the chelate effect, arising from the presence of multiple bonds to the metal center. The chelate effect not only lowers the overall energy of the systems but also increases the kinetic stability of the complex so that it will be less likely to dissociate. Moreover, the steric and electronic effects of accompanying ligands should not be overlooked; bulky ligands can hinder coordination, while strong field ligands can bring significant stabilization through intensified interaction. It is the fine balance of these effects that is most critical in understanding the nature of coordination compounds in bio and industrial applications. Stability effects are beyond simple chemistry, particularly in pharmaceutical applications where coordination compounds are drugs. For instance, coordination compounds play vital roles in drug discovery via enhanced bioavailability and selective action. As indicated in the literature, stability of such complexes should be comprehended for maximizing therapeutic response, particularly for diseases like chronic kidney disease (CKD). For instance, Auryxia®, an iron compound, demonstrates how complex stability control improves iron parameters in patients with CKD and controlled phosphate levels, highlighting the critical complex stability-therapeutic efficacy relationship. The use of these principles in the design of drugs emphasizes the necessity for an interdisciplinarity that combines chemical, biological, and pharmacological insights in designing new and successful drugs, thus underlining the universal relevance of coordination chemistry in health and disease.

REFERENCES

1. Abraha Habtemariam, A., Aird, A., Alama, A., Alama, A., Alberto, R., Allardyce, C., Allen, C., et al. (2012). Designing organometallic compounds for catalysis and therapy. https://core.ac.uk/download/9561100.pdf (accessed on 13th September 2025).
2. Acerete, A., Altenau, T., Amjad, M., Baker, J., Baker, K., Bhasikuttan, A. C., Bhaumik, A., et al. (2014). Extending metal-to-polyoxometalate charge transfer lifetimes: The effect of heterometal location. https://core.ac.uk/download/29109178.pdf (accessed on 13th September 2025).
3. Aelmans, A., Aubert, C., Aubert, T., Aubert, Y., Chantarasiri, N., Chen, Y., Chigrinov, V., et al. (2014). Preparation and characterization of bis-[1,3,5] triazinyl diazenes and their utilization as flame retardants in polypropylene films. https://core.ac.uk/download/42138202.pdf (accessed on 13th September 2025).
4. Albright, T. A., Zhugayevych, A., Armstrong, D. R., Baxter, R. J., Braekken, H., Burdett, J. K., Elliott, S. R., et al. (2010). Electronic structure and the glass transition in pnictide and chalcogenide semiconductor alloys. Part I: The formation of the $pp\sigma$-network. http://arxiv.org/abs/1006.0771 (accessed on 13th September 2025).
5. Ang, W. H., Ang, W. H., Barnard, P. J., Barnes, K., Bednarski, P. J., Bergamo, A., Bregman, H., et al. (2008). New trends for metal complexes with anticancer activity. https://core.ac.uk/download/46362.pdf (accessed on 13th September 2025).
6. Arnold, F. H., Coelho, P. S., & Lewis, J. C. (2011). Enzymatic functionalization of carbon-hydrogen bonds. https://core.ac.uk/download/4888847.pdf (accessed on 13th September 2025).
7. Bailey, N. P., Jacobsen, K. W., & Schiøtz, J. (2003). Simulation of Cu-Mg metallic glass: Thermodynamics and structure. https://core.ac.uk/download/13732598.pdf (accessed on 13th September 2025).
8. Barone, G., Casini, A., De Almeida, A., Kavanagh, G. M., et al. (2017). The mechanism of aquaporin inhibition by gold compounds elucidated by biophysical and computational methods. https://core.ac.uk/download/96660185.pdf (accessed on 13th September 2025).
9. Baulin, V. E., Elshof, J. E. ten, Glebov, A. S., Solovev, S. N., et al. (1992). Complexation of lithium and sodium cations with B-phosphorylate ethers, modelling terminal groups of organophosphorus pod ands. An experimental and theoretical study. https://core.ac.uk/download/pdf/11455068.pdf (accessed on 13th September 2025).

10. Bino, A., Ganz, T., & Salusky, I. B. (2019). Mechanism of action and clinical attributes of Auryxia® (ferric citrate). https://core.ac.uk/download/323076204.pdf (accessed on 13th September 2025).
11. Borodin, O., Cho, S. J., Jang, M., Lee, U., et al. (2020). Nonflammable lithium metal full cells with ultra-high energy density based on coordinated carbonate electrolytes. https://core.ac.uk/download/323383913.pdf (accessed on 13th September 2025).
12. Bruijnincx, P. C. A., & Sadler, P. J. (2009). Controlling platinum, ruthenium, and osmium reactivity for anticancer drug design. https://core.ac.uk/download/44701.pdf (accessed on 13th September 2025).
13. Bryantsev, V. S., de Jong, W. A., Diallo, M. S., Goddard, W. A., et al. (2009). Addition of H_2O and O_2 to acetone and dimethyl sulfoxide ligated uranyl(V) dioxocations. https://core.ac.uk/download/4882289.pdf (accessed on 13th September 2025).
14. Buiarelli, F., Carcassi, E., Gullifa, G., Materazzi, M., et al. (2019). Modeling solid state stability for speciation: A ten-year long study. https://core.ac.uk/download/231747435.pdf (accessed on 13th September 2025).
15. Burr, P. A., Grimes, R. W., Lumley, S. C., Murphy, S. T., et al. (2013). Hydrogen solubility in zirconium intermetallic second phase particles. http://arxiv.org/abs/1307.7617 (accessed on 13th September 2025).
16. Cardin, D., Hartl, F., & Ryan, D. E. (2017). η3-Allyl carbonyl complexes of group 6 metals: Structural aspects, isomerism, dynamic behaviour and reactivity. https://core.ac.uk/download/74207026.pdf (accessed on 13th September 2025).
17. Cheng, J., Haught, M., Li, Z., Tang, W., et al. (2006). Direct methane conversion to methanol by ionic liquid-dissolved platinum catalysts. https://core.ac.uk/download/4874459.pdf (accessed on 13th September 2025).
18. Dengo, N., Gross, S., & Vittadini, A. (2017). Functionalization of colloidal transition metal sulphides nanocrystals: A fascinating and challenging playground for the chemist. https://core.ac.uk/download/141699569.pdf (accessed on 13th September 2025).
19. Dingwell, D. B., Mottana, A., Paris, E., Romano, C., et al. (1994). X-ray absorption study of Ti-bearing silicate glasses. https://core.ac.uk/download/12166895.pdf (accessed on 13th September 2025).
20. Dobrosavljević, V., Lanatà, N., Lee, T. H., Stevanović, V., et al. (2018). Critical role of electronic correlations in determining crystal structure of transition metal compounds. http://arxiv.org/abs/1710.08586 (accessed on 13th September 2025).

21. Eisenberg, R., & Gray, H. B. (2011). Noninnocence in metal complexes: A dithiolene dawn. https://core.ac.uk/download/4892476.pdf (accessed on 13th September 2025).
22. El-Helw, M. A., El-Ayaan, U., Abou El-Reash, Y. G., & Abu El-Reash, G. (2024). Synthesis, characterization, biological potency, and molecular docking of Co^{2+}, Ni^{2+}, and Pd^{2+} complexes derived from carbohydrazone ligand and its application in dye removal. *Applied Organometallic Chemistry*. https://www.semanticscholar.org/paper/d5dfb9ddb9fbbe33c41ca4bc197d3cceb0c79f04 (accessed on 13th September 2025).
23. Enyedy, É. A., Fülöp, F., Gál, G. T., May, N. V., et al. (2018). Impact of copper and iron binding properties on the anticancer activity of 8-hydroxyquinoline derived Mannich bases. https://core.ac.uk/download/213444633.pdf (accessed on 13th September 2025).
24. Golden, J. C., Ho, V., & Lubchenko, V. (2017). The chemical bond as an emergent phenomenon. http://arxiv.org/abs/1705.08429 (accessed on 13th September 2025).
25. Gray, H. B., & Langford, C. H. (1966). Ligand substitution processes. https://core.ac.uk/download/4890295.pdf (accessed on 13th September 2025).
26. Grossman, J. C., & Xie, T. (2018). Hierarchical visualization of materials space with graph convolutional neural networks. http://arxiv.org/abs/1807.03404 (accessed on 13th September 2025).
27. Halicioglu, T. (2025). Simulation studies for surfaces and materials strength. https://core.ac.uk/download/pdf/42834396.pdf (accessed on 13th September 2025).
28. Hoogenboom, R., Jerca, V. V., & Xu, X. (2020). Self-healing metallo-supramolecular hydrogel based on specific Ni^{2+} coordination interactions of poly(ethylene glycol) with bistriazole pyridine ligands in the main chain. https://core.ac.uk/download/299792665.pdf (accessed on 13th September 2025).
29. Jackson, A. J., & Walsh, A. (2013). Oxidation of GaN: An ab initio thermodynamic approach. http://arxiv.org/abs/1309.6232 (accessed on 13th September 2025).
30. Kapilas Das, S., Maji, A., Pal, T., & Pal, T. (2025). Ion-association reaction involving anionic complexes of arsenic (III and V) by monocationic surfactant. *Journal of Surfactants and Detergents*. https://www.semanticscholar.org/paper/043a28d4fa47fa573bfa4e2b66a222352cb3dfba (accessed on 13th September 2025).

31. Krishnaraj, C., Leus, K., Tahir, N., Van Der Voort, P., et al. (2019). Development of covalent triazine frameworks as heterogeneous catalytic supports. https://core.ac.uk/download/227041018.pdf (accessed on 13th September 2025).

32. Kruglov, I. A., Kvashnin, A. G., Oganov, A. R., Semenok, D. V., et al. (2018). Actinium hydrides AcH_{10}, AcH_{12}, AcH_{16} as high-temperature conventional superconductors. http://arxiv.org/abs/1802.05676 (accessed on 13th September 2025).

33. Micoulaut, M. (2016). Relaxation and physical aging in network glasses: A review. http://arxiv.org/abs/1605.07403 (accessed on 13th September 2025).

34. Nováčková, K. (2009). Stability of salts and complexes of humic acids. https://core.ac.uk/download/30278457.pdf (accessed on 13th September 2025).

35. Pizarro, A. M., & Sadler, P. J. (2009). Unusual DNA binding modes for metal anticancer complexes. https://core.ac.uk/download/46353.pdf (accessed on 13th September 2025).

CHAPTER 7

ELECTRONIC SPECTRA AND MAGNETIC PROPERTIES

Contents

INTRODUCTION

In the research of materials having unique electronic spectra and distinct magnetic properties, it is crucial to understand the fundamental definitions and nature distinguishing various solid-state materials. Electronic spectra refer to the structure and energy levels of electrons of a material, which are actually influenced by the nature of the atomic structure and the nature of bonding of the atoms involved. This contact has the potential to alter the light-material interaction, which creates different optical effects. When in relation to magnetic behavior, electron structure, along with potential external parameters such as temperature and pressure, is extremely significant. The interplay of these properties not only defines the behavior of materials but also offers avenues to new applications in electronics and photonics, a topic of significance for research and development in modern materials science. Studies of materials like two-dimensional HfS2 have illuminated our understanding of how dopant atoms can induce dramatic modifications to a material's electronic properties. For instance, incorporation of lanthanide dopants within the HfS2 monolayer exhibits electronic ground state modifications, converting the system from non-magnetic to magnetic behavior. More significantly, research indicated that most lanthanide dopants form defect states, effectively changing the density of states and enabling metallic conductivity, thereby expanding the usability of the materials in various fields of technology such as photovoltaic cells and catalysis (Obodo et al., 2020). It is important to comprehend such transformations at the microscopic scale in order to make use of the potential of two-dimensional materials, leading to new energy conversion and storage technologies. Furthermore, organic solids' characteristics put further richness into the tale of electronic spectra and magnetic properties. These compounds are expected to exhibit unusual excitonic properties depending on their molecular geometry, which allows sophisticated optical processes like exciton formation and mechanisms for energy transfer (Schwoerer et al., 2019). The better scientists understand the innate nature of organic semiconductors, the more the knowledge of how they transport charge becomes important. By examination of the extent to which structural properties affect conductivity, there is evident proof that such properties bear intimate relationships with electronic spectra and magnetic properties. Consequently, this intricate relationship speaks to conducting additional research in a bid to optimize organic materials in applications such as flexible electronics and sensors, which could accelerate the invention of next-generation technological uses.

Definition of Electronic Spectra

The definition of electronic spectra encompasses the forms of energy transitions electrons undergo in atoms and molecules when they absorb or emit electromagnetic radiation. This is, essentially, linked with the quantization of electronic energy states, wherein electrons are present in distinct energy levels only. The resulting electronic spectra appear as discrete lines or bands in a spectrum, and each corresponds to a specific transition between such energy levels. Since the energy of the radiation absorbed or emitted is equal to the energy difference between two quantum states, the spectrum is a representation of the characteristic electronic structure of the considered atoms or molecules. This unique fingerprint is also of central importance to spectroscopy, enabling researchers to be able to characterize materials and investigate their properties within various uses, such as chemistry and materials science. Furthermore, electronic spectra rely on environmental factors, e.g., the presence of magnetic fields. When external magnetic fields are applied to a system, as is the situation in most solid-state materials, degeneracy of electronic states is lifted and effects such as the Zeeman effect can happen. This is an effect that leads to splitting of spectral lines so that interactions between electronic transitions and magnetic properties can be investigated. Realization of this relationship is significant for both simple physics and applications, as it can be applied to design novel devices such as sensors and quantum computing elements. Greater realization of the way electronic transitions respond to magnetic environments is associated with ongoing work in materials science, where the possibility of designing electronic spectra can be applied to optimize material performance. Computational approaches in the guise of software's like Quantum ESPRESSO have transformed the study of electronic spectra by merging solid simulations on density-functional theory with perturbation methods. Such approaches enable scientists to calculate electronic properties and generate spectra in congruence with experimental outcomes. Electronic spectrum simulations based on computation enable one to better understand the nature of material as well as their magnetic nature. By such tools, scientists can explore a wider range of phenomena concerning electron behavior across various fields, including photonics and nanotechnology. Ultimately, convergence of theoretical and computational models enhances our know-how of electronic spectra to enable additional innovations within electronic materials and devices that exploit such interactions to provide specific functional outputs (Giannozzi et al., 2017; Ozawa et al., 2019).

Importance of Studying Electronic Spectra

Understanding electronic spectra is significant to appreciate the underlying interactions among light and matter, especially chemical and physical properties. Electronic spectra of molecules are useful in terms of energy levels for electronic transitions, which may give significant information on molecular bonding and structure. Electronic spectral shifts, for instance, may indicate molecular conformational changes, environmental effects, or the presence of different functional groups. This kind of understanding not only lies at the heart of academic research but also the origin of immense implications in fields such as materials science, where certain electronic properties are essential to the development of solar cells and semiconductors. Furthermore, the ongoing progress of computational methods allows for enhanced prediction of electronic spectra, increasing our ability to design materials with given properties from theoretical modeling. Aside from this, electronic spectra manipulation and interpretation are equally important applications in environmental monitoring and atmospheric science as well. New databases such as HITRAN contain big molecular absorption cross-sections that come into play while processing atmospheric constituents.

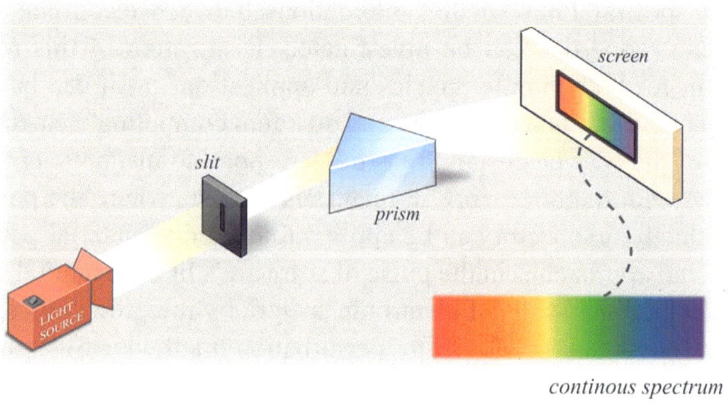

Figure 7.1. *Studying electronic spectra.*

Source: https://chem.libretexts.org/@api/deki/files/345624/1.4.1-1.svg?revision=1&size=bestfit&width=560&height=300.

Electronic spectra interaction with light can provide insights on the dynamics of numerous gases and particles in the atmosphere, which will lead to better predictive models of climate change and air quality. In this context, knowledge of electronic spectra not only serves scientific research but also serves policy-making by providing a critical information to environmental policies and sustainability. As such fields keep on evolving and widening, the importance of electronic spectra will only grow, necessitating ongoing research and funding in this area to address global challenges. In addition, the study of electronic spectra is inherently linked to the study of magnetic properties in materials. Electrons in their excited forms tend to exhibit magnetic behaviors, which may possibly exist in forms that are varied, such as in magnetic resonance imaging (MRI) and the spin nature of electronic materials. By analyzing the electronic spectra, researchers are able to gain insights into fundamental magnetic phenomena like ferromagnetism and paramagnetism, which are utilized in data storage devices and quantum computing. As technology advances further, understanding these characteristics based on electronic spectra is going to be the key to developing innovations in spintronics and other related technologies. Thus, the analysis of electronic spectra is not an academic exercise; it is a doorway to understanding beneficial applications that have the potential to change technology and improve the technologic society (Figure 7.1).

Overview of Magnetic Properties in Materials

The research on magnetic properties in materials is important in the realization of how they conduct themselves when utilized in various uses, particularly in electronics and data storage. Magnetic behavior is primarily attributed to the behavior and alignment of electrons in atoms, particularly unpaired electrons. They can further exhibit different magnetic behaviors, which can broadly be classified as diamagnetic, paramagnetic, ferromagnetic, and antiferromagnetic materials. Diamagnetism is a slight and opposite response to applied magnetic fields, while paramagnetic materials, which have unpaired electrons, become weakly magnetized in the same direction as the applied field. Ferromagnetic substances, on the other hand, possess a lasting, powerful magnetic orientation as a result of cooperative spin interactions at the atomic level between adjacent atoms. This type of magnetic nature indicates the importance of investigating these properties toward future developments in material science, enabling advanced technologies to be created and additional knowledge regarding solid-state systems within a magnetic context. Recent advancements in machine learning (ML) have brought about a paradigm change in materials science, particularly in prediction and understanding magnetic properties. Employing statistical methods, ML has accelerated studies of solid-state systems to make rapid discoveries of stable material as well as associated electronic properties. As highlighted in recent studies, ML techniques allow for detection of quantitative structure-property relationships leading to better understanding of how atomic configuration impacts magnetic behavior. The application of active learning and surrogate-based optimization identifies the power of ML to boost material design processes, particularly for finding new magnetic materials with desired properties for use in technology and energy. This synergy between machine learning and magnetic property science not only streamlines experimental processes but also facilitates better insight into interpretability of materials data and, as such, contributes to the rational design of new materials for application in novel uses (Schmidt et al., 2019). Synthesis and characterization of nanoparticles (NPs) have opened up new avenues to explore magnetic properties at the nanoscale, expanding their applications to diverse fields. Nanoparticles with particle sizes between 1 and 100 nm have new physical and chemical properties that are distinct from their bulk counterparts due to their large surface area and quantum confinement effects. These can be controlled by changing the size, shape, and composition of the NPs such that magnetic responses can be tailored for specific applications, such as biomedical imaging and drug targeting. Recent studies on heavy metal NPs, like lead and mercury species, reveal their high stability and hardness, rendering them a double-edged sword in the context of environmental security. Studying the magnetic properties of such nanoparticles is essential to unlock their potential while neutralizing the hazards involved, further illustrating the intricate relationship between material properties and technological applications in today's rapidly changing world (Khan et al., 2017).

7.1. FUNDAMENTALS OF ELECTRONIC SPECTRA

Electronic spectra analysis, one of the pillars in spectroscopic theory, serves as key importance in obtaining insight into electromagnetic radiation-matter interaction. Electronic spectra result from electron transitions between molecular energy levels and provide significant information on molecular dynamics and structure. These spectra are divided into categories, including absorption, emission, and fluorescence spectra, which exhibit various properties of molecular behavior. Electronic spectra not only aid in identifying chemical compounds, but they also have a critical role in determining the molecular properties, i.e., the presence of some functional groups and electronic structures. Additionally, the advanced techniques and data sets such as HITRAN offer greater accuracy and availability of spectral data, encouraging research and applications in atmospheric science and other fields of chemistry (Gordon et al., 2017). In electronic spectra, quantum mechanical laws are critical to explaining energy transitions and levels. The electronic transitions, typically called upon in the instance of the absorption of ultraviolet or visible light, are quantitatively determined by models of quantum mechanics. For instance, vibronic coupling, the theoretical framework describing how electronic and vibrational states interact, is instrumental in the prediction of the intensity and position of spectral lines. In addition, the solvent environment and molecular interactions are capable of changing these spectra, demonstrating the complex correlation between the molecular properties and outside conditions. Modern advances in computational methods, such as density functional theory (DFT), provide unprecedented capability to model electronic property and spectrum, which can be supplemented with experimental evidence. Techniques such as DFT allow superior understanding of molecular interactions and the resulting spectroscopic behavior, thereby strengthening the field (Akman, 2016). The importance of electronic spectra extends to numerous applications in practice, especially material identification and exploration of magnetic properties. Transition metal complexes, for example, possess complex electronic spectra that are directly associated with their magnetic properties, which have a tendency to inform us about the metal ion oxidation states and ligand surroundings. The understanding of d-d transitions in these complexes is imperative to ascertain the magnetic properties of these complexes, particularly in systems that are paramagnetic or diamagnetic due to unpaired electrons. As the field evolved, advanced spectroscopic techniques combined with computational calculations are transforming the science of materials as well as chemistry. This integration brings to the fore the necessity for a deeper appreciation of electronic spectroscopy, not merely as a diagnostic tool but as a platform for advanced studies on new materials with tailored magnetic properties (Gordon et al., 2017) (Figure 7.2).

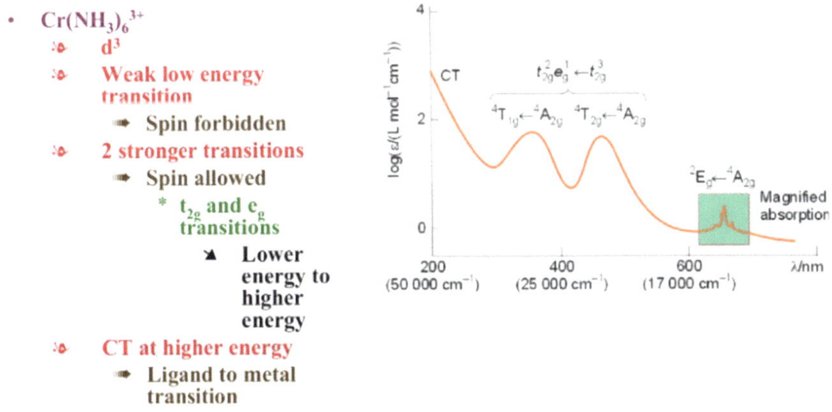

Figure 7.2. *The electronic spectra.*

Source: https://online.anyflip.com/rviq/nink/files/mobile/2.jpg.

7.1.1. Quantum Mechanics and Electronic Transitions

The study of quantum mechanics forms the foundation for the explanation of electronic transitions, which are of central importance in determining electronic spectra and magnetic properties of materials. Electronic transitions occur when electrons acquire or lose energy when hopping from one quantized energy level to another. In this regard, the study of effective mass theory and accompanying models forms a powerful tool for studying confined electron systems, such as those in quantum rings. The relevance of spin–orbit interaction (SOI) and Zeeman splitting reveals itself as these two play a role in light absorption behavior and overall electronic behavior. From the action of electrons interacting with an external magnetic field, derived energy levels and absorption intensities acquire knowledge of material magnetic properties. Thus, by applying quantum mechanical principles to explore these interactions, researchers can obtain useful information on how materials respond to various external conditions and, by extension, directly to technological developments in photonics and spintronics.

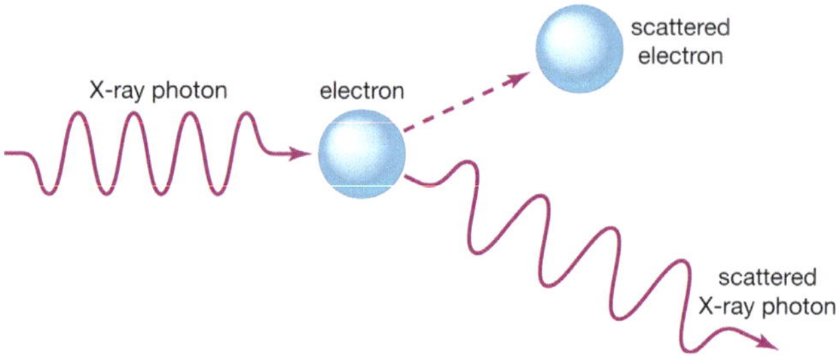

Figure 7.3. *Quantum mechanics and electronic transitions.*

Source: https://cdn.britannica.com/17/96917-050-DD28290F/X-rays-beam-effect-Compton-target-material-some.jpg.

Moreover, the electronic transition dynamics can potentially enlighten mechanisms in quantum memory devices. As has been brought to the forefront in recent times, defects within materials like the hexagonal boron nitride (hBN) offer bright possibilities for efficient quantum memory improvement. This is particularly pertinent since long storage times and strong coupling to other quantum systems are most important for useful quantum networks. The investigation of specific electronic transitions in such defects demonstrates the possibility of achieving high memory efficiency and exploiting favorable electronic structures for quantum applications. Through the use of density functional theory, researchers study a spectrum of spin electronic transitions and reveal complex interplay between electronic properties and potential applications in quantum computing and magnetic resonance. The findings demonstrate the significant role that quantum mechanics has in determining the electronic spectra and magnetic properties that are required to advance quantum technologies. Also, the non-monotonic nature of resonant peaks in various external magnetic fields illustrates the complexity of electronic transitions with quantum mechanics playing its role. Such complexity is once again augmented by structural flaws such as topological defects that not only modify electronic properties and transitions' energy distribution but even change the way the electronic properties get structured. As an illustration, the introduction of topological defects in quantum rings has been shown to alleviate softening of magnetic field crossing energy curves behavior, leading to improved transition energies and dipole moment uniform behaviors. Realization of such subtle interaction helps to improve the understanding of electronic structure and external field interaction and related optical phenomena. Thus, the study of quantum mechanics and electronic transitions is core, both as theory and for use in the development of novel materials and technologies for magnetism and optics (Figure 7.3).

7.1.2. Types of Electronic Spectra: Absorption And Emission

The study of electronic spectra illustrates two fundamental forms: absorption and emission spectra. Absorption spectra describe the process by which electrons in a molecule or an atom absorb specific wavelengths of light and, as a result, transition from lower to higher energy levels. It is a crucial aspect of identifying chemical species because the wavelengths being absorbed uniquely correspond to the electronic structure of the involved atoms. In practical applications, such as spectroscopy, scientists use absorption spectra to determine the concentration of elements in a sample through the measurement of absorbed light intensity. The absorbed wavelengths are unique and important for gaining insight into molecular structure and electronic environment, and further enhancing the understanding of materials properties and interactions in disciplines like chemistry and materials science. In comparison to this, emission spectra occur when electrons that are excited move back to their ground states, releasing energy in the form of light. Emission is possible via some mechanism like thermal excitation or photon absorption and leads to emitting certain spectral lines. The emitted light can be explained in two ways: continuous spectra, which consist of a range of wavelengths, and line spectra, which consist of a line that represents a particular wavelength. Emission spectra play a crucial role in astrophysics because they allow scientists to analyze light from other stars and determine their chemical makeup and temperature. Furthermore, understanding the emission spectra generated from provided materials provides information about their electronic structure and potential applications in luminescence devices such as LEDs and laser technology. The correspondence of absorption and emission spectra is vital in the majority of technological advances, particularly in the production of electronic and optoelectronic devices. For instance, in the case of materials employed in magneto-optical devices, the electronic structure of materials is responsible for the properties of absorption and emission spectra. New developments, such as the synthesis of Al-doped CoxCa(0.90−x) $Ni_0.10Fe_2O_4$ nanoparticles, exhibit improved absorption across several spectral ranges, hence increasing light-matter interactions essential for the efficiency of devices (Billa et al., 2024). Similarly, research on Sr_1-xTbx$Fe_{12}O_{19}$ focuses on the role of variations in electronic structures in changing magnetic and optical properties, contemplating the nature of materials made for specific applications in electronic spectra and magnetic properties (yuelong tang et al., 2024). The complex interaction highlights the importance of understanding the type of electronic spectra in coming up with new technologies.

7.1.3. Role of Energy Levels in Determining Spectra

The intricate interaction of electronic spectra and energy levels is of most significance in the understanding of atomic and molecular behavior. Energy levels are the quantized levels that may be occupied by an electron in an atom or a molecule, and are basic to the establishment of how these objects interact with electromagnetic radiation. Excitation, and electronic transitions therefore occur when an electron is struck by a photon with energy corresponding to the difference between two energy levels. These transitions leave behind distinct absorption or emission spectra that are significant indicators of the electronic structure of a material. Intensive examination of these spectra can reveal information about molecular symmetries, modes of bonding, and even the presence of specific functional groups, highlighting the pivotal role of energy levels in spectroscopic material identification. In addition, the energy level applicability extends to the modulations of magnetic properties within various materials. The atom's electronic structures influence not only the ground state but also the associated energy levels corresponding to the excited states, and the overall magnetic behavior. Transition metal complexes, for example, possess varied magnetic properties due to the different d-orbital occupancies brought about by ligand field effects. The presence of unpaired electrons in the orbitals is responsible for paramagnetism, while paired electrons lead to diamagnetism. The detection of these magnetic properties in the context of electronic spectra requires thorough examination of energy transitions to ascertain the interdependence of electronic structure and magnetism. The spectra provide a point of entry to experiment and predict these properties under various conditions, highlighting the intrinsic role of energy levels in materials science. The sampling and analysis of spectroscopic information, as embodied in such materials as the HITRAN database, also underscore the role of energy levels to specify spectra. The HITRAN compilation contains a large repository of molecular spectroscopic parameters that can account for the absorption characteristics of a vast number of molecules. This is particularly crucial to comprehend phenomena in different atmospheric environments, where variations in energy levels decide the interaction of matter and radiation. Well-established spectroscopic parameters enable not only the study of Earth's atmosphere but also of other planets, providing a broader picture of extraterrestrial environments. Through the provision of an exhaustive data set comprising principal absorption phenomena as well as high-level line-shape formalisms, the database presents the core role of energy levels in building knowledge of electronic spectra as well as their importance in theoretical and practical sciences (Gordon et al., 2017; Mourdikoudis et al., 2018) (Figure 7.4).

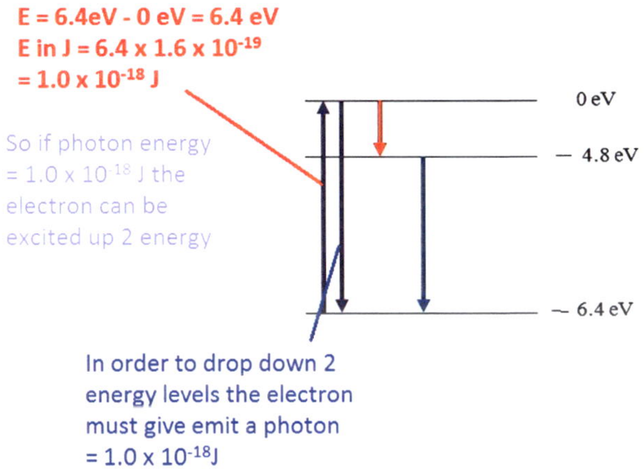

Figure 7.4. *Role of energy levels in determining spectra.*

Source: https://physicsthings.wordpress.com/wp-content/uploads/2014/04/energy-levels2.png.

7.2. THE INTERACTION OF LIGHT WITH MATTER

Matter-light interaction is the key postulate behind the experimental facts of electronic spectra and magnetic properties. When light interacts with a material, it can lead to various possibilities like absorption, emission, or scattering of photons. All these happen under the stringent influence of the electronic structure of the material that determines its energy levels and the allowed transitions between these levels. For example, following the absorption of a photon whose energy is equal to the difference between two states of an electron, an electron is excited into a higher energy level, thus contributing to the electronic spectrum. The ability to measure these spectra provides crucial information on the electronic structure of the material and, as a result, its magnetic behaviors as the electron spin and arrangement dictate its magnetic susceptibility and thus its macroscopic properties. In addition, quantum mechanical origins of light-matter interactions reveal more complex phenomena, for example, the integer quantum Hall effect, which is a useful tool for insights into topological materials. The effect is due to quantization of electronic states in two-dimensional systems under intense magnetic fields, where resulting topological invariants describe the stability of these states towards perturbations. The interaction between light and the quantized states can be used in photonic systems, in which the manipulation of light in periodic media is able to release new behavior, similar to electronic systems. The potential to use photonic metamaterials and waveguide arrays to explore and harness topological effects is an exciting frontier in material science, demonstrating the ways in which fundamental concepts of light-matter interactions can lead to innovative technological uses and advancements in electronic and magnetic materials (Ozawa et al., 2019). Utilizing computational tools such as Quantum ESPRESSO seems to be crucial in advancing the field's understanding of such complex interactions. As an open-source package with a focus on quantum simulations, it has the potential to examine the properties and processes that govern light-matter interactions in electronic systems. With the addition of density-functional theory to Quantum ESPRESSO, it makes it possible for researchers to simulate electronic structures and predict various outcomes, such as electronic spectra and magnetic responses. Enhancements of the codes modularity and parallelization further enhance its ability, allowing scientists to investigate more complex systems and phenomena, such as those encompassing topological properties and quantum Hall effect. With advances in computational methods, they synergistically drive both theory and experiment paradigms, strengthening our understanding of the interactions between light and matter and their relationship with electronic and magnetic properties (Giannozzi et al., 2017).

7.2.1. Photons and Their Energy

Photons and their energy are the main areas of investigation for the understanding of electronic spectra and magnetic properties of materials. Photons, or packets of quantized electromagnetic energy, are at the heart of light-matter interaction. Their energy also depends on the frequency of the electromagnetic wave, as per Planck's equation, $E = hv$, where E is energy, h is Plancks constant, and v is the frequency. This indicates how photons can excite electrons within materials, inducing electronic transitions that are central to the study of electronic spectra. These transitions can be governed by the material structure and composition to produce a broad range of optical properties, which are crucial in the utilization of applications such as photovoltaic and lasers. How the material interacts with photons also calls for an analysis of the material's magnetic properties. For instance, the spectra for the absorption and emission of some magnetic materials can be used to identify the presence of unpaired electrons and their spin configuration. When electrons are struck by photons of given energies, they can travel between energy levels, including the spin states relevant to magnetic properties. For example, the magnetic properties of such compounds as RhNbLi and RhNbAs, as predicted, may be significantly reliant upon their electronic structures and the respective photon energies associated with electronic transitions (Bejo et al., 2025). These connections illustrate the profound effect of photon energy on the vital properties revealed through electronic spectra, furthering our understanding of magnetism in materials. In addition, the investigation of photon energy is extended into its applications in recent work with cosmic rays and the universe as a whole. In ultra-high-energy cosmic ray research, less luminous galaxies emit photons that collide with cosmic backgrounds and contribute substantially to the observed energy spectrum of cosmic rays. Studies on these processes, as in the case of FR0 radio galaxies, indicate that these weaker sources may also contribute substantially to cosmic photon releases, which emit secondary particles, including high-energy photons and neutrinos (Lundquist et al., 2025). This intricate play between photons and energy dynamics has the effect of placing the study of electronic spectra and magnetic properties not just on a terrestrial scale but also on the extended cosmic scale, where the interplay of light and matter transcends the ordinary parameters (Figure 7.5).

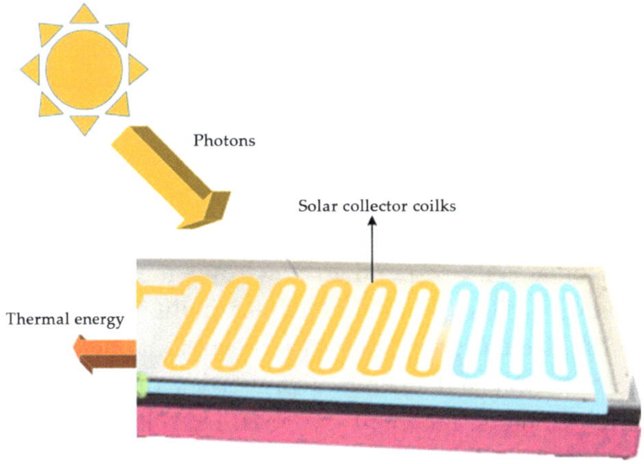

Figure 7.5. *The exploration of photons and their energy.*

Source: https://ars.els-cdn.com/content/image/3-s2.0-B9780128095973003369-f0300336-32-9780128095973.jpg.

7.2.2. Absorption Processes in Different Materials

The intricate dynamics of absorption processes and material electronic spectra play a pivotal role in establishing their magnetic nature. Different materials exhibit different types of absorption according to their unique electronic structures as well as light interactions. When material is subjected to light, it may become absorbed by the electrons, which transfer to higher energy levels, resulting in unique electronic spectra. These spectra also play an important role in determining a material's optical characteristics and influence its magnetic properties as well. Transition metal oxides, for example, are likely to exhibit broad absorption bands corresponding to d-d electron transitions, which is responsible for their magnetism. A better understanding of such processes allows the development of materials with selective electronic and magnetic behavior, which results in breakthroughs in several domains like spintronics and photovoltaics. Nanoparticles (NPs) form an intriguing set of materials whose absorption phenomena are highly sensitive to size and shape. The optical behavior of NPs arises due to their very high surface area to volume ratio, which creates quantum confinement effects that cause the electronic spectra to modify. In particular, as the size of the NPs decreases, the energy levels get quantized, which can cause size-dependent discrete absorption peaks. The diversity makes nanoparticles, classified as metal and ceramic types, applicable in numerous applications like catalysis and imaging in medicine. Their electronic spectra not only define the optical nature but also decide the magnetic nature of such materials, creating abnormal effects, like superparamagnetism, under given situations. Knowledge about these interrelations is necessary to utilize their potential in various technology areas. Furthermore, the role of molecular absorption processes in materials extends beyond traditional solids to encompass gases and liquids, as noted by the HITRAN molecular database. The database provides an enormous dataset of molecular absorption information, illuminating the way various chemicals respond to electromagnetic radiation. Enhanced accuracy and coverage of the HITRAN database facilitate detailed investigations on how different molecular species absorb in different ranges of the spectrum, influencing the emission and magnetic properties of the involved material. The absorption phenomena recorded in the HITRAN compendium, including collision-induced absorption and aerosol indices, reflect the complexity of the absorption processes in different states of matter. Such data are important to atmospheric science and related disciplines, where understanding absorption properties is key to modeling and predicting magnetic behavior across a variety of environmental contexts (Gordon et al., 2017; Khan et al., 2017).

7.2.3. Emission Processes and Fluorescence

The understanding of emission processes is key to the study of electronic spectra and their associated magnetic properties. Emission processes usually occur when an electron within a material acquires energy, transferring it to a higher electronic state. Upon returning to its base state, the electron releases energy in the form of electromagnetic radiation, a phenomenon known as photoluminescence. The change can be in varying efficiency, which influences the characteristics of the released light, including its wavelength and strength. Notably, hybrid lead-free indium perovskites exhibited superior luminescent properties with strong green light emission through self-trapped exciton (STE) processes due to quantum confinement effects. The properties bode well for future potential in creating new optoelectronic devices based on efficient light emission without adverse effects associated with traditional lead-based devices (Sun et al., 2021).

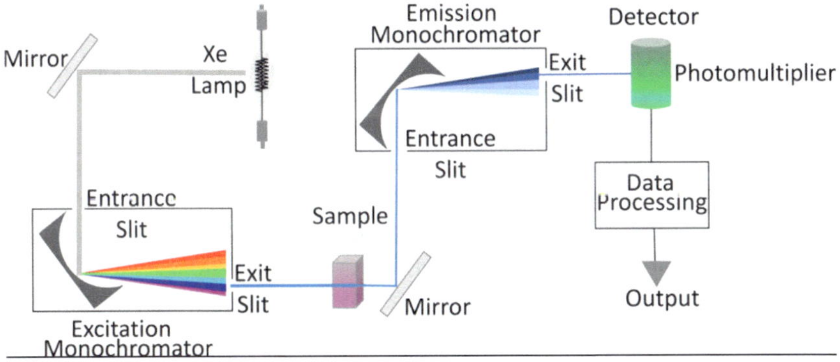

Figure 7.6. *Emission processes and fluorescence.*

Source: https://onlinelibrary.wiley.com/cms/asset/0bfa41e7-5a8e-4521-b235-4ba037d0837c/cjce23506-fig-0001-m.jpg.

Fluorescence, a type of photoluminescence, is where a substance absorbs and emits photons in virtually a flash of time, or in nanoseconds. It's what distinguishes fluorescence from being applied in various different disciplines, such as biological imaging and environmental science, and the intricate relationship between the energy levels involved between exciting and emitting the material. Inside nanomaterials, the inclusion of materials like quantum dots enhances tunability in fluorescence so that accuracy of color emission is possible based on particle size and composition. Quantum dots possess unique properties since they are zero-dimensional systems whose electronic properties are highly conditioned by quantum confinement. These properties lead to very effective emission processes that can be precisely controlled, hence their requirement in biosensors and other applications where specific light-emitting features are required (Naresh et al., 2021). The persisting interest in emission processes and fluorescence in theory and practice continues to highlight their significance in the development of material science and engineering. For instance, the ability to create materials that radiate light effectively directly affects the performance of light-emitting diodes (LEDs) and lasers. Combined with their magnetic character, the electronic spectra of these materials can inform the development of next-generation capabilities in quantum computing and spintronics, opening doors to what is possible in future technologies. Current studies on enhancing emission qualities—e.g., enhanced photoluminescence quantum yields (PLQYs)—are essential for enhanced performance in luminescent materials. Materials like indium halides have yielded discovery in the way forward to acquire high stability and efficiency in emitting light, pointing to continued exploration of their photophysical properties (Sun et al., 2021) (Figure 7.6).

7.3. FACTORS INFLUENCING ELECTRONIC SPECTRA

Electronic spectra give very profound information about a material's electronic structure and magnetic properties. One crucial factor that decides electronic spectra is the nature of chemical bonding in a material. For instance, transition metals usually possess several oxidation states based on the nature of chemical bonding, which in turn affects their d-orbital configurations and leads to different electronic transitions. The ligand field theory explains how the nature and type of ligands that surround a metal ion could lead to degenerate d-orbitals splitting, shifting wavelengths of absorption and, thus, electronic spectra that are observed. Such spectral alteration has relevance in applications from catalysis through understanding electronic character of complex materials, reflecting how critical bonding nature is in deciding electronic spectra. Temperature is also a key parameter in influencing electronic spectra because it has the potential to stimulate populations in electronic states and thus change absorption features. Thermal agitation caused by temperature increases has the ability to initiate electron transitions from the ground state to higher states, thus broadening and shifting spectral peaks. The phenomenon is very prevalent in semiconductors where bandgap energies are capable of altering with changes in temperature to induce extensive changes in their electronic spectra. Further, the electronic transitions at elevated temperatures may be affected by the availability of phonons, thus the measured spectra become complex. Understanding the thermal effects is quite significant in the application of thermoelectrics and semiconductor devices as the performance and efficiency under varying operating conditions can be predicted. Structural and environmental effects cannot be disregarded in any electronic spectra analysis. For example, crystalline symmetry affects the allowed transitions between energy levels; the higher symmetry may result in fewer allowed transitions, altering the spectral behavior observed. The presence of defects may also result in localized states known as trap states, which have the potential to significantly alter the electronic spectra by inducing new absorption bands or altering the ones already present. Environmental parameters, such as solvent interactions within solution-phase systems, can further influence electronic transitions through the modulation of solvation dynamics and coordination environments. Thus, a detailed understanding of these parameters is crucial for scientists to control electronic properties for new material design, such as electronic device applications and magnetic materials (Ozawa et al., 2019; Song et al., 2020).

7.3.1. Molecular Structure and Symmetry

A coherence between molecular structure, symmetry, and the resulting properties of coordination complexes is a coherence that runs through the comprehension of electronic spectra and magnetic properties. Symmetry plays a vital role in coordinating the electronic environment of metal ions, coordinating their stabilization energy and spectroscopic characteristics. C3-point symmetry complexes, for instance, [REIII(Br_4catH)$_3$(tpa)], exhibit characteristic features which can be logically coordinated with different rare earth elements. The long-range hydrogen-bonding features of the coordination environment lead to unusual electronic transitions with distinctive spectral features, thereby allowing researchers to explore the delicate correlation between molecular symmetry and electronic spectra in various complexes. The extremely important role of such symmetries can be appreciated in understanding their physical properties, i.e., magnetism and luminescence, which show the interplay between molecular structure and electronic transition phenomena. Entering the realm of organometallic complexes, particularly Dy(III) ion-containing complexes, adds another level to discussions of molecular structure and symmetry. Complexes such as [Dy(E4)2]- complexes demonstrate how linear to bent geometric shapes have direct effects on magnetic anisotropy and electronic interactions. It has been shown that geometry modifications lead to tunable magnetic relaxation barriers, which are not only important for the definition of intrinsic magnetic behaviors but also hold promise for future data storage technology applications. The findings from simulation, such as by scalar relativistic density functional theory (SR-DFT) and multiconfigurational techniques, describe the manner in which ligand covalency varies with the coordinating atoms and hence influences the overall electronic structure of such complexes. The description enhances the knowledge of how some symmetries relate to the increased magnetic features and warrants the synthesis of materials intended for technological applications. Finally, the interplay between molecular structure, symmetry, and consequent electronic and magnetic properties establishes the applicability of these concepts in coordination chemistry and materials science. The synergy between symmetry and electronic structure not only dictates the spectral properties but also decides how the materials behave under various conditions. Experiments like those of single-molecule magnets and magnetic properties when subjected to magnetic fields indicate that symmetrical structures have the capability to strongly stabilize certain electronic states, which lead to anomalous magnetic properties. As demonstrated by both organometallics and rare earth complexes, the search for symmetry's influence on molecular design continues to give the thrust toward new applications from quantum computing through advanced photonics, eventually guaranteeing the need for knowledge of molecular structure and symmetry to electronic spectra and magnetic properties in modern science.

7.3.2. Solvent Effects on Spectra

The solvent effect on electronic spectra is an important area of research in photochemistry and materials science, particularly concerning the magnetic properties of materials. Spectral properties may profoundly be affected by solvents, generally by processes like solvatochromism, in which electronic behavior in transitions changes according to solvent polarity. Polar solvents, for instance, stabilize charge-separated forms and affect electronic distribution within molecules. This effect can be seen in red-shift of absorption and emission spectra, such as in experiments involving hybridized local and charge-transfer (HLCT) materials. In such cases, the dielectric constant of the solvents is linearly correlated to changes realized in the energy levels of the excited state, demonstrating that care must be taken while selecting solvent environments during spectral recording to make accurate interpretations of electronic features (Chen et al., 2021). Apart from that, solvent dynamics are also involved in the time-dependent features of electronic transitions, influencing both absorption and fluorescence lifetimes. Solvent-molecule and solute-molecule interactions can cause the phenomenon of solvent reorganization or relaxation, where the solvents structure "solvation shell" reaction to the solutes charge state. This can aid or hinder transitions, leading to modifications in oscillator strengths and quantum yields. For HLCT compounds, it is crucial that such dynamics are combined with torsional motions of acceptor and donor subunits since conformational change on coupling can enable efficiencies in charge transfer to varying degrees. Hence, it is important to have a good understanding of solvent effects to optimize the design of high-performance light-emitting diodes and other optoelectronic devices where magnetic properties are critical (Unke et al., 2021). Furthermore, studies have emphasized the necessity of estimating the solvent effects through spectroscopic methods and periodically utilizing solvatochromic correlations in the study of spectral shifts as a function of solvent polarity. Through approaches like the Lippert–Mataga equation, researchers can evaluate the correlation between the observed spectral change and solvent properties to aid the basic understanding of charge transfer processes. Through detailed studies of various solvent surroundings, it is evident that electronic material properties targeted for OLED application are closely related to solvent interactions. Such understanding not only contributes to a deeper insight into electronic spectra but also promises the possibility of designed solvent systems for optimization of desired features, resulting in major breakthroughs for efficient and effective electronic devices (Chen et al., 2021) (Figure 7.7).

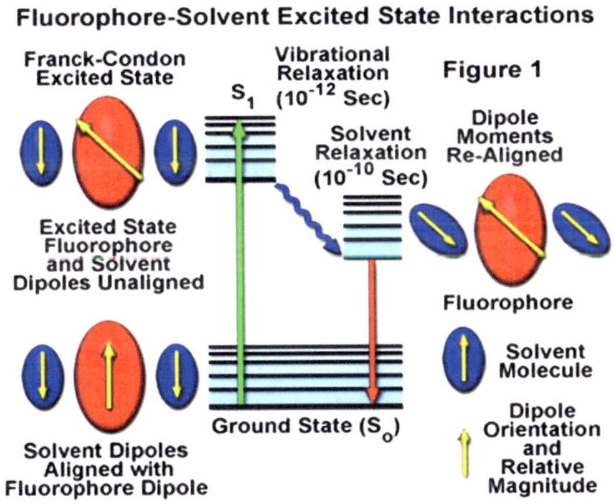

Figure 7.7. *The influence of solvent on electronic spectra.*

Source: https://evidentscientific.com/en/microscope-resource/tutorials/jablonski/media_16f796edadafb274dd509f955685dffebc66ed907.jpeg?width=750&format=jpeg&optimize=medium.

7.3.3. Temperature Dependence of Electronic Transitions

Temperature dependence of electronic transitions is extremely significant in the study of electronic spectra and magnetic materials' properties. As the temperature increases, the lattice vibrations become enhanced, influencing the energy states of electronic states by coupling effects. This phenomenon is predominantly manifested in the case of materials that have strong electronic correlations, and the changes induced by temperature give rise to shifts in spectral features. For instance, recent works that have applied the supercell method in order to study the optical and the magneto-optical properties have revealed that optical spectrum signals rise significantly with lower photon energies as temperature increases. Intraband transition involving phonon and magnon cause the development, highlighting the delicate interaction between electronic dynamics and heat energy, ultimately leading to measurable electronic spectra changes required for application in different fields, including magnetic refrigeration and spintronics (Kang et al., 2024). Additionally, the interconnection between temperature and magnetic states exists in accordance with electronic transitions, particularly in ferromagnetic materials. In materials like $MnFeP_1$−xAsx, temperature variations significantly affect magnetic properties of ions, supported by soft x-ray magnetic circular dichroism (XMCD) data. The observations from the above research suggest that with varying temperature, magnetic moments of Mn and Fe ions respond differently, notably exhibiting paramagnetic to ferromagnetic state transitions. The concentrations of different oxidation states such as Mn_2^+ and Fe_2^+ and their spin configurations are crucial in determining the magnetic character of the system. This interplay of electronic configuration and temperature is decisive for establishing the stability and functionality of magnetic materials. The temperature dependence of the XMCD spectra under various conditions indicates the stability of the electronic configurations, highlighting the significance of investigating these transitions further in relation to thermal effects and magnetic ordering (Takeda et al., 2010). In general terms, temperature dependence implications are important beyond academic science to applications in magnetic refrigeration as well as in optoelectronic devices. Understanding achieved through the analysis of electronic transitions at various temperatures enables the material to be engineered with improved thermal management properties. For instance, electronic spectra-changing materials dramatically can be designed for energy-conserving applications. Gradual evolution of electronic properties with temperature can lead to new concepts in magnetic cooling systems, particularly through the utilization of magnetic entropy change due to thermal fluctuations. The present studies on these phenomena foreshadow a good future for technological advancement through more comprehensive knowledge of electronic and magnetic interactions at both temperature extremes, focusing on the interdependent relationship of temperature, electronic transitions, and performance of composite materials (Kang et al., 2024; Takeda et al., 2010).

7.4. MAGNETIC PROPERTIES OF MATERIALS

Material magnetic properties are intrinsically tied to their electronic structures, which dictate how such materials respond to magnetic fields that are applied to them. At the atomic level, the existence of unpaired electrons also contributes significantly to a material's magnetism. Transition metals, for example, are likely to have varied magnetic properties because of the arrangement of their d-orbitals. When taking into consideration the case of chromium-doped zinc sulfide and selenide, the concern is evident that substituting chromium for zinc not only impacts the electronic properties but leads to varying effects on magnetism as well. This is particularly observed in the form that both substances, if one carries out the spin-polarized calculation, display semiconducting character for the spin-down states but metallic character for the spin-up states. This two-faced nature enhances the significance of these materials in spintronic devices and hence testifies to the central role played by an understanding of their magnetic properties combined with their electronic spectra (Syrotyuk et al., 2025). The development of computational and theoretical methods has significantly contributed to increased understanding of magnetic materials, particularly in two-dimensional (2D) materials. The most recent research has focused on Janus MXenes, such as $MnCrNO_2$ monolayers, which are now promising candidates for new electronic and spintronic devices. Based on density functional theory (DFT) calculations, researchers have described how such materials are ferromagnetic in nature and are resistant to large strain without destabilization. The unique interaction between manganese and chromium within the electronic structure provides a great influence on magnetic anisotropy, reshaping the type of magnetism from out-of-plane to in-plane easy axis with applied strain. This is an understanding that highlights the complex relationship between electronic and magnetic behavior, and thus the need for an integrated approach towards optimizing these materials' use (Yue et al., 2024). Magnetic property studies in materials science are not just limited to their potential applications in devices, but extend to the fundamental nature of magnetism. Investigation of how some impurities, such as chromium, affect semiconductor electronic spectra extends new possibilities for designing desired magnetic properties into materials. The addition of some dopants can dramatically alter local electron configurations, forming new magnetic phases and overall enhanced performance in devices like spintronic devices. With further research coming to light that details the complexity of these interactions, the role of magnetic properties in dictating a material's usefulness in at-the-cutting-edge technologies increasingly comes into focus. This dynamic relationship between electronic spectra and magnetic properties will be of paramount importance in the creation of future materials, finally expanding the potential of electronic and spintronic technologies in our world of technological saturation (Syrotyuk et al., 2025; Yue et al., 2024).

7.4.1. Types Of Magnetism: Diamagnetism, Paramagnetism, and Ferromagnetism

Magnetism is caused by electron movement in matter, which can exhibit various types of magnetic response based on their electronic structures. The primary types of magnetism are diamagnetism, paramagnetism, and ferromagnetism, and each refers to how materials respond to an external magnetic field. Diamagnetic materials, such as copper and bismuth, consist of paired electrons, which make them develop a weak, opposing magnetic response. This is due to the currents induced that counteract the applied field of magnetism, thus causing the materials to be repelled.

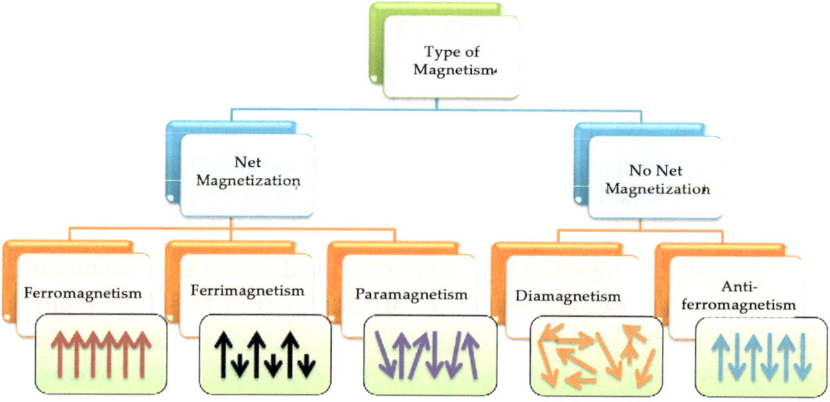

Figure 7.8. *Types of magnetism: Diamagnetism, paramagnetism, and ferromagnetism.*

Source: https://www.researchgate.net/profile/Masoomeh-Yari-Kalashgrani/publication/361666216/figure/fig1/AS:1172988229623808@1656673595467/Types-of-magnetism-arrows-indicate-the-orientation-of-magnetic-moments-35.ppm.

These features are essential in situations where magnetic levitation is needed, highlighting the significance of diamagnetism subtleties in electronic spectra and magnetic response. In contrast to diamagnetism, paramagnetism occurs with materials that contain unpaired electrons, which result in a weak, positive magnetic response to external fields. These substances such as aluminum and transition metals exhibit such an effect where the magnetic moments align with the presence of an external magnetic field, even for a short time. This alignment reduces once the external field is removed because thermal motion results in the direction of the magnetic moments to be random. Paramagnetic materials have interesting electronic properties at varying temperatures, wherein their magnetic susceptibility increases very significantly at lower temperatures and hence are suitable for studying applications in magnetic resonance imaging and electronic sensors. Ferromagnetism exhibits a more intense form of magnetism wherein materials find it difficult to lose their magnetization even without an external magnetic field. Common ferromagnetic materials like nickel and iron have domains where the spins align along the direction of the magnet field, and this results in a net magnetic moment. This process is crucial in making permanent magnets and magnetic storage devices. Recent discoveries emphasize the fascinating potential of half-metallic ferromagnets (HMF), which are metallic conductive for one spin orientation and insulating for the other. Synthesis of materials like FeMnVAl has proven to exhibit a ferromagnetic state, which has prompted research on their electronic behavior and defect-driven effect on magnetism, thereby advancing the field of mobile charge carriers in spintronic devices (Gupta et al., 2022; Yang et al., 2023) (Figure 7.8).

7.4.2. Role of Unpaired Electrons in Magnetism

It is essential to know how unpaired electrons are responsible for magnetism in order to account for the magnetic characteristics of various materials. Unpaired electrons create magnetic moments due to the intrinsic spin that generates magnetic fields. Magnetic moments align in some directions based on interactions between electrons, ultimately determining the magnetic property of the material. For instance, in transition elements, the presence of unpaired d-electrons has significance for ferromagnetism and antiferromagnetism. Alignment of these unpaired electrons by external magnetic fields is capable of generating high magnetic response, making material useful in a variety of applications ranging from storage medium to high-end electronics. Recent studies reveal how defects in crystalline lattices can affect unpaired electron distribution and behavior, thus influencing magnetic properties. For example, certain material defects like cubic boron nitride are capable of creating localized states in the band gap that bear unpaired electrons. The localized states enhance the magnetic property of the material, as demonstrated through the influence of vacancy and interstitial defects on both structural stability and magnetism. The existence of these defects modifies the local electronic environments, which allows for novel spin configurations that result in drastically better magnetic performance, including greater stability and higher sensitivity to magnetic fields, as revealed by recent research on defect engineering approaches in new materials (Nguyen et al., 2024). In addition, the interaction between isolated electrons and catalytic additives is gaining significance in areas of energy storage. In lithium–sulfur batteries, for instance, defect engineering to maximize the spin configuration of additives can greatly influence the performance. Defects, for instance, vacancies, can create unpaired electrons that provide greater spin polarization, enhancing catalytic activity by improving the adsorption of polysulfide and enhancing electrochemical reactions. This effect leads to greater charge distribution and overall enhanced lithium-sulfur redox reaction kinetics, which is crucial for maximizing efficiency and lifespan in the battery. These findings underscore the crucial role of unpaired electrons in manipulating not just providing magnetism, but also enabling critical advances in energy technologies (Yu et al., 2024).

7.4.3. Measurement Techniques for Magnetic Properties

The investigation of measurement techniques of magnetic properties is at the core of understanding many materials, particularly in electronic spectra context. Techniques such as nuclear magnetic resonance (NMR) play a key role in probing the microscopic properties of materials such as Weyl semimetals. As an instance, NMR has been used to investigate the angular dependence of the ^{93}Nb-NMR spectra in NbP, and this has served to elucidate important parameters for both electric quadrupole interactions and magnetic hyperfine interactions. Insight into the Knight shift, one of the important NMR parameters, is the revelation not only of the electronic structure of the materials but also the dynamics of almost massless fermions that are responsible for the overall magnetic behavior. Such minute studies give an idea about quantum oscillations in correlation with band structure, enabling researchers to establish comparisons with previously documented bulk measurements. This correlation points to the importance of NMR techniques in advancing additional understanding of magnetic properties at the microscopic level (Kubo et al., 2024). Another important technique applied in the quantification of magnetic properties is X-ray absorption spectroscopy (XAS), which is capable of differentiating local structural features and oxidation states in magnetic compounds. The application of XAS is present even in the study of $LaFeO_3$ ceramics doped with alkaline-earth metals, where the technique confirmed La^{3+} and Fe^{3+} oxidation states and identified local structural distortions. These measurements are extremely crucial in the revelation of electronic state interaction with magnetic properties, as indicated in the varied magnetic responses observed between differing substitutions in $LaFeO_3$. For example, La0.8Mg0.2FeO3−δ was found to exhibit significant magnetization because of changed hybridization among Fe(3d) and O(2p) orbitals. The fine detail of the structural investigation that arose from XAS ultimately improves our understanding of how electronic configurations are connected with intrinsic magnetic properties (Nandy et al., 2024). The convergence of disparate measurement approaches is crucial for a thorough assessment of magnetic properties in various materials. Techniques such as electron spin resonance (ESR) complement techniques already mentioned in that they provide independent verification of magnetic coupling and measurements of spin dynamics. ESR can generate valuable information about the presence and activity of unpaired electrons, and its use for complex oxides can yield valuable information about their magnetic ordering and electronic spectra. Additionally, the application of a number of methods, including XAS, NMR, and ESR, not only confirms results but also provides scientists with a three-dimensional view of magnetic behavior. This holistic comprehension enhances advancements in material design with certain magnetic properties, enhancing their application in electronic devices and spintronics technologies, bridging the gap between prediction and observation in electronic spectra and magnetic properties research (Figure 7.9).

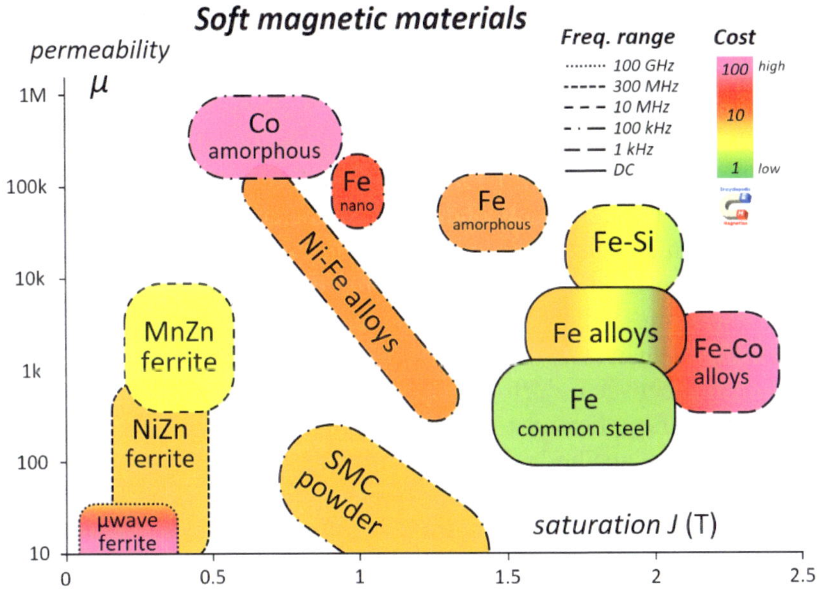

Figure 7.9. *Measurement techniques for magnetic properties.*

Source: https://www.e-magnetica.pl/lib/exe/fetch.php/soft_magnetic_materials_magnetica.png.

7.5. RELATIONSHIP BETWEEN ELECTRONIC SPECTRA AND MAGNETIC PROPERTIES

Magnetic behavior and electronic spectra are an important area of investigation that reveals underlying mechanism for material response. The interaction between the electronic atom configurations and their corresponding magnetic moments plays the key role in regulating how materials respond to a magnetic field. For instance, the participation of unpaired electrons in transition metals is crucial in determining magnetic susceptibility, which may be directly assessed from electronic spectral investigations. Consequently, techniques like electron paramagnetic resonance (EPR) provide information regarding the spin state of such free electrons. Advances in EPR technology, such as EPR-on-a-chip (EPRoC), have enhanced the ability to probe such relationships in semiconductor materials, emphasizing the importance of correlating electronic and magnetic characterization methods to attain comprehensive understanding of material properties (Lettens et al., 2025). Furthermore, the geometries of the materials' structures directly influence their electronic spectra and, by extension, magnetic properties. Spinel ferrites are a case in point, with complexity in their crystal structure and magnetic properties being masterfully intertwined. The arrangements of cations in tetrahedral and octahedral sites give rise to varying magnetic interactions that dictate their electronic states. Inferences from studies incorporating sophisticated computational methods, such as the B3LYP and HSE06 hybrid functionals, indicate that spin configuration changes result in different semiconducting characteristics as well as magnetic orderings in the materials. Their intricate dependence of vibrational modes and spin configurations also reflects how electronic spectra could be an image of magnetic states, stressing the need to couple these features in order to achieve accurate material characterization (Almutairi, 2025). The influence of knowing the relationship between electronic spectra and magnetic properties extends to a broad set of applications from spintronics to new magnetic recording technologies. New methods like EDMR offer valuable means to probe electrically active defects, which can have a determining effect on electronic and magnetic properties in devices. By connecting the electronic spectral properties with magnetic resonance data, researchers are able to obtain critical information on defect dynamics and recombination processes in materials. The continuous developments in detection methods hold promise to unveil even more subtle interrelations, thus further refining predictability and material performance in real-world applications. The synergistic research path will certainly advance our understanding of electronic and magnetic properties, enabling the creation of next-generation materials with designed functionalities (Lettens et al., 2025).

7.5.1. Spin States and Their Influence on Spectra

Spin state and electronic spectra interaction is a key factor in understanding magnetic properties within materials. Spin states refer to the intrinsic angular momentum of electrons, which can be in different arrangements such as singlets or triplets. Spin arrangements significantly influence the electronic transitions within the material, which form an observable function in spectra. For instance, singlet-triplet state distinction can lead to varying absorption and emission spectra, and this affects the way materials absorb light. How material and light interact informs us about the electronic structure and the magnetic characteristics of the material, from quantum computing to photovoltaic devices. Thus, the investigation of material interactions with spin states and electronic spectra is crucial in learning magnetic materials' intrinsic behavior. Spin state influence extends beyond mere electronic transition to impact intrinsic magnetic material properties. Knowledge of these spin states makes it possible to predict such phenomena as ferromagnetism, antiferromagnetism, and paramagnetism. For example, in certain transition metal complexes, the presence of unpaired electrons in certain spin states can greatly enhance magnetic susceptibility. This phenomenon finds significance while creating next-generation magnetic sensing and data storage materials. Understanding the influence of spin state configurations on electronic spectra allows researchers to engineer materials with desired magnetic properties, thus triggering material science breakthroughs. This understanding translated into reality is significant considering the fact that next-generation technologies rely more and more on manipulating magnetic properties at atomic scales. Computational tools play a central role in spin state analysis and their contribution to electronic spectra. Advanced software platforms such as CP2K enable atomistic simulations that aid researchers in the analysis of electronic structure, as well as magnetic properties, of different systems. Employing density functional theory and various post-Hartree–Fock methods, these tools provide a robust platform for the analysis of spin interactions and their spectral fingerprints. High-performance computing developments make highly accurate simulations possible to represent the complexity of spin states, leading to a clearer picture of material behaviors. Moreover, such computational studies can offer insights into the influence of topological effects and spin effects on electronic properties, thereby being more extensive in studying materials both in fundamental studies and real applications (Kühne et al., 2020; Ozawa et al., 2019).

7.5.2. Correlation Between Electronic Transitions and Magnetic Behavior

The relation between magnetic properties and electronic transitions is fundamentally rooted in interactions that exist on the atomic and molecular levels. Electronic transitions, which are the phenomenon of exciting electrons from the lower energy to higher energy levels, are influenced by the magnetic surroundings of the material. In the majority of magnetic systems, such as ferromagnets, transitions are strongly correlated with unpaired electron distributions, which are responsible for the net magnetic moment. Owing to the fact that the materials possess different electron configurations, the electronic transition character can lead to various magnetic properties, ranging from paramagnetic to ferromagnetic phases. Understandings of such correlations reveal the subtle interplay of spin and charge that governs the properties of modern functional materials, offering potential for new uses in electronics and magnetism. In exploring how the addition of transition metals impacts electronic and magnetic properties, study has referenced the role that electron-

electron interactions play on both. For instance, incorporation of transition metals such as Cu, Ag, or Au into non-magnetic host compounds such as magnesium oxide (MgO) results in the formation of new energy levels as a consequence of d-f hybridization of the electronic states, which produces well-marked ferromagnetic characteristics. Specifically, in dilute magnetic phases of doped materials, the richness of electronic structure and magnetism is enhanced by the dopants present that can stabilize the ferromagnetic phase through spin polarization.

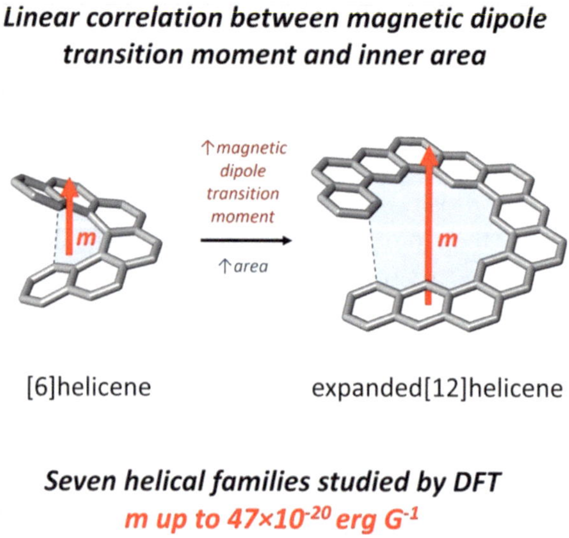

Figure 7.10. *The relationship between electronic transitions and magnetic behavior.*

Source: https://onlinelibrary.wiley.com/cms/asset/b3875ef5-1e4f-48db-83bf-11a094a9d6ab/anie202316696-toc-0001-m.jpg.

This advance underscores the importance of electronic transitions in deciding magnetic behavior, as confirmed by density functional theory calculations that reveal the ensuing energy levels, magnetic moments, and stability of such new systems as represented in the findings on the formation energy values of co-doped MgO (Halais et al., 2024). As we examine further the processes of electronic transitions, it is evident that spectral properties of such changes can be rich sources of information on magnetic material behavior. Systems with incoherent spectral properties, often associated with quantum excitations, pose fascinating opportunities and challenges for the understanding of magnetic processes. Experimental techniques such as angle-resolved photoemission spectroscopy (ARPES) allow scientists to probe the impact of electronic states on spin dynamics, once again affirming the connection between magnetic properties and spectral response. Interestingly, such research has identified such phenomena as magnon softening and anomalous decoherence displayed by spin excitations in compound systems. The realization of a uniform framework correlating electronic transitions and magnetic properties not only enhances our understanding of basic interactions but also holds profound implications for the development of new materials with tailored electronic and magnetic properties, as evidenced by the multi-band systems exhibiting multifaceted behavior in the presence of localized spins (Moreo et al., 2025) (Figure 7.10).

7.5.3. Case Studies of Materials Exhibiting Both Properties

The study of materials that exhibit both electronic spectra and magnetic behavior has made tremendous advancement in understanding their intrinsic interactions. One very interesting case study is $CrSiTe_3$, a two-dimensional material that exhibits an abundant mixture of ferromagnetic and antiferromagnetic interactions. Experimental investigation exhibits antiferromagnetism emerging below 1 kOe whereas ferromagnetic order is observed at 33 K. Interestingly, these magnetic states coexist in a given range of temperatures, from 15 K–33 K, suggesting complex intra-layer and inter-layer interactions at play. In addition, specific heat and Raman response measurements also suggest alterations in magnetic order under the effect of external magnetic perturbation, with substantial alterations in the Eg and Ag modes at lower temperatures. This spin-lattice dynamics coupling illustrates the need to examine further the magnetic ground state properties that have increasing relevance for technological use in two-dimensional materials (Gohil et al., 2025). Another good example is in the examination of the Kondo lattice compounds $CeCuSb_2$ and $CeCuAs_2$, which provides significant information about their electronic structure and magnetism. $CeCuSb_2$ has antiferromagnetic ordering at a Neel temperature of approximately 6.9 K, but $CeCuAs_2$ has no long-range magnetic order down to the lowest observed temperatures. High-resolution depth-resolved photoemission spectroscopy enabled researchers to separate electronic behavior at surface and bulk scales. Notably, the variations in the spectral signatures across the Ce 3d spectra of the two compounds, the presence of a low binding energy peak in $CeCuAs_2$ as those which occur in cuprates, reflect variations in their hybridization effects. The variations in the electronic behaviors highlight complexity in Kondo systems and illustrate mechanisms through which such factors affect overall magnetic properties observed (Datta et al., 2023). The comparison and synthesis of these case studies reveal the richness of electronic and magnetic property interactions in materials science. The establishment of significant spin-lattice interaction in $CrSiTe_3$, and hybridization disparity for $CeCuSb_2$ and

$CeCuAs_2$, show the complexity of these materials. Each case study illustrates how theoretical predictions may not match experimental results and thus why it is important to keep learning in order to understand the mechanisms behind these occurrences. With such materials showing such bimodal characteristics of interest in emerging electronic technologies, a continuous push to probe their intrinsic tendencies will provide the keys to understanding the mysteries of complex magnetism and its technological applications. Such understanding is at the heart of both elementary physics and the creation of future materials with potential for spintronic devices and beyond (Datta et al., 2023; Gohil et al., 2025).

7.6. APPLICATIONS OF ELECTRONIC SPECTRA AND MAGNETIC PROPERTIES

Applications of electronic spectra are extensive and apply to numerous fields such as chemistry, materials science, and environmental monitoring. By studying electronic spectra, scientists can extract vital information regarding the electronic structure of molecules, and in doing so, gain insights into their chemical properties. UV-Vis spectroscopy, for instance, is widely applied for studying the electronic transitions of molecules, providing insight into their bonding behavior and energy levels. The ability of this technique to explain organic molecules makes it particularly valuable in synthetic chemistry, where knowing the purity of a compound or reaction mechanisms is important. Beyond the laboratory, electronic spectra also have critical applications in remote sensing and atmospheric sciences, where studying the spectra can inform scientists about the composition of distant worlds or monitor air quality indicators on our own world. Magnetic characteristics also enhance the practicality of electronic spectra. Values such as magnetic susceptibility, a measure of a material's response to a magnetic field, can be related to its electronic structure. Their knowledge is very important in the creation of advanced materials, such as those in magnetic resonance imaging (MRI) and other medical devices. The magnetic field-electron spin interaction provides the means by which substances that are ferromagnetic or antiferromagnetic can be recognized and delineated, an aspect of significance in magnetic device design. As electronic and magnetic properties become increasingly entwined in emerging devices, methods such as electron paramagnetic resonance (EPR) spectroscopy are all-important instrumentalities in allowing scientists to untangle these complex interconnections so that materials can be tailored for application in various technological uses. Electronic spectrum and magnetic property incorporation is demonstrated in computational techniques such as those offered by packages like Quantum ESPRESSO, enabling researchers to model the properties at the quantum level. The package enables one to calculate a broad range of physical properties and processes with the help of sophisticated techniques like density-functional theory. The ability to predict electronic and magnetic behavior with high accuracy allows for novel material design with specific characteristics before laboratory synthesis. Coupled with resources like the HITRAN molecular spectroscopic database, holding large amounts of molecular transition and absorption rate information, scientists can achieve high levels of accuracy in predicting material behavior under different conditions. Thus, the correlation of electronic spectra with magnetism remains at the center of advances in materials science and technology.

7.6.1. Use in Material Science and Engineering

In material science and engineering, the investigation of electronic spectra is very important for determining properties and performance of materials. Electronic structure control enables researchers to predict and tailor magnetic properties of materials, which is crucial in designing next-generation materials. Transition metal oxides, for instance, have a rich variety of electronic configurations that play a critical role in influencing magnetic properties. Tailoring these properties through doping or alloying has far-reaching implications in applications ranging from magnetic storage disks to superconductors. This capability to fine-tune materials makes it possible to selectively design materials that satisfy certain engineering requirements, thereby maximizing the performance of technology and industrial equipment. Ultimately, the in-depth knowledge of electronic spectra will guarantee innovation in material sciences because it offers a foundation for the rational design of novel materials with characteristics of interest.

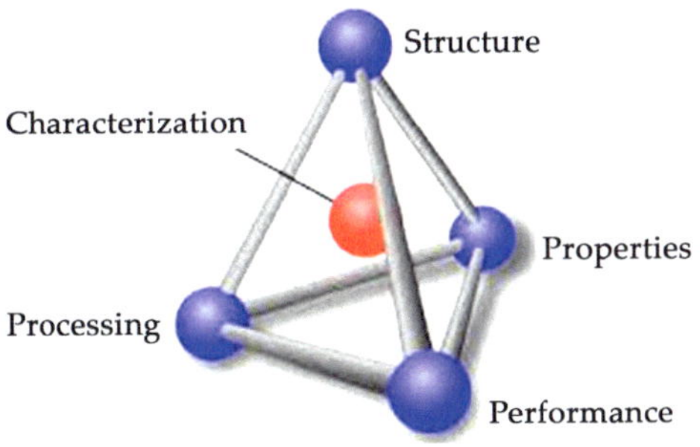

Figure 7.11. *Use in material science and engineering.*

Source: https://encrypted-tbn0.gstatic.com/images?q=tbn:ANd9GcRXpWGvyC3M9a6yosvdre3jEwZ7NNVrTViQdQ&s.

Besides that, magnetic behavior resulting from electronic structures is of the greatest interest in engineering. Materials with specially engineered magnetic behaviors, such as ferromagnets or ferrimagnets, are vital for technological advancements in spintronics and magnetic resonance imaging (MRI). Whether or not the electronic states interact with the magnetic moments will define the usability of the materials in actual applications on a fundamental level. For instance, the processing of high-coercivity magnets is guided by an overall understanding of the electronic spectra of the constituent atoms and of the interactions between them. This serves to optimize material behavior by aligning the electronic structure with that desired magnetic property. Thus, the relationship between electronic spectra and magnetic properties not only advances material science but also enhances engineering practice by providing the tools for designing high-performance materials for particular applications. In addition, the integration of advanced characterization techniques in material science enables the detailed examination of electronic and magnetic properties. Techniques such as electron paramagnetic resonance (EPR) and X-ray absorption spectroscopy (XAS) provide essential information on the local electronic environments and how they influence magnetic properties. These methods allow electronic states to be correlated with macroscopic magnetic properties, resulting in a deeper understanding of material functionality. By employing these techniques, researchers are able to get a clearer picture of how atomic-level changes are likely to impact overall material performance. Through this information, materials are continually optimized for applications in future technologies such as quantum computing and green energy systems. Lastly, the growing intersection between electronic and magnetic property measurement in material science is unleashing new solutions to complex engineering challenges, with profound implications for technological progress in the future (Figure 7.11).

7.6.2. Applications in Medical Imaging and Diagnostics

The integration of electronic spectra and magnetic properties into medical diagnostics and imaging has ushered in a new age in medicine. With nanomaterials, in nanoparticle form, center stage, the field will stand to benefit much from their unique electronic and magnetic properties. Nanoparticles possess high surface-area-to-volume ratios and improved interactions with biological tissues. This characteristic is highly beneficial to imaging processes such as magnetic resonance imaging (MRI) and computed tomography (CT), as they enhance contrast and allow for clarity of images obtained. Moreover, the particle-based contrast agents make it easier to more precisely localize tumors and other abnormalities, leading to better diagnosis and treatment eventually. Thus, the use of such properties is opening the doors for better imaging modalities and setting the stage for more targeted medical intervention, which is the foundation for better patient treatment. Also, the functioning of biosensors, which convert biological information into an electrical signal, relies greatly on nanomaterials unique properties. Due to the use of devices at the nanoscale, one can make biosensors extremely sensitive that play a vital role in diagnostics such that the detection of disease is possible at very early stages. For instance, integrating nanoparticles in biosensors has greatly improved the efficiency of these devices, giving them fast response rates and the ability to detect single biomolecules. This is particularly critical in cases such as the monitoring of chronic and infectious diseases, where timely intervention could mean life. Besides, the flexibility of nanomaterials encourages the ability to design unique diagnostic agents that are disease-specific, thus propelling personalized medicine initiatives and improving overall healthcare outcomes, as reported in their extensive uses in medical diagnostics (Naresh et al., 2021). Furthermore, the use of electronic spectra is indispensible in anticipating the action and interaction of nanomaterials within biological systems. Their optical characteristics can be optimized to interact effectively with a variety of biological probes, and disease development and treatment response can be monitored in real time. This makes it possible to tune their properties and design targeted imaging probes that enhance diagnostic technique specificity by limiting background interference and providing clearer insights into disease conditions. As science evolves, the potential to merge more than a single imaging modality with enhanced nanotechnology brings about the potential to create hybrid systems that take advantage of complementary information to deliver more comprehensive diagnostics. Research into these frameworks not only describes biological phenomena but also illustrates the continuous development of medical imaging technologies, so that electronic spectra and magnetic properties continue to be at the forefront of innovative applications in medical imaging and diagnostics (Joudeh et al., 2022).

7.6.3. Role in Developing New Technologies and Materials

The development of new materials and technologies heavily relies on the synergy between electronic spectra and magnetic properties. Progress in two-dimensional materials, for instance, illustrates the ways in which electronic functionality can be engineered and exploited for cutting-edge applications. As in the instance of the g-C_4N_3/MoS_2 van der Waals heterostructure, the marriage of ferromagnetic half-metallicity with exceptional gas adsorption capability is an intriguing path towards enhancing spintronic devices. These developments not only serve to illustrate the potential of such materials towards multifunctional device enhancements but also suggest the physics of function that makes them so compelling. With the study of spin-dependent electronic transport processes, researchers can manipulate configurations such as parallel and anti-parallel spins to make breakthroughs in energy-efficient electronics and sensing performance enhancement in sensor technologies (Dong et al., 2022). Beyond their current applications, ongoing research on electronic spectra has profound implications for the manufacture of next-generation materials. Application of comprehensive databases, such as HITRAN, is an oriented attempt to understand molecular interaction and spectral features relevant for the design of new compounds. The inclusion of infrared cross-sections determined by experiment for more molecules enhances the community's capacity to predict behavior in atmospheric and materials science applications. This degree of specificity allows for advanced modeling and simulation, which is essential in the discovery of photonic materials and sensors that can operate under diverse and extreme environmental conditions. Such innovations are immediate results of the understanding provided by molecular transitions and energy levels, which underscore the intrinsic worth of spectral information in contemporary materials science (Gordon et al., 2017). Furthermore, electronic spectra and magnetic property research are paramount in leading the way for sustainability in technological innovations. Material scientists are increasingly interested in producing materials that meet functional specifications and are at the same time environmental-friendly. The investigation of new magnetic materials that are capable of efficient energy conversion and storage points to this direction. For example, the material spintronic devices like g-$C4N_3/MoS_2$, their ability to operate with single-spin transport and sensitivity to gas molecules, indicate the twin interest in performance and environmental impact. Because the materials offer high transport anisotropy, selective interactions of the materials lay the foundation for effective sensors that can identify harmful gases while, at the same time, motivating cleaner technology. This union of material innovation and environmental sustainability is an ideal representation of the significant role that electronic spectra and magnetic properties play in mapping the destiny of most technological areas (Dong et al., 2022).

SUMMARY

In Summary, the intricate correlation between electronic spectra and magnetic properties sheds light on a background knowledge of material behavior, particularly in catalysis. The research on electronic transitions gives critical insights into the behavior of materials under various conditions and is thus vital for their effectiveness as catalysts. For instance, oxygen evolution reaction conditions and catalyst design both accentuate the demand for particular materials that will support optimal efficiencies and stability in catalytic processes. These findings stress the need to study electronic spectra and magnetic properties simultaneously because they both contribute to the overall activity of catalytic systems (Song et al., 2020). Moreover, the utilization of advanced computational softwares such as Quantum ESPRESSO enables in-depth study of electronic structures and their correspondence to magnetic properties. The ability of this softwares to simulate a broad range of material properties enables researchers to predict and optimize physical system phenomena. By using density-functional theory and many-body perturbation theory, as shown by Quantum ESPRESSO, researchers are able to readily explore new techniques for enhancing our information regarding materials at a quantum level (Giannozzi et al., 2017). These advances are seminal to future research towards the development of new materials with enhanced electronic and magnetic properties. Ultimately, the adoption of these complex interdependencies promises the potential for innovative material design and breakthrough applications across a range of industries. Since electronic spectra and magnetic properties are entwined, gaining understanding into their dynamics not only advances theoretical systems but also develops practical applications. With continued breakthroughs in computational and experimental methods to explore these phenomena, the scientific community is able to work toward designing novel materials that are capable of meeting the requirements of future technological needs. This concerted process provides a sound foundation for future research into the rich interplay among electronic structures, magnetic properties, and their applications within practical contexts.

Electronic spectra and magnetic behaviors have been significantly advanced with the advent of new materials and new methods of research. Perhaps the most fascinating discipline is the field of magnonics, an emerging discipline that utilizes spin waves (magnons) for information storage, processing, and transmission. New advancement in magnonics has brought to the forefront the possibility of lower energy consumption and improved device integration with CMOS architecture, and the design of sub-micrometer devices. Of special interest is the challenge of circumventing difficulties involved in nanoscale magnon manipulation and the production of low-Gilbert damping materials. The advancement emphasized within this area is a very significant advance in the comprehension of magnetic properties at the nanolevel, providing the possibility of new knowledge within information technology and other areas. Through the resolution of these problems, researchers are continually striving for more efficient and compact electronic devices, echoing the dynamic relationship between electronic spectra and magnetic properties within modern physics (Barman et al., 2021).

Simultaneously, the advent of two-dimensional (2D) materials has catalyzed groundbreaking research in condensed matter physics and material science. Initially focused on graphene, 2D materials have expanded to a vast family of materials, each with distinctive structural features and physicochemical properties. Recent breakthroughs have propelled the advancement of novel synthetic methods, including liquid and vapor phase deposition, which form the backbone of engineering the materials' properties for desired purposes. The finding of superconducting, optical, and magnetic properties in 2D materials has revealed unparalleled opportunities for electronics, optoelectronics, and other applications. As syntheses and characterizations of 2D materials become better, researchers are poised to unlock new paradigms in device fabrication and operation, thereby highlighting the imperative interaction between electronic spectra and magnetic phenomena (Chang et al., 2021). The overlap of electronic spectra and magnetic properties is a fruitful field in material science, particularly as researchers seek to push past existing boundaries. Such grand debates highlight the need for new methods of dealing with quantum materials, expanding their functionality, and integrating them into existing technologies. With ongoing research in fields like magnonics and 2D materials, new developments are being made, and the potential exists for future breakthroughs that could change electronic device technology. Material synthesis with custom-tailored properties could also lead to greatly efficient energy consumption and device performance. Lastly, manipulation, synthesis, and application of materials in the context of electronic spectra and magnetism are not just promising but also crucial for developing next-generation technologies sensitive to the needs of contemporary times for function and efficiency (Barman et al., 2021; Chang et al., 2021).

REFERENCES

1. Akman, F. (2016). Spectroscopic investigation, HOMO–LUMO energies, natural bond orbital (NBO) analysis and thermodynamic properties of two-armed macroinitiator containing coumarin with DFT quantum chemical calculations. *Canadian Journal of Physics, 94*, 583–593. https://doi.org/10.1139/cjp-2016-0041 (accessed on 13th September 2025).
2. Almutairi, T. (2025). Cation and spin interactions in cadmium ferrite: A quantum mechanical study. *International Journal of Molecular Sciences, 26.* https://www.semanticscholar.org/paper/db142509ca15c082bdc6380f1a7d1f863c3d6012 (accessed on 13th September 2025).
3. Bagio, S., Gonzalez, J., Gable, R., Hall, C. R., Boskovic, C., & Giansiracusa, M. (2025). Redox, spectroscopic and magnetic properties of C3-symmetric rare earth complexes featuring atypical ortho-dioxolene binding. *Dalton Transactions.* https://www.semanticscholar.org/paper/9d47a6cd921c12748ec0123adacfae72e87e4b87 (accessed on 13th September 2025).
4. Barman, A., Gubbiotti, G., Ladak, S., Adeyeye, A. O., Krawczyk, M., Gräfe, J., Adelmann, C., et al. (2021). The 2021 magnonics roadmap. *Journal of Physics: Condensed Matter, 33*, 413001–413001. https://doi.org/10.1088/1361-648x/abec1a (accessed on 13th September 2025).
5. Bejo, A. B., Shura, M. W., Tadele, K., Afrassa, M. A., & Maremi, F. T. (2025). Structural, electronic, mechanical, optical and magnetic properties of RhNbZ (Z = Li, Si, As) Half-Heusler compounds: A first-principles study. *Materials Research Express, 12.* https://www.semanticscholar.org/paper/b3945304da64053cac45ac888edc01945f317f8a (accessed on 13th September 2025).
6. Billa, M., Islam, M. T., Alam, T., Islam, M., Moubark, A. M., Alsaif, H., Albadran, S., et al. (2024). Structural, morphological, optical, electrical, and magnetic properties of aluminum-doped CoxCa(0.90−x)Ni0.10Fe_2O_4 flexible substrate for visible to NIR spectra applications. *APL Materials.* https://www.semanticscholar.org/paper/b63d2db0e61013f1e4408252e185bc6a6ad2ad3f (accessed on 13th September 2025).
7. Chang, C., Chen, W., Chen, Y., Chen, Y., Chen, Y., Ding, F., Fan, C., et al. (2021). Recent progress on two-dimensional materials. *Acta Physico-Chimica Sinica, 0*, 2108017–0. https://doi.org/10.3866/pku.whxb202108017 (accessed on 13th September 2025).
8. Chen, X., Ma, D., Liu, T., Zhu, C., Yang, Z., Zhao, J., Yang, Z., et al. (2021). Hybridized local and charge-transfer excited-state fluorophores

through the regulation of the donor–acceptor torsional angle for highly efficient organic light-emitting diodes. *CCS Chemistry, 4*, 1284–1294. https://doi.org/10.31635/ccschem.021.202100900 (accessed on 13th September 2025).

9. Cholsuk, C., Çakan, A., Suwanna, S., & Vogl, T. (2023). Identifying electronic transitions of defects in hexagonal boron nitride for quantum memories. *Advanced Optical Materials, 12*. https://www.semanticscholar.org/paper/e2ac81ee292ff37ecf17956c80351f0c72d98dcd (accessed on 13th September 2025).
10. Datta, S., Pandeya, R., Dey, A., Gloskovskii, A., Schlueter, C., Peixoto, T., Singh, A., et al. (2023). Complexity in the hybridization physics revealed by depth-resolved photoemission spectroscopy of single crystalline novel Kondo lattice systems, $CeCuX_2$ (X = As/Sb). https://www.semanticscholar.org/paper/d2a247168f0ee5751a0228860133ca0008ea2469 (accessed on 13th September 2025).
11. Dong, X., Chen, T., Liu, G., Xie, L., Zhou, G., & Long, M. (2022). Multifunctional 2D g-C_4N_3/MoS_2 vdW heterostructure-based nanodevices: Spin filtering and gas sensing properties. *ACS Sensors, 7*, 3450–3460. https://doi.org/10.1021/acssensors.2c01785 (accessed on 13th September 2025).
12. Giannozzi, P., Andreussi, O., Brumme, T., Bunău, O., Buongiorno Nardelli, M., Calandra, M., Car, R., et al. (2017). Advanced capabilities for materials modelling with Quantum ESPRESSO. *Journal of Physics: Condensed Matter, 29*, 465901–465901. https://doi.org/10.1088/1361-648x/aa8f79 (accessed on 13th September 2025).
13. Gohil, S., Halder, S., Iyer, K. K., Ghosh, S., Thamizhavel, A., & Maiti, K. (2025). Unusual magnetic order, field induced melting and role of spin-lattice coupling in 2D Van der Waals materials: A case study of $CrSiTe_3$. https://www.semanticscholar.org/paper/1cfb4ddf0c8df5a10fee3ec43c28bbebc125275d (accessed on 13th September 2025).
14. Gordon, I. E., Rothman, L. S., Hill, C., Kochanov, R. V., Tan, Y., Bernath, P. F., Birk, M., et al. (2017). The HITRAN2016 molecular spectroscopic database. *Journal of Quantitative Spectroscopy and Radiative Transfer, 203*, 3–69. https://doi.org/10.1016/j.jqsrt.2017.06.038 (accessed on 13th September 2025).
15. Gupta, S., Chakraborty, S., Pakhira, S., Barreteau, C., Crivello, J. C., Bandyopadhyay, B., Grenèche, J. M., et al. (2022). Coexisting structural disorder and robust spin-polarization in half-metallic FeMnVAl. *Physical Review B, 106*. https://doi.org/10.1103/physrevb.106.115148 (accessed on 13th September 2025).

16. Halais, W. T., Doghmane, M., Bouhlala, A., & Chettibi, S. (2024). Structural, magnetic, and optoelectronic properties of rock salt MgO co-doped with Eu-TM (TM = Cu, Ag, and Au) for spintronic and UV detector applications: SP-DFT investigations. *Modern Physics Letters B*. https://www.semanticscholar.org/paper/a12c12292b5d44ba2e56600beed4be7fac6be618 (accessed on 13th September 2025).

17. Joudeh, N., & Linke, D. (2022). Nanoparticle classification, physicochemical properties, characterization, and applications: A comprehensive review for biologists. *Journal of Nanobiotechnology, 20*. https://doi.org/10.1186/s12951-022-01477-8 (accessed on 13th September 2025).

18. Kang, K., Cahill, D. G., & Schleife, A. (2024). Temperature-dependent optical and magneto-optical spectra of ferromagnetic BCC Fe. https://www.semanticscholar.org/paper/d1fee8195a861a02921b5cb177a5a653e0b663ff (accessed on 13th September 2025).

19. Khan, I., Saeed, K., & Khan, I. (2017). Nanoparticles: Properties, applications and toxicities. *Arabian Journal of Chemistry, 12*, 908–931. https://doi.org/10.1016/j.arabjc.2017.05.011 (accessed on 13th September 2025).

20. Kubo, T., Yasuoka, H., Kasinathan, D., Ranjith, K. M., Schmidt, M., & Baenitz, M. (2024). Static and dynamic electronic properties of Weyl semimetal NbP–A single crystal ^{93}Nb-NMR study. *Journal of the Physical Society of Japan*. https://www.semanticscholar.org/paper/c13b36e45641f0a373d4b205904aa7f9faf5c861 (accessed on 13th September 2025).

21. Kühne, T. D., Iannuzzi, M., Del Ben, M., Rybkin, V. V., Seewald, P., Stein, F., Laino, T., et al. (2020). CP2K: An electronic structure and molecular dynamics software package–Quickstep: Efficient and accurate electronic structure calculations. *The Journal of Chemical Physics, 152*. https://doi.org/10.1063/5.0007045 (accessed on 13th September 2025).

22. León-González, J. C., Toscano-Negrette, R. G., Morales, Á., Vinasco, J. A., Yücel, M. B., Sari, H., Kasapoglu, E., et al. (2023). Spin–orbit and Zeeman effects on the electronic properties of single quantum rings: Applied magnetic field and topological defects. *Nanomaterials, 13*. https://www.semanticscholar.org/paper/6d9f1e047151a2a93a1a7c54755768945b35a849 (accessed on 13th September 2025).

23. Lettens, J., Avramenko, M., Vandevenne, I., Chu, A., Hengel, P., Kern, M., Anders, J., et al. (2025). Integration of electrically detected magnetic resonance

on a chip (EDMRoC) with charge pumping for low-cost and sensitive defect characterization *in silicon* carbide metal–oxide–semiconductor field-effect transistors. *Journal of Applied Physics*. https://www.semanticscholar.org/paper/f6f32eefa3f47831051daef786618d63ba679834 (accessed on 13th September 2025).

24. Lundquist, J., Vorobiov, S., Merten, L., Reimer, A., Boughelilba, M., Da Vela, P., Tavecchio, F., et al. (2025). Multi-messenger insights into ultra-high-energy cosmic rays from FR0 radio galaxies: Emission spectrum, composition, and secondary photons and neutrinos. *Proceedings of 7th International Symposium on Ultra High Energy Cosmic Rays–PoS(UHECR2024)*. https://www.semanticscholar.org/paper/75eff478905898ab787cb3e3a7dbee0f44658992 (accessed on 13th September 2025).

25. Moreo, E., Dagotto, E., Alvarez, G., Tohyama, T., Mierzejewski, M., & Herbrych, J. (2025). Magnon damping and mode softening in quantum double-exchange ferromagnets. *Reports on Progress in Physics, 88*. https://www.semanticscholar.org/paper/42b657a54bc789863caed412f9cf973b56250e21 (accessed on 13th September 2025).

26. Mourdikoudis, S., Pallares, R. M., & Thanh, N. T. K. (2018). Characterization techniques for nanoparticles: Comparison and complementarity upon studying nanoparticle properties. *Nanoscale, 10*, 12871–12934. https://doi.org/10.1039/c8nr02278j (accessed on 13th September 2025).

27. *N/A*. (2023). Retracted: Heating properties of graphene oxide nanosheets and their application in clothing design. *Advances in Materials Science and Engineering*. https://www.semanticscholar.org/paper/a4d94550310b6f78596eb5711c2fc87d988c9a89 (accessed on 13th September 2025).

28. *N/A*. (2023). Retracted: Studying the effect of metallic precursor concentration on the structural, optical, and morphological properties of zinc sulfide thin films in photovoltaic cell applications. *Advances in Materials Science and Engineering*. https://www.semanticscholar.org/paper/6ab52870db5de9532566a992a55473283ae91149 (accessed on 13th September 2025).

29. Nandy, S., Theingi, M., Ghosh, S., Chae, K. H., & Sudakar, C. (2024). Influence of local structure and metal-oxygen hybridization on the electrical and magnetic properties of alkaline earth metal (Mg^{2+}, Ca^{2+}, Sr^{2+}) substituted $LaFeO_3$ ceramics. *Journal of Applied Physics*. https://www.semanticscholar.org/paper/b16aa32fe4ac829aeb8501a3ba419b9e39cebc8c (accessed on 13th September 2025).

30. Naresh, V., & Lee, N. (2021). A review on biosensors and recent development of nanostructured materials-enabled biosensors. *Sensors, 21*, 1109–1109. https://doi.org/10.3390/s21041109 (accessed on 13th September 2025).

31. Nguyen, N. L., Dang, H. T., Pham, T. L., & Nghiem, T. M. H. (2024). Computationally predicted electronic properties and energetics of native defects in cubic boron nitride. *Journal of Electronic Materials*. https://www.semanticscholar.org/paper/685921b8eeb0f09273a642c3f351812c97fcfec5 (accessed on 13th September 2025).

32. Obodo, K. O., Gebreyesus, G., Ouma, C. N. M., Obodo, J. T., Ezeonu, S. O., Raics, P., & Bouhafs, B. (2020). Controlling the electronic and optical properties of HfS_2 monolayers *via* lanthanide substitutional doping: A DFT+*U* study. *RSC Advances, 10*, 15670–15676. https://doi.org/10.1039/d0ra02464c (accessed on 13th September 2025).

33. Ozawa, T., Price, H. M., Amo, A., Goldman, N., Hafezi, M., Lü, L., Rechtsman, M. C., et al. (2019). Topological photonics. *Reviews of Modern Physics, 91*. https://doi.org/10.1103/revmodphys.91.015006 (accessed on 13th September 2025).

34. Schmidt, J., Marques, M. R. G., Botti, S., & Marques, M. A. L. (2019). Recent advances and applications of machine learning in solid-state materials science. *npj Computational Materials, 5*. https://doi.org/10.1038/s41524-019-0221-0 (accessed on 13th September 2025).

35. Schwoerer, M., & Wolf, H. (2019). *Organic Molecular Solids*. CRC Press eBooks. https://doi.org/10.1201/9780429126086 (accessed on 13th September 2025).

36. Song, J., Wei, C., Huang, Z. F., Liu, C., Zeng, L., Wang, X., & Xu, Z. J. (2020). A review on fundamentals for designing oxygen evolution electrocatalysts. *Chemical Society Reviews, 49*, 2196–2214. https://doi.org/10.1039/c9cs00607a (accessed on 13th September 2025).

37. Sun, C., Zang, J. P., Liu, Y., Zhong, Q. Q., Xing, X., Li, J., Yue, C. Y., et al. (2021). Lead-free hybrid indium perovskites with highly efficient and stable green light emissions. *CCS Chemistry, 4*, 3106–3121. https://doi.org/10.31635/ccschem.021.202101092 (accessed on 13th September 2025).

38. Syrotyuk, S., Hussain, M. K., & Nakonechnyi, R. (2025). Electronic and magnetic properties of hexagonal ZnSeS solid solutions modified by Cr impurity and the anion concentration. *Physics and Chemistry of Solid State*. https://www.semanticscholar.org/paper/1034aa14b98fdfde2a9f1bd5ebcadffadc0c9af6 (accessed on 13th September 2025).

39. Takeda, Y., Okane, T., Ohkochi, T., Fujimori, S., Saitoh, Y., Yamagami, H., Yabuta, H., et al. (2010). Electronic states of magnetic refrigerator materials $Mn_{0.9}Fe_{1.1}P_{0.55}As_{0.45}$ using soft X-ray magnetic circular dichroism. *Volume, 200*, 012199. https://www.semanticscholar.org/paper/19956e3b28ed05e43f4ddc56d5596b98a34baea8 (accessed on 13th September 2025).

40. Tang, Y., Wang, C., Tang, J., Luo, L., & Wang, C. (2024). Study on structure and magnetic properties of rare earth terbium-doped ferrite $Sr_{1-x}Tb_xFe_{12}O_{19}$. *Physica Scripta, 100*. https://www.semanticscholar.org/paper/fc3bfffab65398b6d346ffd32206fbe0c12cf4a8 (accessed on 13th September 2025).

41. Tarannum, I., & Singh, S. (2024). Unravelling the electronic structure, bonding, and magnetic properties of inorganic dysprosocene analogues $[Dy(E_4)_2]^-$ (E = N, P, As, CH). *Physical Chemistry Chemical Physics: PCCP*. https://www.semanticscholar.org/paper/2a8ed72f017eeb67ee1446cdc9a77e9d2ccbcc31 (accessed on 13th September 2025).

42. Unke, O. T., Chmiela, S., Sauceda, H. E., Gastegger, M., Poltavsky, I., Schütt, K. T., Tkatchenko, A., et al. (2021). Machine learning force fields. *Chemical Reviews, 121*, 10142–10186. https://doi.org/10.1021/acs.chemrev.0c01111 (accessed on 13th September 2025).

43. Yang, F., Hu, P., Yang, F. F., Chen, B., Fei, Y., Sun, R., Hao, K., et al. (2023). Emerging enhancement and regulation strategies for ferromagnetic 2D transition metal dichalcogenides. *Advanced Science, 10*. https://doi.org/10.1002/advs.202300952 (accessed on 13th September 2025).

44. Yu, J., Huang, C., Usoltsev, O., Black, A. P., Gupta, K., Spadaro, M. C., Pinto-Huguet, I., et al. (2024). Promoting polysulfide redox reactions through electronic spin manipulation. *ACS Nano, 18*, 19268–19282. https://www.semanticscholar.org/paper/ac5667d961e1ff24d4b688da4ba9a6ab92579f30 (accessed on 13th September 2025).

45. Yue, W., Shan, J., Jiao, R., Zhang, L., Chen, Y., & Hao, D. (2024). Strain-controlled electronic and magnetic properties of Janus nitride MXene monolayer $MnCrNO_2$. *Applied Sciences*. https://www.semanticscholar.org/paper/4a37bf1601e7ba9a5973e28738c71f7392207bc9 (accessed on 13th September 2025).

CHAPTER 8

REACTIONS OF COORDINATION COMPOUNDS

Contents

INTRODUCTION

Coordination compound study is a major aspect of the overall branch of inorganic chemistry. Coordination compounds, a central metal atom surrounded by ligands, possess a diverse range of properties and reactivity patterns that are of fundamental and practical importance to sciences. These compounds have advanced interactions that regulate their reactivity, geometry, and stability, which can be determined by a number of factors such as metal oxidation states and ligand properties. Further, the reactivity of these compounds is commonly used in catalysis, biological organisms, and materials science, hence contributing to a fundamental knowledge necessary to continue chemical research. This prelude leads to a comprehensive discussion of how the inherent properties of coordination compounds control the nature of their reactions and implications in diverse fields of science. With scientists investigating the chemistry of coordination compounds more, they appreciate that their reactivity is based on various factors. These can include like the electronic and steric characteristics of ligands, coordination number, and the oxidation state of the central metal. Furthermore, more recent advances in coordination chemistry also point towards the importance of novel and innovative synthesis methods that disclose novel coordination modes and functional architectures. For example, it has been demonstrated through studies that the targeted modification of ligands can significantly influence reaction pathways as well as product stability. The relevance of knowing these parameters cannot be overstressed, as they have direct implications for applications as diverse as drug design to catalysis. Identification of these reactivity trends places new opportunities in areas such as material sciences and bioinorganic chemistry, thereby emphasizing the multidisciplinary nature of reactions in coordination compounds. The scope of influence from coordination compounds extends beyond simplistic theoretical concepts and intrudes into applied science applications that resonate within the scientific community. Studies on these compounds have led to significant progress toward understanding catalysts in chemical industry processes, drugs, and other technologies. For instance, the preparation of targeted delivery systems of drugs on a routine basis relies on the properties of coordination complexes, which can be designed to react preferentially to biological targets. Additionally, their susceptibility to radical reactions indicates their versatility and applicability in organic synthesis, especially in green operations where water acts as solvent or medium. This increasing body of knowledge emphasizes the crucial significance coordination compounds have in understanding fundamental concepts of chemistry as well as in solving existing problems in every branch of science, finally culminating in future advances in chemistry (Biso et al., 2025; Chatgilialoglu et al., 2024).

Definition of Coordination Compounds

Coordination compounds is a term for a collection of compounds that have a central metal atom or ion surrounded by a collection of molecules or ions called ligands. Coordinate covalent bonds are the interactions between the metal center and the ligands, with both electrons in the bonding being supplied by the ligand. This new architecture is accountable for the compounds' diverse chemical properties and reactivity, thus qualifying them as a valuable area of study in coordination chemistry. The core metal is generally a transition metals member, which is accountable for the compounds' diverse oxidation states and coordination numbers, resulting in complicated geometries. These characteristics are particularly important in understanding the way coordination compounds carry out chemical reactions, aspects like reactivity, stability, and alteration of electronic states in these reactions, which are key concepts in the broader field of inorganic chemistry. Another key characteristic of coordination compounds is their classification according to the nature of ligands involved. Ligands can also be monodentate, bidentate, or polydentate, in reference to the number of donor atoms they bring to the metal center. Monodentates bind via a single atom, whereas bidentates have two binding sites with which to form chelate complexes. The capacity of polydentates to wrap around the central atom stabilizes them even more. This built-in diversity in ligand coordination not only enhances the stability of coordination compounds but also plays an important role in their reactivity in chemical reactions. Ligand exchange reactions, which are very frequent in these compounds, are the cause of their involvement in biological and industrial processes, such as hemoglobin and chlorophyll reactions, as well as catalysis and extraction, to affirm the functional relevance of learning about these definitions in coordination compound reactions. The electronic structure and the spatial geometry of the ligands have an important influence on the reactivity of coordination compounds. The geometry of the coordination complex—whether it adopts octahedral, tetrahedral, or square planar geometries—is often very influential in determining the electronic properties of the compound. For instance, changes in geometry can lead to variations in ligand field splitting and consequently the stability of different oxidation states of the central metal ion. Coordination number and geometry may also control the kinetics of ligand exchange reactions, an important process in bringing about chemical change. Knowledge of these electronic and geometrical principles is critical to anticipating the behavior of coordination compounds in some reactions because it has a direct relationship with their reactivity profiles. Consequently, the behavior and definition of coordination compounds give a necessary basis for the investigation of reactions, justifying relationships throughout the coordination chemistry discipline, which plays a pivotal role in material science, catalysis, and bioinorganic chemistry developments (Diercks et al., 2017; Hoffmann et al., 2016).

Importance in Chemistry and Industry

In industry and chemistry, coordination compounds have a paramount role that extends far beyond standard synthesis methods. Their unique ability to stabilize complexes with metal ions is responsible for the synthesis of various catalysts needed for optimizing chemical reactions. This is particularly true in aromatic hydroxylation reactions where reactions catalyzed by transition metals have proven to be highly efficient. The regioselective ortho-hydroxylation of arenes serves to showcase this, with transition metal catalysts like palladium and rhodium allowing the conversion of C–H bonds into more active C–O bonds and resultant phenolic compounds of value. These processes are significant in the drugs sector, where demand for specific phenols as intermediates in drug production is high.

Figure 8.1. *Importance in chemistry and industry.*

Source: https://etimg.etb2bimg.com/photo/91267079.cms.

The advancements in such catalytic reactions highlight not just the magical potential of coordination compounds but also their necessity for the enhancement of production procedures in industries (Iqbal et al., 2020). Additionally, the synthesis of nitrogen-containing organic compounds is a classic instance of how coordination chemistry solves major industrial problems. The quest to discover means of synthesizing useful organic compounds directly from dinitrogen (N_2) rather than from ammonia (NH_3) is a paradigm shift in industrial chemistry. Complexes of transition metals offer a new frontier for the initiation of N_2 functionalization, enabling the establishment of N–C bonds essential to the synthesis of a broad spectrum of chemical products. These reactions, encompassing different coordination modes of [N_2-M] complexes, not only have deepened our understanding of nitrogen chemistry but also opened up new grounds for synthesizing agrochemicals, medicinals, and materials. The ramifications of these reactions are far-reaching in that they balance both economic and environmental sustainability goals by potentially facilitating the greener synthesis of essential compounds (Ze-Jie et al., 2020). Another essential characteristic of coordination compounds in industry is their responsiveness and flexibility to different chemical environments. Advancements in the formation of these compounds have led to catalysis technologies that have revolutionized standard operating procedures for chemical manufacture. Different transition metals can participate in the same reactions but produce distinctive catalytic products, highlighting the importance of metal choice in reactions. For instance, metal complexes' differential reactivity enables them to be applied in selective reactions that limit the formation of byproducts, hence enhancing overall yield and reducing waste. With industries seeking to comply with stringent regulatory regimes and sustainability needs, proper use of coordination compounds presents effective pathways toward such ends. Lastly, their role in chemistry and industry emphasizes the all-important link between theory and application, propelling the topic towards solution-based paradigms of chemical challenges of today (Figure 8.1).

Overview of Reactions Involving Coordination Compounds

Coordination compounds are the subject of investigation in the study of the diverse reactions they undergo, which constitute an integral component of various fields, including catalysis and biochemistry. One interesting set of reactions are the electron transfer reactions catalyzed by these compounds, whereby central metal ions, such as transition metal ions, may be oxidized or reduced. The unique electronic structures of these metal ions enable them to react via redox, bearing testimony to their versatility. For instance, the capacity for ligands to stabilize multiple oxidation states allows for the wide range of possible reactions that can be designed for syntheses or catalysis applications. Such electron transfer reactions are not only significant in synthetic processes but also replicate biologically pertinent systems, highlighting the importance of coordination complexes in life sciences and inorganic chemistry. Besides, coordination compounds are highly reactive in catalytic reactions and often serve as catalysts in organic conversions. Artificial hydrogenase mimetics—coordination compounds that duplicate the catalytic activity of natural hydrogenases—are one of the new approaches. Hydrogenase mimics catalyze the hydrogen evolution reaction effectively at mild conditions, delivering sustainable methods for hydrogen production (Leone et al., 2023). Synthesis of such mimetics is a synthesis of inorganic coordination chemistry and green chemistry protocol, with a focus on coordination compounds as playing a central role in green energy technology. Structural optimization of these catalysts through organic chemist refinement can enhance their efficacy, stability, and selectivity, thereby widening their application to synthetic organic processes and energy conversion technology. Furthermore, the light photoreaction with coordination complexes brings a fascinating aspect into their reactivity, particularly within photoredox catalysis. Cyclometalated iridium(III) complexes have been found to be highly efficient photocatalysts for a variety of applications, ranging from organic light-emitting diodes to energy conversion in solar energies (Schreier et al., 2022). If the compounds are specifically designed to be soluble in water, then they can catalyze reaction which is commonly less efficient under conditions in aqueous media. This versatility opens up new avenues for biochemical and synthetic chemistry, allowing coordination complexes to participate in complex networks of biocatalytic reactions. With continuing work making water-soluble coordination complexes with functionalized characteristics, the marriage of coordination chemistry and photophysics has the ability to produce new, groundbreaking achievements in sustainable chemistry and energy use.

8.1. TYPES OF COORDINATION COMPOUNDS

The enormity of coordination compounds is largely a result of their varied structure and stability depending on the nature of their ligands and metal centers. Ligands can be classified into two categories: monodentate and polydentate. Monodentate ligands bind to a central metal atom at one point of attachment, for example, ammonia or water, while polydentate ligands, for example, ethylenediamine or EDTA, establish multiple bonds, thus resulting in more stable complexes. This capacity to make multiple bonds is due to the chelate effect, whereby ring formation enhances the overall stability. Understanding about the classes of coordination compounds is extremely crucial in fields such as bioinorganic chemistry, where the coordination compounds play a critical role in biological mechanisms, including oxygen transport and electron transfer processes that are essential for life. One of the most important classes of coordination compounds are those containing transition metals, which possess unusual and multilateral patterns of reactivity. Transition metal complexes often engage in redox reactions in which they accept or donate electrons depending on their oxidation state. This feature renders them valuable in a broad variety of catalytic reactions and electron transfer mechanisms. For instance, the copper-iron interaction in heme/Cu model complexes accounts for the mechanistic dynamics of dioxygen reduction and nitrogen oxides. As revealed in recent studies, reduced heme/Cu complexes show marked reactivity with dioxygen to form μ-oxo heme-FeIII–O–CuII(L) products, thus showing a crucial interplay between biological activity and coordination chemistry (Hematian et al., 2015). Furthermore, the role of coordination compounds is not only stability and redox behavior. Their actions are typically complicated mechanisms that are vital to biological conditions. An example is the redox interconversion of nitric oxide and nitrite and biochemical signaling processes where heme/Cu assemblies play a central role. The cupric center within the complexes is a Lewis acid and the heme a redox-active center, while electron transfer and nitrite binding is coordinated (Hematian et al., 2015). Advanced mechanistic studies have explained how ligands are positioned during reactions, shedding more light on the ways coordination compounds enable essential processes in living organisms. Coordination compound chemistry and reaction chemistry thus hold the solution to advancing synthetic and biochemistry.

8.1.1. Homoleptic vs. Heteroleptic Complexes

In coordination chemistry studies, the distinction between homoleptic and heteroleptic complexes is of foundational significance because it directly influences the stability and reactivity of the formed compounds. Homoleptic complexes are distinguished by having a central metal ion bonded solely to equivalent ligands, hence creating a homogeneous structure that tends to lead to chemically expected behavior. Complexes like [Cu(NN)2]+ are a good example that shows a homogeneous coordination environment, leading to molecular stability. In contrast, heteroleptic complexes are composed of greater than one ligand type bonded to the central metal ion. This diversity can have a drastic effect on the chemical behavior and stability of the complex, as established in the dynamic coordination/decoordination processes described in the study on dinuclear copper(I) complexes with a series of 1,10-phenanthroline ligands, where entropic considerations control the balance between homoleptic and heteroleptic species (Jean-François Nierengarten et al., 2019). The unique characteristics of heteroleptic complexes also extend their application in other fields such as catalysis and materials science. Through the introduction of different metal ions, as with mixed-metal metal-organic frameworks (MM-MOFs), these complexes can leverage the synergistic effect due to the presence of more than one metal center. This is particularly beneficial to catalytic applications, where metal-mischmetal juxtapositions result in increased stability and reactivity compared to homometallic. Such features have been observed in recent studies to show the enhanced performance of MM-MOFs in tandem reactions and gas separations (Masoomi et al., 2019). Therefore, researchers are increasingly focusing on synthesizing heteroleptic complexes in a strategic manner to leverage their multi-faceted advantages in practical applications. The implication of homoleptic versus heteroleptic complexes extends far beyond mere chemical stability; they significantly influence the physical properties of materials as well. For example, in the above copper(I) complexes, rigidification of certain architectures by bulky substituents can affect photophysical characteristics such as emission quantum yield. In some cases, addition of other types of ligands can lead to enhancement of luminescence, enhancing the suitability of these complexes for optoelectronic devices. The ability to modulate properties at the molecular level underscores the importance of selection of a suitable coordination environment and, as such, comparative study of homoleptic and heteroleptic complexes an extremely important area of research in the overall background of coordination compound reactions. It is, as such, the understanding of these differences that not only aids in predicting reactivity but also in furthering technology and material science (Figure 8.2).

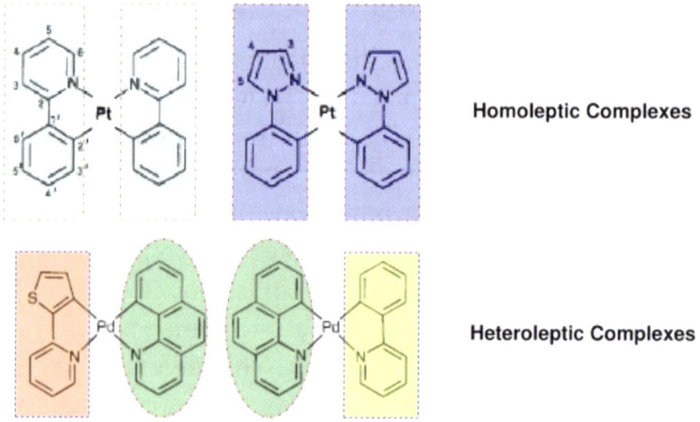

Figure 8.2. *Homoleptic vs. heteroleptic complexes.*

Source: https://image1.slideserve.com/1742292/slide5-l.jpg.

8.1.2. Coordination Number and Its Significance

The coordination number of a metal in a coordination compound is a key characteristic that decides its geometric shape, stability, and reactivity. It is the number of ligand atoms bound directly to the central metal ion. The coordination number is a unifying concept in forming an understanding of how these compounds behave in various chemical environments. For instance, coordination numbers such as four, six, and eight correspond to specific geometric figures, such as tetrahedral, octahedral, and cubic structures, respectively. These geometries not only define the physical properties of the compounds but also their mode of action in chemical reactions. The coordination number ultimately determines electron density distribution and total stabilization of the metal ion, which are both essential parameters defining reactivity and catalytic potential of coordination compounds towards various chemical reactions. Coordination number becomes even more significant in biological and medicinal chemistry, particularly in the design of pharmaceuticals. For example, the four-coordination number platinum-based drugs like cisplatin are mainly implicated in cancer treatment through covalent DNA binding, preventing cancer cell replication of DNA. There has been a recent drift towards the synthesis of nonclassical platinum complexes with various coordination numbers to enhance activity and minimize side effects. It also allows for the exploration of new delivery routes, such as the incorporation of platinum complexes within nanoparticles to be optimized for targeted treatment. All these advances underscore the importance of coordination number knowledge in developing new drug development and therapeutic strategies (Johnstone et al., 2016). Moreover, coordination number is the key in metal-organic framework (MOF) design and use, which is increasingly appreciated for its amenability in various fields ranging from gas storage to catalysis and sensing. Stability of the frameworks, porosity, and functionality in MOFs are all dependent on the coordination number. By varying coordination number through the selection of metal cations or organic ligands, researchers can engineer such materials with specified properties for specific uses, such as chemosensing luminescent sensors. Recent findings highlight the potential of sensing and detection of luminescent MOFs due to their unique photoluminescence properties, which are developed by their coordination environment. These new applications demonstrate the real interaction of coordination chemistry and modern technological advancements (Hu et al., 2014).

8.1.3. Examples of Common Coordination Compounds

In the learning of coordination compounds, several important examples depict the diversity of such compounds and their reactions. Heme is one such important coordination complex that is a necessary component found in hemoglobin. Heme, characterized by its iron coordination complex, coordinates oxygen molecules to facilitate their transport via blood. This molecule demonstrates the coordination compounds' very important biological role in that not only does it coordinate with other ligands but also participates in redox reactions necessary for cellular respiration. The reactivity of heme illustrates the major role played by coordination chemistry in biological phenomena, and a complete knowledge of such interactions is required when exploring potential medicinal applications. A further convincing illustration is the coordination compound ethylenediaminetetraacetic acid (EDTA). EDTA is widely utilized in industry and medicine due to its ability to form strong complexes with metal ions. In metal ion sequestration, the chelation of polyvalent metal ions such as lead and mercury by EDTAs is an expression of its applicability in environmental chemistry. With the coordination of the metal ions by EDTA into stable coordination compounds, stability guarantees their removal from mediums, thus inhibiting toxic effects. The action of EDTA not only highlights the structural intricacies of coordination compounds but also highlights the utility of such chemistry for problem-solving in reality, from pollution control to advanced medicinal interventions. Furthermore, metal-organic frameworks (MOFs) highlight an interesting field of investigation for coordination compounds. MOFs are constructed by reticular synthesis, wherein metals are coordinated with organic ligands, resulting in extremely porous materials. The unique properties of MOFs, like the chemical content that can be tuned and stability, enable applications like gas storage and separation, catalysis, and drug delivery. It is possible to tailor their structure for specific intentions, which demonstrates the versatility of coordination compounds for technological applications. As the search for MOFs continues, the potential for them extends to areas like energy storage, much like what is seen in sodium-ion battery technology, which signals a lucrative area for new uses (Furukawa et al., 2013; Slater et al., 2012).

8.2. LIGAND TYPES AND THEIR INFLUENCE

The impact of the types of ligands on the reactivity of coordination compounds is a crucial topic to be familiar with in understanding their reactions. Ligands can be classified based on their donor properties, charge, and steric bulk. Bidentate ligands such as ethylenediamine, for instance, form stable chelate complexes due to the fact that they are capable of coordination through more than one site, which significantly increases the stability of the resulting coordination compound. This increased stability permits a variety of reactions, including substitution and redox reactions, through permitting reagents a more convenient pathway. Besides, the electronic character of ligands, e.g., their ability to donate or withdraw electron density, has immediate consequences for catalytic efficiency and selectivity in processes like homogeneous catalysis, which requires ascertaining the connection between ligand types and reaction outcomes (Cheng et al., 2021). Besides, the geometrical arrangement of ligands around a metal center can dictate the rate and mechanism of reaction with coordination compounds. Octahedral complexes, for instance, commonly formed from six ligands, can have varying reactivity due to variation in axial and equatorial ligand orientation compared to tetrahedral complexes. Orientation dictates sterics and electronics that determine transition state formation when reactions are occurring. Sterically encumbered ligands impose severe constraints, and bulkiness may preclude access to the metal center, influencing pathways of reaction. Choice of ligand therefore not only alters electronic interaction but also steric accessability, thereby leading to differing reaction features. An understanding of such geometric features is vital in designing efficient catalytic systems and in optimizing reaction conditions (Liang et al.,

2020). The influence of ligands also extends into the thermodynamic and kinetic domains of coordination compound reactions. Ligands that stabilize a given oxidation state of metal can lower activation barriers significantly, allowing reactions that otherwise are unfavorable to occur. This stabilization widens the chemical landscape for electrochemical and photochemical reactions to occur, enabling a variety of applications in fields such as organic synthesis and materials chemistry. The design of novel ligands capable of supporting multiple coordination modes or having distinctive electronic properties holds exciting prospects for the synthesis of next-generation catalysts. This technology has potential applications in the production of more efficient and selective catalysts, highlighting the vital role ligand chemistry plays in controlling the reactivity profiles of coordination compounds (Cheng et al., 2021) (Figure 8.3).

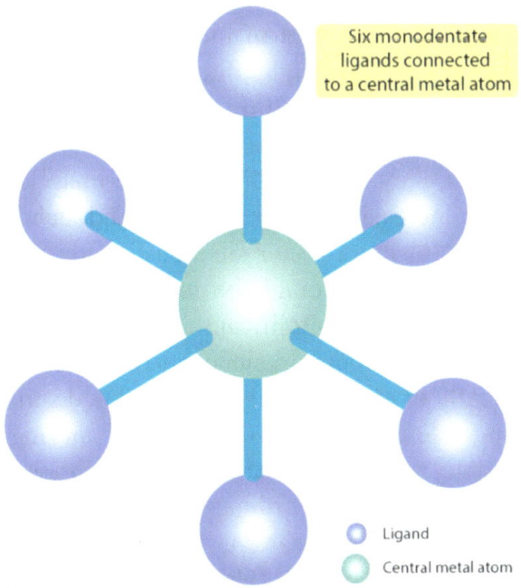

Figure 8.3. *The influence of ligand types on the reactivity of coordination comp.*

Source: https://www.chemistrylearner.com/wp-content/uploads/2023/07/Ligand.jpg.

8.2.1. Classification of Ligands (Monodentate, Bidentate, etc.)

Classification of ligands into classes in coordination chemistry is necessary in order to understand various reactions involving coordination compounds. Ligands can be classified based on the number of donor atoms they possess and the type of bonding. Monodentate ligands have one donor atom that coordinates with the metal ion and creates a single site of attachment. Water, ammonia, and chloride ions are common examples of monodentate ligands. Their ease of coordination allows them to stabilize metal ions effectively and initiate a series of chemical reactions. A study of monodentate ligands thus becomes the foundation of conducting further in-depth investigation into more complex ligand systems that exhibit different reactivity and nature towards coordination compounds. In contrast, bidentate ligands have two donor atoms that can bind to a metal center at the same time with two coordination positions. This two-binding capability makes bidentate ligands, such as ethylenediamine and oxalate, produce stable chelate complexes. The ability of the ligands to form ring conformations with the metal ion significantly enhances the thermodynamic stability of the resulting complexes, which is very important when handling reaction mechanisms and reaction kinetics in coordination chemistry. Understanding the nature of bidentate ligands and their chelating effect on metal ions provides insight into how ligand structure influences coordination compound reactions and directs the synthesis of new ligands to achieve desired chemical characteristics. Moreover, there are some ligands that are polydentate, that is, they can coordinate to the metal center by more than one donor atom, thus increasing the complexity and stability of coordination compounds. With three or more donor sites on ligands, e.g., tripodal ligands, complex spatial arrangements around the metal become increasingly likely. Configurational freedom plays a significant role in the characteristic reactivity patterns of coordination compounds since polydentate ligands can generate more crowded sterically or electronically differentiated surroundings about the metal cation. Taking into account the influence of varying ligand types, like monodentate, bidentate, and polydentate ligands, is important to predict and control the chemical behavior of coordination compounds in a wide variety of reactions, the foundation for the greater understanding of their significance in both biological and synthetic contexts (Ho et al., 2020; Newton et al., 2017).

8.2.2. Chelation and Its Effects on Stability

The chelation process is a key concept to enhance coordination complex stability. Chelating agents, typically consisting of multidentate ligands, coordinate to metal ions effectively to form stable ring-shaped complexes called chelates. This complex structure significantly reduces the reactivity of metal ions, which is extremely useful in most chemical and biological reactions. For example, in the field of medicine, chelation is employed in treating heavy metal poisoning, where chelating compounds such as EDTA sequester poisonous metals, preventing them from exerting their harmful effects on living organisms. Therefore, stability afforded by chelation translates into safer and more effective medical interventions. The study of coordination complexes thus highlights the essential liaison between ligand design and stability, indicating how molecular design has the potential to influence chemical behavior and curative action, and paving the way for future avenues in reticular chemistry as well as material sciences as witnessed in Freund et al. (2021). Besides biomedical uses, the effect of chelation is experienced beyond biological conditions and extends into industrial processes as well. In metallurgy, chelating agents serve to enhance recovery of metal from ores to a maximum, making leaching operations more efficient and selective. Stability of the metal-chelate complex inhibits loss of metal ions into unwanted side reactions, thus maximizing the yield. Stability also allows processing conditions, making it possible to keep metals in solution when extraction and purification are done. The strategy of design of chelating agents to maintain such stability has resulted in metal recovery and recycling processes that have little environmental impact with regards to mining and metal manufacturing. The principle that governs chelation and stability is thus essential in seeking sustainable use of resources, bridging the microscopic world of coordination compounds to issues of industrial-level sustainability, which aligns with innovations under discussion by Jean-François Gaucher et al. (2016). Moreover, learning about chelation and its effect on stability reveals more general chemical concepts that are the basis of coordination chemistry. Kinetics of metal-ligand interactions and the thermodynamics of stability of chelate complexes are universal in many fields of science, for example, catalysis and material science. Chelation stabilizes certain coordination compounds that are catalysts in chemical reactions to make them more efficient and longer lasting, for example. Understanding how various ligands influence metal stability allows for the design of tailor-made catalysts that operate under specified conditions, thereby optimizing reaction efficiencies. This subtle balance between chelation and stability not only emphasizes the importance of coordination compounds but also increases the demand for out-of-the-box strategies in chemical synthesis and applications. Therefore, chelation continues to be an active field of research, spanning advanced theory and practice in addressing chemical challenges of sophistication in numerous disciplines.

8.2.3. Role of Ligands in Determining Reactivity

Ligands take center stage in regulating the reactivity of metal complexes in coordination chemistry. Ligands act as electron-pair donors and stabilize metal cations through a variety of modes of bonding interaction that ultimately regulate the thermodynamics of chemical reactions. The ligands' electronic properties, including their electron-donating and withdrawing capabilities, largely influence the metal's oxidation state as well as coordination environment. For example, electron-donating ligands can make the metals' lower oxidation states stable to a high degree, which would make them more reactive toward redox processes. This is particularly important in transition metal complexes, where the ligand nature can affect reaction courses by stabilizing particular transition states and intermediates, hence validating the crucial role ligands have in modulating the reactivity of coordination compounds (Cheng et al., 2021). Apart from that, ligand stereochemistry is also an important factor that determines the reactivity of coordination compounds.

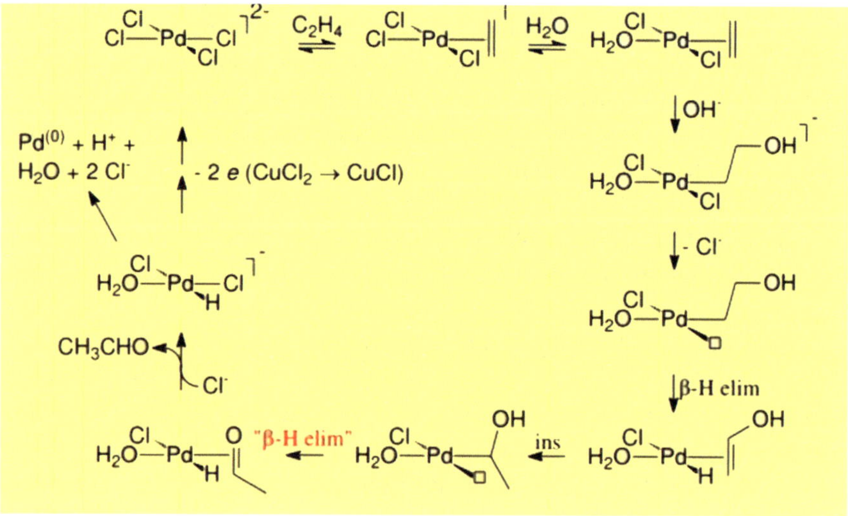

Figure 8.4. *Role of ligands in determining reactivity.*

Source: https://cdn5.slideserve.com/9392193/ligand-substitution-reactivity-of-co-ordinated-ligands-l.jpg.

Different geometrical disposition of ligands around the central metal ion, depending on which steric accessibility and hence reactivity for ligand substitution or catalysis can be influenced. Bidentate ligands, for example, would normally show five- or six-membered chelate rings that impart extra stability to metal complexes and therefore lowered tendency for dissociation during reaction. Sterically hindered ligands, however, would shield substrates from approaching, thereby reducing the reaction rate. In addition, the specific ligand coordination has the potential to facilitate some forms of reactions such as nucleophilic attacks or oxidative additions and make them more applicable in catalysis and synthetic chemistry (Liu et al., 2020). Ligand electronic and steric effects are propagated to their influence on the overall stability of metal coordination complexes under the conditions of reaction. Certain ligands can provide protective coordination environments that will stabilize sensitive metal states against oxidation or reduction, and permit desired reactions to go on without metal center decomposition. Furthermore, a ligands ability to participate in secondary bonding interactions can provide more than one reaction pathway so that more than one reaction path is open, which makes the richness and complexity of coordination chemistry more. This multi-faced interaction underscores the importance of selecting appropriate ligands for specific reactions, as they are the deciding factors for influencing reactivity as well as optimizing yield and selectivity in most coordination chemistry processes (Cheng et al., 2021) (Figure 8.4).

8.3. MECHANISMS OF REACTIONS

Coordination compound reaction mechanisms are central to their reactivity as well as roles in various chemical reactions. Coordination compounds, which are characterized by coordination of metal ions and organic or inorganic ligands, possess unique characteristics based on their geometries. The geometry of the metal center is particularly critical in determining reaction pathways toward differentiated reaction rates and product outcomes. For instance, scientists have identified coordination bond templates as playing a central role in facilitating reactions such as olefin [2+2] photocycloaddition. By investigating olefin pair disposition in coordination polymers, academics explain the photoreactivity of these compounds and how structural modifications lead to different cyclobutane products, thereby illustrating the complexity of reaction mechanisms in coordination chemistry (Zhang et al., 2024). The probing of electron transfer reaction mechanisms by metalloproteins serves to further highlight the significance of oxidation states of metal ions. Metals serve as crucial cofactors in biological machinery that alter their oxidation state to be effective catalysts. Advanced techniques, including cryogenic electron microscopy and X-ray crystallography, expose the protein's structure dynamics and the matching reaction mechanisms. However, experimental outcome interpretation can be complicated by conditions such as radiation damage and crystallization problems. Recent advances in machine learning models have even addressed these issues by providing accurate predictions of metal cluster oxidation states in metalloproteins, which has also boosted our understanding of their reactivity patterns and catalytic potential (Kapuścińska et al., 2023). The marriage of computational techniques with traditional experimental technique is gaining more popularity in the study of reaction mechanisms in coordination compounds. Machine learning algorithms not only offer a tool for the analysis of already proven data but also for predicting new outcomes based on well-proven principles of chemistry. This intersection of disciplines sheds light on complex mechanisms, such as regulating the reactivity of coordination compounds and metalloproteins, in which collections of variables compete against one another to determine reaction paths. Such interdisciplinary research uncovers the complex nature of chemical reactivity and opens new avenues for inquiry, eventually leading to new applications in materials science and biochemical catalysis. As researchers go further in refining these techniques, the possibility of applying such information in numerous fields increases more favorably, highlighting the importance of understanding reaction mechanisms in coordination chemistry.

8.3.1. Substitution Reactions in Coordination Compounds

Substitution reactions in coordination compounds are simple reactions that exert a strong influence on the reactivity and character of such complexes. These replacement reactions typically involve the displacement of one ligand from a coordination sphere by another and may be broadly categorized into two mechanisms: associative and dissociative mechanisms. The substitution occurs via a transition state in which the entering ligand makes a temporary bond to the metal center in associative mechanisms, while in the dissociative mechanism a leaving ligand leaves prior to the attachment of the entering one. The electronic character of the metal ion, the electronic nature of the ligand, and steric factors are inherent features in determining the reaction mechanism. Moreover, these substitutions can bring about variations in the physical and chemical characteristics of resultant complexes, reflecting the importance of investigating such mechanisms in coordination chemistry. The susceptibility of coordination compounds to undergo substitution reactions is generally governed by the electronic nature of the ligands involved. For example, ligands which are found to possess high field strengths, such as π-acceptors or those that are composed of strong electronegative atoms, will stabilize the metal-ligand bond and consequently affect the rate of substitution. Weaker field ligands, however, are able to facilitate faster substitution reactions via easier dissociation of the ligand in place. Studies have shown that substituent change on ligands may have a significant influence on both the strength of bonding and overall reactivity in the course of substitution reactions. As an example, urea–fluoride-based complexes have exhibited substituent-dependent tunable reactivity, illustrating that the architecture design of ligands can optimize desirable effects of substitution in catalysis, material science, and elsewhere (Pfeifer et al., 2016). In applications involving synthesis, substitution reactions are crucial for the development of new coordination compounds with desirable characteristics. Coordination cluster chemistry, represented by cluster assembly from N-substituted diethanolamine ligands with transition metals and lanthanides, demonstrates the versatility of substitution reactions to achieve complex architectures with preselected functionalities. X-ray crystal structure analysis reveals that varying ionic size and pairs of ligands leads to differences in geometric organization, thereby enhancing chemical versatility of the compounds. The resulting coordination clusters can exhibit new electronic and magnetic properties and are thus useful for application in the fields of magneto-optics and catalysis (Rinck et al., 2015). With scientists still exploring various substitution pathways, complex coordination compounds are being synthesized more and more comprehensively, with the possible lead to major advances in both fundamental chemistry and applied science.

8.3.2. Redox Reactions Involving Coordination Complexes

Redox reactions of coordination complexes are an important part of the versatility and functionality of metal complexes in a broad variety of chemical situations. Redox reactions are usually based on the ability of metal centers for electron transfer processes, which can potentially greatly alter the electronic and structural characteristics of the complex systems involved. In coordination compounds, the oxidation state of metal ions can vary, leading to a range of coordination geometries and ligand bonding interactions. The dynamic dependence of oxidation and reduction reactions allows for the production of materials with desired properties, and redox-active coordination compounds are invaluable for catalysis and material science. Through the modulation of redox states, researchers can control reaction pathways and improve the efficiency of various processes, including energy storage and conversion systems. Redox reaction research in recent times has focused on the significance of structural differences in coordination complexes. For instance, synthesis of nanoparticles composed of zerovalent metals such as zirconium, hafnium, and scandium has provided new synthetic methodologies based on redox chemistry. Of interest here is also the preparation of new crown-ether coordination compounds from these nanoparticles, which was achieved through taking advantage of their unique redox properties. As a result, different structural structures have been attained, including layered and chiral chain-type compounds, indicating the variety that redox reactions can cause in coordination chemistry. These findings not only extend the fundamental understanding of metal-ligand interactions but also present pathways for the synthesis of advanced materials with catalytic and molecular electronic uses, further showing the potential of redox processes in coordination chemistry (Faden et al., 2025). The biological potential of redox processes in coordination compounds cannot be overemphasized, particularly with regards to the drug applications of metal-based drugs. The metal complexes have become the core of the study of pharmacology due to the phenomenal success of cisplatin, a metal-containing drug employed in cancer treatment. The ability of these complexes to engage in redox-dependent reactions allows one to have targeted therapeutic interventions, for instance, apoptosis in cancer cells by selective metal-ligand interactions. Research today is still discovering therapeutic applications of diverse metal centers, like manganese and others, which exhibit potential redox characteristics. These investigations serve to establish novel therapies for cancer and other diseases based on the specificity of reactivity characteristic of redox-active coordination complexes. Consequently, the exploration of redox processes in coordination compounds not only enhances our knowledge of basic chemistry but also creates new possibilities for innovative medicinal uses (Kozie et al., 2024) (Figure 8.5).

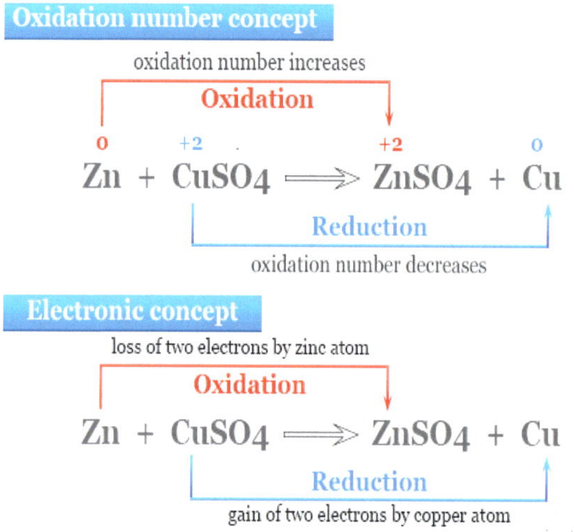

Figure 8.5. *Redox reactions involving coordination complexes.*

Source: https://www.priyamstudycentre.com/wp-content/uploads/2019/01/Redox-reactions-example.png.

8.3.3. Ligand Exchange Mechanisms

Ligand exchange reactions are the focus of the physical properties of coordination compounds, such as stability, reactivity, and utility in function. Reactions include replacement of one ligand within a complex by another and can have a significant impact on the physical properties of the compounds. Ligand exchange kinetics depend on a myriad of factors, including the type of ligands, steric reasons, and electronic features of the metal center. For instance, chelating and bidentate ligands are generally more binding-affinity-rich since they can afford to form more than one bond with the metal center, which consequently enhances the stability of the resultant complex. The coordination sphere of the metal may also be responsible for controlling the activation energy that goes into the reaction of ligand exchange, leading to varying reaction rates among various coordination complexes. The study of ligand exchange reactions is critically important in many fields, including medicinal chemistry and materials science. A prime example is reflected in studies involving paddlewheel diruthenium complexes, where ligand exchange is responsible for their capacity for therapeutic activity. Specifically, the complexes could prevent amyloid fibrillation in Alzheimer's disease. By the employment of different ligands, such as the bulky DPhF groups, the complexes revealed differing extents of interaction with the Aβ1–42 peptide. The ability to modulate ligand exchange by structural variation not only provides greater understanding of these intricate processes but also introduces the possibility of using such interactions to design new metallodrugs for degenerative disease treatment, thus verifying the close connection between ligand exchange and biological activity (Manna et al., 2024). Further, the recent advancements in inorganic-organic hybrid material synthesis place emphasis on the role of ligand exchange mechanisms in enhancing ionic conductivity for solid-state electrolytes. Investigations have demonstrated that coordination compounds can be good Li^+ ion conductors by design. Copper maleate hydrate nanoflakes synthesized, for example, exhibited 1D channels possessing *in situ* Cu^2+/Cu^+ nodes and functional groups facilitating rapid ionic transport through coupled pathways. Such an approach is an exemplary reflection of how specially engineered ligand environments can optimize ion hopping and mobility to offer optimal performance in devices such as solid-state batteries. The interplay between material functionality and ligand exchange offers promising directions of research aimed at creating electrochemical devices, illustrating the broader importance of unraveling ligand exchange processes in coordination chemistry (Zhan et al., 2024).

8.4. FACTORS AFFECTING REACTION RATES

In coordination chemistry, reaction rates can be regulated by a variety of significant factors, including concentration, temperature, and the use of catalysts. Concentration is a significant factor because higher concentrations of the reactants are found to lead to more collision of reactant molecules, thereby augmenting the rates of the reactions. In coordination compounds, the ligand concentration can have a considerable impact on the formation and stability of complexes, and this can be observed in experiments involving ligands such as pyrazolylpyrimidines. As an example, in studies of LH and LMe compounds, variations in the ligand concentration were reported to affect not only the rate of complex formation but also the resulting photophysical properties of the final coordination compounds, with small concentration variations leading to observable differences in luminescence behavior (Skvortsova et al., 2025). Temperature is another important factor in controlling reaction rates in coordination chemistry. In general, an increase in temperature increases the kinetic energy of the molecules, thereby guaranteeing higher frequency and effectiveness of collisions between reactants. The heat energy can supply sufficient energy to overcome the activation energy barrier for reactions to proceed, particularly in thermally-sensitive reactions that are found in coordination complexes. Changes in temperature can also influence the stabilization of coordinated ligands and the resulting complex geometric structure, hence affecting the yield and reaction rate. This phenomenon was illustrated in the copper(I) complex research with changes in temperature coming with changes in Cu□Cu distances and thus influencing the luminescent nature of the complexes and their quantum yield for photoluminescence (Skvortsova et al., 2025). The role of catalysts in regulating the speed of the reaction cannot be overlooked, particularly on the catalysis of coordination compound reaction efficiency. Catalysts provide an alternate reaction route which is lower in activation energy, and hence the reaction is faster and more selective. In zinc coordination complexes used in energy storage devices, for instance, aqueous zinc (Zn) metal batteries, addition of designed surface modifications can have a significant influence on reaction kinetics, particularly during the designing of dendrite-free anodes. By preventing unwanted side reactions and optimizing Zn^{2+} deposition, the changes lead to longer battery cycle life and stability, which demonstrates how effectively judicious chemical interventions can be employed to manage reaction rates. In conclusion, the relationship between concentration, temperature, and catalysts in coordination compounds demonstrates the subtlety in working with and optimizing reaction kinetics.

8.4.1. Concentration of Reactants

The concentration of reactants plays a crucial role in organizing compound reactions based on kinetics and thermodynamics. With higher concentration, there is greater likelihood of successful collision of molecules as well, which has a tendency to enhance the rate of product formation significantly. The impact can be analyzed quantitatively by applying rate equations that consider reaction rate in terms of concentration. For instance, in ruthenium(II) tris-heteroleptic compound preparation, the concentration of the ligand affects reaction rates directly, as in the sequential synthesis protocol where varying concentrations yielded varying rates and product distribution. How various concentrations affect the reaction mechanism is core business for chemists who want to optimize synthesis routes and tailor product yields according to their requirements, hence reiterating concentration as a critical factor in coordination chemistry. In determining the concentration of the reactants, it is important to respect the manner in which the presence of multiple substituents on ligands makes the topography of the reaction even more complicated. The diversification of ligands brings with it different electronic and steric properties to the coordination sphere, controlling the kinetics of formation and stability of the intermediates through the reaction. For example, control over the partitioning of linkage isomers is exerted by electron-donating or electron-withdrawing groups on the ligands as seen in the complexation of cis-[Ru(R-phen)(dcbH$_2$)(NCS)$_2$]. Such precise control of reaction conditions based on substrate concentration will allow coordination compounds of interest to be prepared with greater control. Consequently, modifying reactant concentrations and ligand characteristics provides a strategic avenue for chemists interested in probing the diversity and character of coordination chemistry. Furthermore, mechanisms governing how concentration of reactants influences these reactions can often be uncovered through studies of catalysis. Cerium dioxide (CeO_2), for example, has emerged as a key component in various catalytic systems owing partly to its efficacy under different reactant concentrations. The catalytic behavior of CeO2 in applications as varied as auto converters and photocatalysis shows the function of reactant concentration in encouraging chemical reactions to advance. The maximization of concentration in such situations not only accelerates reaction rates but also enhances the selectivity of products formed. Hence, not only do reactant concentrations dictate the short-term outcomes of coordination compound reactions, but also the scale of greater applications in catalysis and material science, and are a fundamental area of study in chemical science (Figure 8.6).

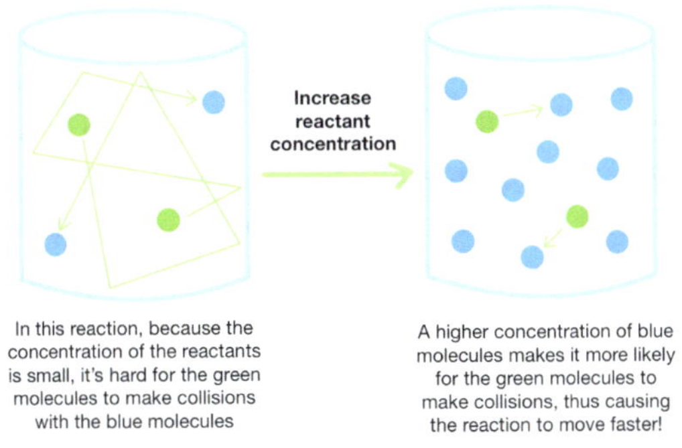

Figure 8.6. *Concentration of reactants.*

Source: https://d20khd7ddkh5ls.cloudfront.net/concentration_affects_reaction_rate.jpeg.

8.4.2. Temperature and Its Impact on Reaction Kinetics

The role played by temperature to impact reaction kinetics is the key to understanding coordination compound behavior. When the temperature goes up, so does the energy of the molecules, meaning there are more energetic collisions between reactants. This greater activity of molecules leads to greater reaction rates, something that can exert a significant influence on stability and reactivity of coordination complexes. For energetic coordination compounds (ECCs) synthesized with transition metal ions, the ignition and explosion limits are extremely sensitive to the changes in temperature. For example, establishing ignition/explosion temperatures through testing is critical to obtain secure handling and application, especially in detonators that utilize such compounds as alternatives to traditional explosives.

This kind of study emphasizes the importance of aggressively learning how temperature is influencing the reaction environment as well as the potential effects of coordination compound reactions, discussed in ECC synthesis and testing (Pawlus et al., 2024). Similarly, temperature also plays a critical role in determining the efficiency of coordination compound and related ligand reaction. In reactions such as in the production of bioethanol, optimizing temperature is crucial to attaining product yields. Literature indicates that changing reaction parameters such as temperature can have a critical effect on kinetics of reaction pathway and influencing factors such as final concentration of the product and reaction rates. The temperature sensitivity of productivity factors emphasizes controlling it within a specific optimal range. For instance, maintaining a specific temperature can induce the functioning of anaerobic fermentation processes in a direct correlation with the thermal conditions. Thus, the interaction between temperature and reaction rates is used to develop more efficient chemical processes based on coordination compounds, a best example of a critical understanding of reaction mechanisms in various applications (Aslan et al., 2024). In addition, interaction between temperature, reaction kinetics, and stability of coordination compounds is of interest for researchers to develop more efficient industrial and scientific applications. Temperature not only regulates the reaction rates but also states of coordination complexes. For instance, increased temperature can lead to higher thermal agitation, which can destabilize certain coordination bonds and thus their overall reaction dynamics and thermodynamic stability of the resulting compounds. This is particularly relevant when looking for new compounds for specific applications, where their reactivity can vary significantly with temperature fluctuations. Understanding such nuances allows chemists to condition reactions in an efficient manner to achieve desired effects, either in the synthesis of new coordination compounds or optimizing previous reactions to be used industrially. With advancements in material science, the role of temperatures in the kinetics of reactions will continue to be a vital area of study (Pawlus et al., 2024).

8.4.3. Solvent Effects on Coordination Reactions

The solvent environment plays a dominant role in controlling the kinetics and thermodynamics of coordination reactions, i.e., coordination reaction between metal ions and ligands of different nature. Solvents can influence rates of reactions, selectivity, and the nature of coordination complexes formed. Ionic intermediates, for instance, are stabilized by polar solvents and their reaction rates are accelerated with ionic species. Conversely, nonpolar solvents may selectively stabilize nonpolar species and lower ionic reactant solubilities, potentially resulting in decreased reaction efficiencies. Solvent influences can affect the electronic nature of the metal ion, impacting metal-ion d-orbital energy levels. These changes can impact equilibrium between possible coordination states, highlighting the essential importance of solvent choice in designing and optimizing coordination reactions to achieve specified goals in synthesis processes. In coordination chemistry, the molecules of the solvent are not a mere passive presence; they play an active role in influencing the coordination environment, particularly in solvent-solute interactions. Such interactions can create a dynamic environment where the coordination sphere of the ligand is altered, resulting in changes in the reactivity and stability of the obtained complexes. For example, the reaction dynamics of coordination reactions can be controlled to a great extent by dipole and hydrogen bonding with solvent molecules. This can lead to non-observable transient phases under different conditions of the solvent. Moreover, polarity and the viscosities of solvents can affect the solvent-solute interactions governing the thermodynamics and kinetics of ligand substitution reactions. Thus, an understanding of the way various solvents influence such interactions is needed to predict the outcome of coordination reactions both in theory and in experiment. The exploration of solvent influences on coordination reactions involves more than the study of straightforward solvation; it yields information regarding complex reaction mechanisms and can be utilized to manipulate the creation of new catalysts and processes. Advances in spectroscopic techniques, such as ultrafast laser spectroscopy, allow researchers to probe the rapid evolution of solute-solvent interactions during the course of reaction events and to disentangle the fundamental processes controlling such reactions. For instance, specific solvent environments can facilitate single reaction manifolds through solvation dynamics, dramatically modifying the rate constants and equilibrium values of coordination reactions. This can particularly be achieved in photochemical processes, where solvent influences can lead to altered excitations and deactivation processes of coordination compounds. A close understanding of solvent effects interaction with coordination chemistry thus aids in the development of more efficient coordination compounds and catalysts with special applicability, indicating the importance of solvent selection in chemical studies and industry (Liberto et al., 2023; Venkatraman et al., 2021) (Figure 8.7).

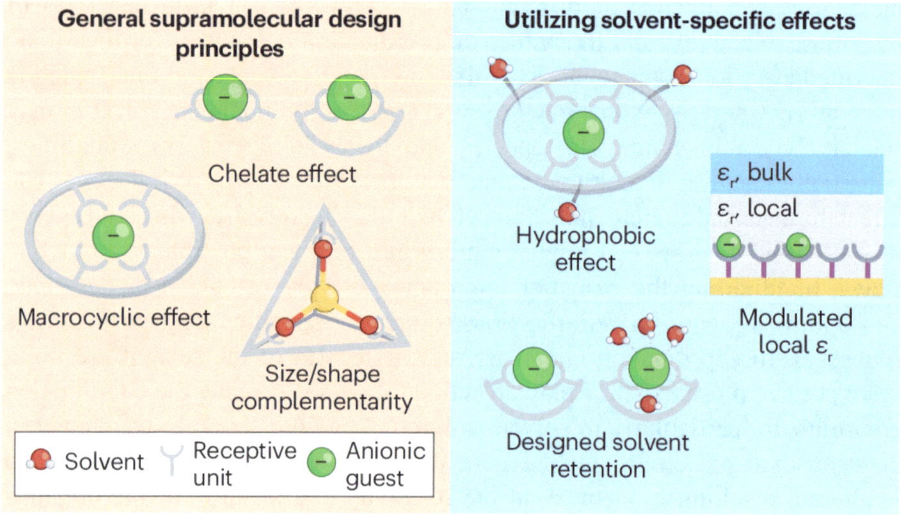

Figure 8.7. *The solvent environment plays a pivotal role in shaping the kinetics and thermodynamics of coordination reactions.*

Source: https://media.springernature.com/full/springer-static/image/art%3A10.1038%2Fs41570-024-00584-4/MediaObjects/41570_2024_584_Figa_HTML.png.

8.5. APPLICATIONS OF COORDINATION COMPOUND REACTIONS

Coordination compound reaction applications far exceed simple chemistry, extending to other areas of science like medicinal chemistry, material science, and catalysis. One of them includes synthesis of bioactive N-heterocyclic compounds, imidazo/benzimidazo[1,2-c]quinazolines, etc. These compounds have risen to prominence as drugs discovery agents due to their profound pharmacological activity, which can be finely tuned by adjusting the positioning of the nitrogen atom and functional groups around their molecular architecture. These heterocycles are also vital in building coordination complexes and polymers, which are crucial towards enhancing electrical conductivity and sensory functions. Thus, the intricate nature of coordination compound reactions is not only the basis for the synthesis of novel drugs therapeutically but also provides the impetus for environmentally friendly material synthesis strategies in consonance with global objectives for greener chemistry methods (Bairy et al., 2024). In catalysis, coordination compounds are the basis for facilitating a vast variety of chemical reactions, particularly in organic substrate conversion. For example, rhodium(III) derivative complexation reactions, such as {Rh[PyCH=C(O)Rn]3} complexes, display excellent catalytic activity in the hydrogenation of ketones. The intelligent selection and adjustment of support ligands allow higher yields of the product and effectiveness in the reaction. Advanced characterization techniques like FT-IR and NMR are routinely applied to comprehend the structures and mechanisms of coordination complexes. The turnover numbers (TON) and turnover frequencies (TOF) of high values observed in catalytic reactions are a testament to the extensive role played by coordination compounds in organic synthesis, as evident from their proven worth in academia and industry (Moniriyan et al., 2024). Besides medicinal and catalytic applications, coordination compound reaction is also vital in environmental applications, particularly in sensing and remediation technology. Metal ion and organic ligand-based coordination polymers can have unique properties enabling the identification of hazardous substances and the elimination of pollutants from various matrices. The materials have selective binding and are stable, enabling them to act as targeted environmental monitoring agents. Moreover, the presence of functional groups in coordination compounds makes them more effective in sensing ions and the trapping of target pollutants, therefore offering new solutions to key environmental issues. The capability of coordination compounds to address both synthesis and utilitarian problems attests to their significance in advancing sustainable approaches in most scientific disciplines (Bairy et al., 2024).

8.5.1. Role in Catalysis

In the context of coordination compounds, catalysts are at the very center of accelerating reaction rates and selectivity. Catalysis, which is fundamentally an acceleration of chemical reactions by adding a substance that does not get consumed following the reaction, facilitates a range of chemical conversions required in industrial manufacturing and in the laboratory as well. Coordination compounds can serve as extremely effective catalysts since they can stabilize transition states and reveal special active sites to reactants. For instance, transition metal complexes enable electron transfer processes, which are crucial for many reactions, including organic synthesis. Recent work has highlighted the way in which homogeneous catalysis—when the catalyst is in the same phase as the substrates—can optimize the conditions of reactions by controlling various selectivity factors such as chemo-, regio-, and enantioselectivity via delicately adjusted metal-ligand interactions and coordination geometries, thereby its sophistication and worth for catalysis (Cheng et al., 2021). Moreover, the hybridization of electrochemical approaches has also expanded the realm of coordination compound uses in catalysis. The microchemical reactions employ redox properties of metal sites in coordination complexes towards dynamic control over the reaction path. For example, transition metals like cobalt and nickel have been effectively employed in electrochemical reactions, building functional organic compounds via catalyzed hydrogen evolution and functionalization reactions. The kinectics of electron transfer and structural stability of catalysts are highly significant as they regulate the efficacy and effectiveness of these reactions. Notably, advances in electrochemical homogeneous catalysis have enabled one to initiate new strategies of synthesis that require less toxic reagents and milder conditions for the reaction, thereby adhering to the concepts of green chemistry. This advancement points to the significance of coordination compounds in meeting the current synthetic challenges by way of innovative catalytic strategies (Valekar et al., 2020). The practical implications of these kinds of catalytic processes are profound, particularly in constructing environmentally friendly syntheses that diminish waste and energy consumption. By modifying the coordination sphere of the metal-ligand, reaction rates can be conveniently improved, whereas the formation of unwanted byproducts can be minimized. The coordinative dynamic chemistry of these compounds, owing to the structural and electronic versatility, opens up new realms for the design of next-generation catalysts with unique transformational attributes. Moreover, the unparalleled flexibility of coordination compounds renders them adaptable to applications in many industries, ranging from pharmacy to materials science, where catalytic reactions are critical for innovation. With advancing chemistry research, a solid understanding of coordination chemistry will remain crucial to the design of catalytic systems that drive efficient, economically sound, and sustainable chemical transformations (Cheng et al., 2021).

8.5.2. Use in Medicinal Chemistry

Medicinal chemistry has increasingly recognized the significance of coordination compounds, particularly in drug development and therapeutic applications. The bimodal functional nature of the compounds provides the ability to incorporate varying metal ions, which can contribute to their efficacy and bioactivity. Coordination complexes, such as transition metal-containing ones, display unique biological properties that can be utilized in drug development. For instance, the interactions of metal ions with biologically active molecules can regulate numerous physiological functions and influence the pharmacodynamics of lead compounds. This metal-organic ligand interaction emphasizes the need for more research on coordination reactions that lead to the creation of new compounds with greater therapeutic activity. Nitroso compounds have proven to be a core feature of medicinal chemistry as they offer a rich source of chemical reactivity leading to the formation of various biologically active entities. These compounds are extremely useful due to their ability to both function as nucleophiles and electrophiles in organic reactions, reflecting their application in synthetic processes.

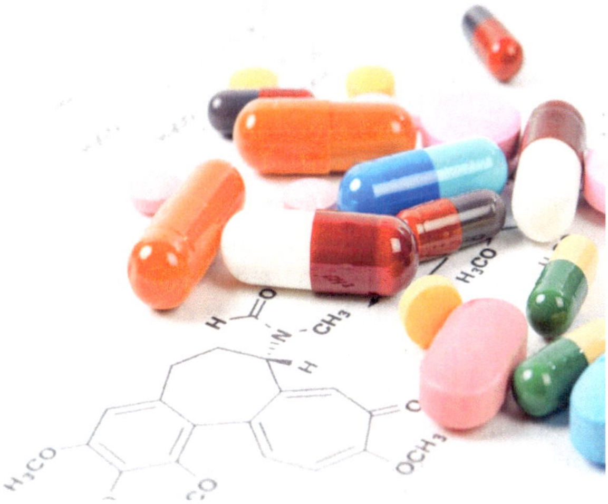

Figure 8.8. *Medicinal chemistry has increasingly recognized the significance of coordination comp/*

Source: https://chemistry.csuci.edu/career-pathways/images/medicinal.jpg.

Nitrosoarenes, a very important subclass, provide an enormous contribution to the development of drugs as a result of being involved in coordination chemistry. They have been discovered to act as drug building blocks for drugs of diverse nature, especially in terms of antimicrobial and anticancer activities. Reaction path exploration with nitroso compounds not only enhances the tools of medicinal chemists but also suggests promising therapeutic avenues for future research, emphasizing the need to pursue research in this area (Lin et al., 2024). Schiff bases and metal complexes are yet another important class of coordination compounds in medicinal chemistry. They are synthesized via condensation reaction between amino and carbonyl compounds to give products having some marvelous biological activities. The coordination ability of Schiff bases with various metal ions makes them exhibit catalytic activity as well as improve therapeutic efficacy. Recent studies recognize them as potential drugs for infections, cancer, and other conditions due to their characteristic interaction patterns. An exploration of structure-activity relationship of Schiff base metal complexes can ultimately lead to the development of more effective chemotherapeutic drugs. While researchers and researchers engage their zeal in synthesizing such new drugs, their adoption in the field of coordination chemistry appears to be very promising in addressing unserved medical needs (Bhowmik, 2024) (Figure 8.8).

8.5.3. Applications in Materials Science.

The catholicity of coordination complexes has led to significant advancements in materials science, particularly in catalysis, sensor fabrication, and energy storage. Transition metal complexes are efficient catalysts in many reactions owing to their unique electronic and structural properties, fostering green synthetic methods and enhancing reaction efficiencies. Current studies have portrayed the potential of such compounds in electrochemical applications, with non-noble-metal and noble-metal catalysts used therein. This application is specific to facilitate key conversions with lower use of destructive reagents, in harmony with green chemistry and sustainable development protocols (Cheng et al., 2021). Besides, optimization of catalyst structures allows for higher selectivity and activity for various chemical reactions to allow further exploration of more efficient and sustainable material synthesis processes. In material science, hydrogen isotope exchange (HIE) reactions have been found to be important in the synthesis of deuterium- and tritium-labeled compounds. These compounds are of further utility in biological studies, radioligand binding assays, and optimization of the pharmacokinetics of established drugs. HIE reactions offer latitude for late-stage introduction of isotopes to streamline synthetic procedures that previously utilized cumbersome multistep methodology. The built-in selectivity of metal-catalyzed HIE reactions enhances the prospects for achieving high isotope incorporation long sought after in medicinal chemistry. Recent findings have brought to the fore the efficacy of nanocatalysts, i.e., Ru and Ir, with a high degree of functional group tolerance and regioselectivity, thereby broadening the scope of application for tagged compounds in drug development (Lepron et al., 2021). The interface of coordination chemistry with materials science embodies the unrelenting quest for synthetic progress in order to respond to changing demands from different fields. Yet another significant application is the preparation of high-performance materials that utilize the unique properties of coordination compounds, particularly in nanotechnology. Metal nanoparticles exhibit a dynamic reactivity, which results from their large surface area and the heterogeneous coordination environments at nanostructured materials. These have been used to create novel functional materials for many applications, including high-performance catalysts, sensors, and drug delivery systems. For example, recent studies highlight the application of metallic nanoclusters to achieve chemoselective and enantiospecific labeling processes, a pursuit of vital relevance to pharmaceuticals and peptides. Additionally, advances in coordination compound science have provided new tools that cover homogeneous and heterogeneous catalysis, thereby propelling the disciplines of material science and the construction of intricate chemical systems towards practical uses (Cheng et al., 2021). The importance of the developments highlights the central role played by coordination compounds in determining future technologies and materials.

8.6. EXPERIMENTAL TECHNIQUES FOR STUDYING REACTIONS

Studies of reaction with coordination compounds need the application of various experimental techniques through which chemists are aware of the mechanism of the reactions in detail. Among the most popular techniques are spectroscopic techniques such as UV-Vis and NMR spectroscopy. UV-Vis spectroscopy is particularly valuable in the observation of electronic transitions in coordination compounds so that ligand field strengths may be calculated and potential electronic arrangements predicted. NMR spectroscopy, on the other hand, provides data about local structure around a single nucleus in a coordination complex, reporting ligand coordination and metal-ligand motion. With such techniques, scientists can disentangle coordination chemistry's complexities and arrive at a clearer picture of reaction mechanisms and thermodynamic properties that are critical to advance the field. Another primary experimental technique for the study of coordination compound reactions is mass spectrometry, electrospray ionization mass spectrometry (ESI-MS). This method is especially suited to large inorganic complexes, with highly precise mass measurements that render coordination number and molecular structure determination straightforward. ESI-MS is underutilized within coordination chemistry, however, perhaps due to the perceived limitations in the characterization of labile inorganic complexes. Advances in mass spectrometry in recent years, for example, in ionization sources and high-resolution mass analyzers, are gradually beginning to fill this gap. By challenging mass spectrometry's capability, researchers can significantly enhance coordination compound study weaponry, increasing the accuracy and range of analyses of their reactivity and stability in differing conditions, according to Geue (2024). Furthermore, identification of redox reactions in coordination compounds is specially reliant on procedures for tracing alterations in electron transfer processes. For instance, cyclic voltammetry is a powerful electrochemical method for the exploration of redox behavior in metal complexes. The technique provides information on electron capacity as well as mechanistic pathways of redox reactions, which are crucial to the reactivity of coordination compounds in biological and synthetic processes. When combined with studies of thiol-disulfide metabolism, for example, with copper coordination complexes, such electrochemical methods can reveal dynamic redox fluctuations that influence biological functions Pantea et al. (2025). Such cooperation of experimental techniques like cyclic voltammetry with mass spectrometry and spectroscopic analysis provides an integrated picture of the reactivity pattern characteristic of coordination compounds in different milieus.

8.6.1. Spectroscopic Methods (UV-Visible (UV-Vis) and Nuclear Magnetic Resonance (NMR))

The investigation of coordination compounds remains a core branch of modern chemistry, particularly when using spectroscopic methods such as UV-Vis and NMR spectroscopy. UV-Vis spectroscopy holds a dominant role in coordination complex electronic transition studies. This technique enables one to analyze the structural features and the environment of metal ions because it provides information regarding their d-orbitals. Absorption spectra obtained from UV-Vis can also describe the nature of ligand interactions and confirm the formation of the coordination compound. By studying the wavelength at which there is maximum absorbance, many things can be concluded by scientists, such as the ligand field splitting and degree of coordination, which are important for the explanation of chemical species involved in different reactions in coordination chemistry. Also, NMR spectroscopy complements UV-Vis by providing molecular-level explanations about the dynamics and the environment of atoms in coordination compounds. Notably, NMR is required for the determination of complex molecules' structures and the connectivity of the atoms. NMR provides us with detailed information on hydrogen environments (1H NMR) and carbon skeletons (13C NMR) and demonstrates dynamic behavior of coordination reactions. The capability of NMR is elaborated in the context of Schiff-based compounds, where data regarding enol-imine and keto-amine conformations were elucidated by spectroscopic studies, thereby emphasizing structural nuances central to coordination chemistry (Güzel et al., 2024). The coupling of these techniques furnishes an holistic picture of static and dynamic properties of coordination compounds, testifying to their significance in synthetic and analytical chemistry. Additionally, the application of UV-Vis and NMR spectroscopy can be seen in the analysis of natural dyes as sensitizers in photovoltaic devices, for instance, Dye-Sensitized Solar Cells

(DSSC). Both these spectroscopic methods are applied in current research to establish the stability and photodegradation kinetics of natural dyes purified from the sources Medina Lawsonia Inermis and others (Sukardi et al., 2025). By tracking spectral changes in UV-Vis data, researchers can determine the efficacy of photodegradation reactions and compare for the influence of different dye structures on long-term stability. The insights that are gained through these processes not only push the boundaries of knowledge in the field of coordination reactions but also pave the way to new breakthroughs in the field of materials science and environmental chemistry, highlighting the significance of spectroscopic techniques in inter-disciplinary research (Figure 8.9).

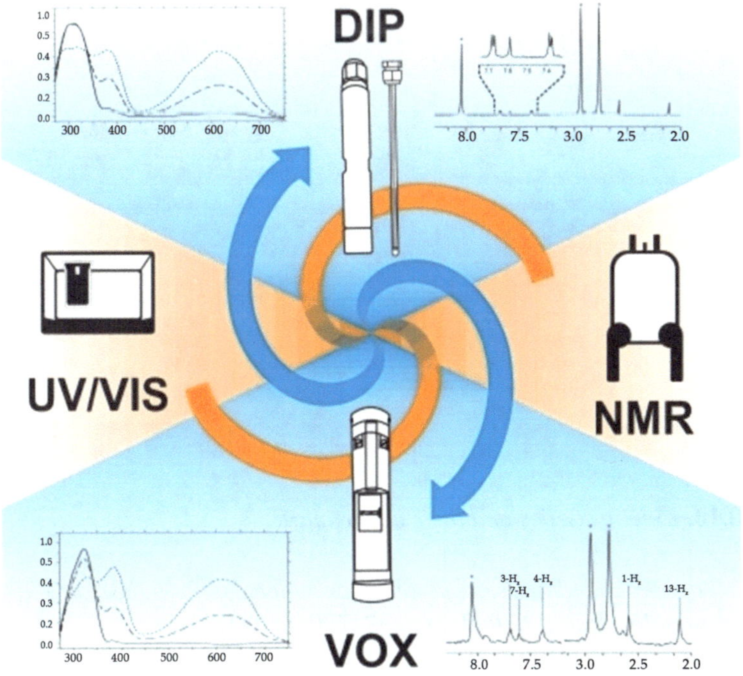

Figure 8.9. *Spectroscopic methods (UV-visible (UV-Vis) and nuclear magnetic resonance (NMR))*

Source: https://chemistry-europe.onlinelibrary.wiley.com/cms/asset/ad3387d2-78c8-498b-8f13-a7f4a153d6f3/chem202304016-toc-0001-m.jpg.

8.6.2. Kinetic Studies and Their Importance

The study of reaction kinetics is a fundamental component of understanding the dynamic process that leads to coordination compound reactions. Kinetic study provides extremely valuable data regarding the speed at which these reactions occur, enabling chemists to identify what mechanisms are involved. By analyzing the rates of reaction and the influence of numerous factors such as temperature, concentration, and catalysts, scientists are able to determine how various coordination compounds behave in specific conditions. This information comes in handy when creating effective strategies for the synthesis of novel materials and in maximizing industrial processes. Secondly, kinetic information is used to predict reaction outcomes, towards an improved understanding of coordination chemistry in terms of numerous applications, some of which are drug design and catalysis.

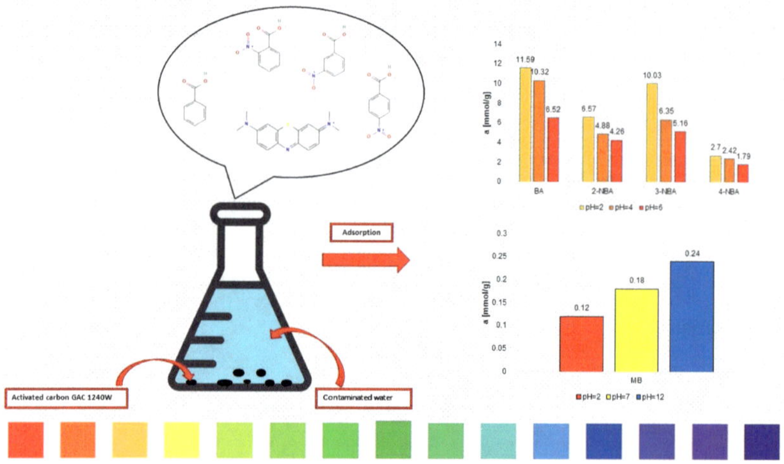

Figure 8.10. *Kinetic studies and their importance.*

Source: https://pub.mdpi-res.com/molecules/molecules-29-03032/article_deploy/html/images/molecules-29-03032-ag.png?1720515085.

Lastly, kinetic studies are crucial towards understanding electron transfer processes in coordination compounds. These processes need to be understood to apply to electrochemistry, particularly in the formation of energy storage cells like zinc-ion batteries, which are enhanced by efficient electron transfer reactions (Song et al., 2018). These investigations aid in reaction pathway elucidation and identification of the intermediate species formed from the course of the reaction. Through the use of various kinetic models, scientists can also identify the rate-determining steps, thus having an understanding of how electronic and structural changes of coordination compounds can influence their reactivity. This piece of information is essential in advancing electrochemical technology and contributes significantly to the field of green chemistry. Finally, another significant role of the kinetics studies research is in designing efficient catalysts for reactions involving coordination compounds. Catalysts can accelerate reaction rates without being consumed in the process themselves, and their operation often hinges on awareness of the kinetics of the reactions themselves. In oxygen electrocatalysis, the central theme of metal–air batteries, kinetic studies can unravel the advanced dynamics of the oxygen-reduction and evolution reactions (Yan et al., 2017). Investigation of the effect of various parameters, i.e., structure and composition of the catalyst surface, aids in optimizing the performance of the catalyst. Further understanding of how defects on a catalyst affect its activity will reveal strategies for maximizing catalytic efficiency, eventually leading to the conception of energy conversion technology and the capability of producing sustainable energy solutions (Figure 8.10).

8.6.3. Computational Chemistry Approaches

In the nascent field of computational chemistry, a number of methods are very useful in accounting for the complexity of reactions of coordination compounds. Quantum mechanical computations, especially density functional theory (DFT), allow researchers to predict the electronic structure and thus the resulting reactivity of coordination complexes successfully. Such computational insight allows the study of ligand field effects, geometric shapes, and electronic transitions, which are all important in determining the stability and nature of metal complexes. DFT calculations have been invaluable to include in the study of the coordination modes of ligands, helping immensely in the interpretation of experimental data. For instance, it has been demonstrated by studies that ligand interactions with metal sites by DFT provide a closer approximation of the steric and electronic forces that govern coordination chemistry and therefore facilitate the design of new, more effective metal-based compounds. Computational methods also include the prediction of thermodynamic stability and reactivity of metal-ligand complexes. Through theoretical calculation, researchers can analyze the stability and formation constants of various coordination compounds and establish their potential application. Using Jobs method alongside computational modeling makes it possible to identify the optimal metal-to-ligand ratios, which are vital for achieving necessary chemical properties. Recent studies indicate that computational techniques are capable of computing stability constants and determining the most stable complex geometries, which is critical for catalysis and materials science applications. For example, relative comparison of complexes of Fe(III), Zn(II), and Cu(II), as in experimental outcomes, not only validated DFT predictions but also presented the efficiency of computational chemistry in guiding experimental designs and interpretations. Additionally, the combination of computational chemistry with experimental approaches has a profound impact on our knowledge of coordination compounds. The complementarity between DFT calculations and *in vitro* experiments enables the determination of biological activities of metal complexes prepared by synthesis, contributing to an understanding of their potential as medicines. For instance, studies utilizing molecular docking simulations may provide a deeper understanding of metal complex-biological target interactions that dictate their effectiveness as antimicrobial or anticancer drugs. As in the synthesis of novel metal-ligand pairs, computational methods allow scientists to make educated hypotheses on reactivity in the biological and stability in the environmental context, which highlights the relevance of theoretical models in predicting empirical evidence. Therefore, the combination of computational and experimental approaches remains a robust interface for the decoding of the complexity involved in coordination compound reactions and further advances in fundamental and applied chemistry.

SUMMARY

The examination of the reactions of coordination compounds illustrates their complexity and multi-dimensionality in various chemical processes. More specifically, these reactions are the benchmark of the intricate relationship between coordination chemistry and its medicinal applications, as exemplified in the setting of pharmacogenomics. Findings from research that incorporate pharmacometabolomics highlight the importance of understanding how various compounds, even ones as well-defined as drugs like cyclosporine, act within biological systems. These actions highlight the importance of coordination compounds in drug metabolism and again highlight their core role in personalized medicine by tweaking approaches in relation to individuals' metabolic profiles. By combining extended analyses, researchers are able to uncover hitherto unreported metabolites that can have significant effects on therapeutic outcomes, paving the way for more studies in this field to ensure maximum drug efficacy and safety in therapeutic uses. Moreover, the intensive investigation of coordination compounds transcends their biological significance and also encompasses their environmental and occupational implications. Because sensitization to chemical substances like lanolin mirrors industrial exposure so closely, awareness of the nature of such chemicals can be the key to establishing means to prevent the danger of contact dermatitis. The observed disparity in the frequency of sensitization highlights the imperative for ongoing investigation along these lines, with the appeal to clinicians to consider these factors when evaluating dermatological responses in patients. Coordination of the comprehension of both chemical reactions and human consequences of such reactions can make our knowledge of such reactions more comprehensive and hence drive future prevention in workplaces or risk patients of occupational exposure. In conclusion, the detailed study of coordination compounds reasserts their high relevance in practical as well as theoretical chemistry. The reactivity patterns and interaction patterns disclosed by the most recent studies highlight the necessity of having thorough knowledge of these compounds in drug development and treatment of patients. As illustrated within the studies on both pharmacometabolomics and lanolin sensitization, the knowledge generated from familiarity with coordination compounds can potentially enhance safety protocols and drug therapies. These findings not only impact drug safety and health but also pave the way for drug formulation innovations, culminating in the creation of personalized medicine and improved patient care in the clinic.

The intricate coordination compound world boasts an enormous collection of reaction mechanisms, each contributing to its role in the complete function of these complexes. Among the fore lights in this discussion is the role of high-valent metal oxo species, particularly Fe(IV)-oxo complexes. The species have been shown to disrupt the typical activity-selectivity trade-off present in reactive oxygen species and thereby enhance the efficiency of catalytic processes. From studies, it has been shown that through the coordination engineering of single-atom catalysts, including Fe-Nx SACs, heterogeneous FeIV=O oxidation behaviors can be precisely regulated. This advancement not only optimizes peroxymonosulfate (PMS) activation but also enables the selective removal of pollutants by specific reactions, with the emphasis on the driving importance of coordination milieus to chemical selectivity and reactivity (Lin et al., 2024). Oligomeric and polymeric coordination complex syntheses also indicate the dynamic nature of coordination chemistry. Self-assembly procedures have emerged as a viable method of assembling complex molecular architectures, particularly via photochemical reaction. For instance, ruthenium complexes are synthesized by photosubstitution reactions using bridging ligands that have pyrazole groups. The photochemical reactions produce a number of oligomeric structures with novel structural and functional properties. Intramolecular hydrogen bonding in the complexes, which accounts for their stability and orientation, enhances the capability to form one-dimensional nanostructures. Notably, these multinuclear complexes also exhibit excellent promise for applications in materials science and nanotechnology, highlighting the versatility and practical importance of state-of-the-art coordination compound synthesis (Honda et al., 2024). Beyond this, coordination compound science encompasses interaction between their reactive nature and structural characteristics under various conditions. The ability to regulate these compounds through external agents, such as pH adjustments, but again shows their adaptability to diverse environments. The photoinduced assembly and disassembly displayed in certain coordination networks show the potential for dynamic applications, where the material properties can be switched back and forth. This feature enables responsive systems to be constructed that can be tailored for specific applications, ranging from drug delivery to the cleanup of the environment. Hence, understanding coordination compound reactions and complexation processes holds great potential for future scientific exploration and useful applications in various science disciplines (Honda et al., 2024; Lin et al., 2024).

REFERENCES

1. Aslan, C., Devianto, H., Wonoputri, V., & Harimawan, A. (2024). Novel study of reaction kinetics and mass transfer in bioreactor modelling: Prediction of bioethanol fermentation performance by *Saccharomyces cerevisiae* on continuous fixed bed biofilm plug flow reactor. *Bulletin of Chemical Reaction Engineering & Catalysis*. https://www.semanticscholar.org/paper/15c812943f8aba533e8670fe63683772e3f9537e (accessed on 13th September 2025).
2. Bairy, G., & Sinha, C. (2024). An extensive analysis of current synthetic methodologies and applications of imidazo/benzimidazo[1,2-c]quinazolines. *ChemistrySelect*. https://www.semanticscholar.org/paper/12416938298aef460717c1bf3013571a855b283c (accessed on 13th September 2025).
3. Bhowmik, K. R. N. (2024). Biomedical applications of some Schiff bases and their transition metal complexes: A review. *Asian Journal of Chemistry*. https://www.semanticscholar.org/paper/6337b8105a022189674f5cc571dac5e6a7b3db81 (accessed on 13th September 2025).
4. Biso, L., Carli, M., Scarselli, M., & Longoni, B. (2025). Overview of novel antipsychotic drugs: State of the art, new mechanisms, and clinical aspects of promising compounds. *Biomedicines, 13*. https://www.semanticscholar.org/paper/01f22eb1974736b7a261739245a88c59758aca04 (accessed on 13th September 2025).
5. Cegolon, L., & Larese Filon, F., et al. (2024). Sensitization to lanolin in North-Eastern Italy, 1997–2021: Prevalence, risk factors and the impact of occupation. *Life, 14*. https://www.semanticscholar.org/paper/090e3d2be77e94eb4bd163591eb9db4fdc1a7c2f (accessed on 13th September 2025).
6. Chatgilialoglu, C., Barata-Vallejo, S., & Gimisis, T. (2024). Radical reactions in organic synthesis: Exploring in-, on-, and with-water methods. *Molecules, 29*. https://www.semanticscholar.org/paper/2b588c0300f865df72276fe54abc0b3403eb250d (accessed on 13th September 2025).
7. Cheng, X., Lei, A., Mei, T. S., Xu, H. C., Xu, K., & Zeng, C. C. (2021). Recent applications of homogeneous catalysis in electrochemical organic synthesis. *CCS Chemistry, 4*, 1120–1152. https://doi.org/10.31635/ccschem.021.202101451 (accessed on 13th September 2025).
8. Di Liberto, G., & Pacchioni, G. (2023). Modeling single-atom catalysis. *Advanced Materials, 35*. https://doi.org/10.1002/adma.202307150 (accessed on 13th September 2025).
9. Diercks, C. S., & Yaghi, O. M. (2017). The atom, the molecule, and the covalent organic framework. *Science, 355*. https://doi.org/10.1126/science.aal1585 (accessed on 13th September 2025).

10. Faden, L. P., Bayarjargal, L., Milman, V., Winkler, B., & Feldmann, C. (2025). Crown-ether coordination compounds of zirconium, hafnium and scandium using metal nanoparticles. *Dalton Transactions*. https://www.semanticscholar.org/paper/e76cf4f519bd06b6ab60a708869863025ad3dbb3 (accessed on 13th September 2025).

11. Ferrier, M. G., Childs, B. C., Silva, C. M., Greenough, M. M., Moore, E. E., Erickson, K. A., Monreal, M., et al. (2024). Laser-induced thermal decomposition of uranium coordination compounds with non-oxidic ligands to produce nitride and carbide materials. *Inorganic Chemistry*. https://www.semanticscholar.org/paper/227a671862aa5f39a700e4c3af908e614a6b6864 (accessed on 13th September 2025).

12. Freund, R., Zaremba, O., Arnauts, G., Ameloot, R., Skorupskii, G., Dincă, M., Bavykina, A., et al. (2021). The current status of MOF and COF applications. *Angewandte Chemie International Edition, 60*, 23975–24001. https://doi.org/10.1002/anie.202106259 (accessed on 13th September 2025).

13. Furukawa, H., Cordova, K. E., O'Keeffe, M., & Yaghi, O. M. (2013). The chemistry and applications of metal-organic frameworks. *Science, 341*. https://doi.org/10.1126/science.1230444 (accessed on 13th September 2025).

14. Gaucher, J. F., Reille-Seroussi, M., Gagey-Eilstein, N., Broussy, S., Coric, P., Seijo, B., Lascombe, M. B., et al. (2016). Biophysical studies of the induced dimerization of human VEGF receptor 1 binding domain by divalent metals competing with VEGF-A. *PLoS ONE, 11*, e0167755–e0167755. https://doi.org/10.1371/journal.pone.0167755 (accessed on 13th September 2025).

15. Geue, N. (2024). Modern electrospray ionization mass spectrometry techniques for the characterization of supramolecules and coordination compounds. *Analytical Chemistry, 96*, 7332–7341. https://www.semanticscholar.org/paper/1f519816781a6cc0cec5a88fdcff7fab04081bde (accessed on 13th September 2025).

16. Güzel, E., Çoruh, U., Duteanu, N., Cıcek, C., Ağar, E., Vazquez-Lopez, E., & Yavuz, M. (2024). Experimental studies of Schiff base (E)-4-methoxy-2-(((2-phenoxyphenyl)imino)methyl) phenol compound with various spectroscopic methods. *Журнал структурной химии*. https://www.semanticscholar.org/paper/3d33584c8795111a70348188755e68310c7c553f (accessed on 13th September 2025).

17. Hematian, S., Garcia-Bosch, I., & Karlin, K. D. (2015). Synthetic heme/copper assemblies: Toward an understanding of cytochrome *c* oxidase interactions with dioxygen and nitrogen oxides. *Accounts of Chemical Research, 48*, 2462–2474. https://doi.org/10.1021/acs.accounts.5b00265 (accessed on 13th September 2025).

18. Ho, T. C. S., Chan, A. H. Y., & Ganesan, A. (2020). Thirty years of HDAC inhibitors: 2020 insight and hindsight. *Journal of Medicinal Chemistry, 63*, 12460–12484. https://doi.org/10.1021/acs.jmedchem.0c00830 (accessed on 13th September 2025).

19. Hoffmann, R., Kabanov, A. A., Golov, A. A., & Proserpio, D. M. (2016). *Homo citans* and carbon allotropes: For an ethics of citation. *Angewandte Chemie International Edition, 55*, 10962–10976. https://doi.org/10.1002/anie.201600655 (accessed on 13th September 2025).

20. Honda, T., Hirake, T., Kondo, M., Masaoka, S., Fujimori, K., Moriuchi-Kawakami, T., & Hirahara, M. (2024). Photoinduced self-assembly: An alternative strategy for the construction of coordination oligomers and polymers. *Chemistry*. https://www.semanticscholar.org/paper/9ffbe9016383bb768bf220bf96de684a540f0ca2 (accessed on 13th September 2025).

21. Hu, Z., Deibert, B. J., & Li, J. (2014). Luminescent metal–organic frameworks for chemical sensing and explosive detection. *Chemical Society Reviews, 43*, 5815–5840. https://doi.org/10.1039/c4cs00010b (accessed on 13th September 2025).

22. Inam Omar, M., Alharas, M., Feizi-Dehnayebi, M., Alharbi, S. K., Abo-Dief, H., Qasem, H. A., El-Khatib, R., et al. (2025). Design, synthesis, physico-chemical characterization, stability determination, and biomedical applications of some novel tetra-dentate imine metal chelates supported by theoretical approaches: Bridging coordination chemistry and life sciences. *Applied Organometallic Chemistry*. https://www.semanticscholar.org/paper/236e7237fefb3a995330784d1f84e84bca3c709c (accessed on 13th September 2025).

23. Iqbal, Z., Joshi, A., & De, S. R. (2020). Recent advancements on transition-metal-catalyzed, chelation-induced *ortho*-hydroxylation of arenes. *Advanced Synthesis & Catalysis, 362*, 5301–5351. https://doi.org/10.1002/adsc.202000762 (accessed on 13th September 2025).

24. Johnstone, T. C., Suntharalingam, K., & Lippard, S. J. (2016). The next generation of platinum drugs: Targeted Pt(II) agents, nanoparticle delivery, and Pt(IV) prodrugs. *Chemical Reviews, 116*, 3436–3486. https://doi.org/10.1021/acs.chemrev.5b00597 (accessed on 13th September 2025).

25. Kapuścińska, K., Dukała, Z., Doha, M., Ansari, E., Wang, J., Brudvig, G., Brooks, B., et al. (2023). Bridging the coordination chemistry of small compounds and metalloproteins using machine learning. *Journal of Chemical Information and Modeling*. https://www.semanticscholar.org/paper/0df34e819995086f7953065f941552b981cbde5b (accessed on 13th September 2025).

26. Kozieł, S., Wojtala, D., Szmitka, M., Sawka, J., & Komarnicka, U. K. (2024). Can Mn coordination compounds be good candidates for medical applications? *Frontiers in Chemical Biology*. https://www.semanticscholar.org/paper/7d18acfc5bd7876c9e2a77441259e210a2765adb (accessed on 13th September 2025).

27. La Manna, S., Panzetta, V., Di Natale, C., Cipollone, I., Monti, M., Netti, P., Terán, A., et al. (2024). Comparative analysis of the inhibitory mechanism of Aβ1-42 aggregation by diruthenium complexes. *Inorganic Chemistry*. https://www.semanticscholar.org/paper/502b2e5e2b254d805c75b00a1070b2caa68fb30e (accessed on 13th September 2025).

28. Leone, L., Sgueglia, G., La Gatta, S., Chino, M., Nastri, F., & Lombardi, A. (2023). Enzymatic and bioinspired systems for hydrogen production. *International Journal of Molecular Sciences, 24*, 8605–8605. https://doi.org/10.3390/ijms24108605 (accessed on 13th September 2025).

29. Lepron, M., Daniel-Bertrand, M., Mencia, G., Chaudret, B., Feuillastre, S., & Pieters, G. (2021). Nanocatalyzed hydrogen isotope exchange. *Accounts of Chemical Research, 54*, 1465–1480. https://doi.org/10.1021/acs.accounts.0c00721 (accessed on 13th September 2025).

30. Li, J., He, B., Zhang, Y., Cheng, Z., Yuan, L., Huang, Y., & Li, Z. (2022). *In situ* constructing coordination compounds interphase to stabilize Zn metal anode for high-performance aqueous Zn-SeS_2 batteries. *Small*. https://www.semanticscholar.org/paper/fb98d68241208081f9d275a0654f316a296dc139 (accessed on 13th September 2025).

31. Liang, X. Q., Gupta, R. K., Li, Y. W., Ma, H., Gao, L., Tung, C. H., & Sun, D. (2020). Structural diversity of copper(I) cluster-based coordination polymers with pyrazine-2-thiol ligand. *Inorganic Chemistry, 59*, 2680–2688. https://doi.org/10.1021/acs.inorgchem.9b02919 (accessed on 13th September 2025).

32. Lin, F., Tang, R., Liu, S., & Tan, Y. (2024). Recent advances in the synthetic applications of nitroso arene chemistry. *Organic & Biomolecular Chemistry*. https://www.semanticscholar.org/paper/a14869ba3010d140eeef35e88489c116ca68aec9 (accessed on 13th September 2025).

33. Lin, Y., Wang, Y., Weng, Z., Zhou, Y., Liu, S., Ou, X., Xu, X., et al. (2024). Coordination engineering of heterogeneous high-valent Fe(IV)-oxo for safe removal of pollutants via powerful Fenton-like reactions. *Nature Communications, 15*. https://www.semanticscholar.org/paper/b3ebae40c9e148cec7e3c586bd23497e30e852fa (accessed on 13th September 2025).

34. Liu, L., & Corma, A. (2020). Evolution of isolated atoms and clusters in catalysis. *Trends in Chemistry, 2*, 383–400. https://doi.org/10.1016/j.trechm.2020.02.003 (accessed on 13th September 2025).

35. Lv, Z. J., Wei, J., Zhang, W. X., Chen, P., Deng, D., Shi, Z. J., & Xi, Z. (2020). Direct transformation of dinitrogen: Synthesis of *N*-containing organic compounds via N−C bond formation. *National Science Review, 7*, 1564–1583. https://doi.org/10.1093/nsr/nwaa142 (accessed on 13th September 2025).

36. Masoomi, M. Y., Morsali, A., Dhakshinamoorthy, A., & García, H. (2019). Mixed-metal MOFs: Unique opportunities in metal–organic framework (MOF) functionality and design. *Angewandte Chemie International Edition, 58*, 15188–15205. https://doi.org/10.1002/anie.201902229 (accessed on 13th September 2025).

37. Moniriyan, F., Sabounchei, S., & Sedghi, A. (2024). Synthesis, characterization and theoretical studies of tris-cyclometalated rhodium(III) complexes: Applications to asymmetric transfer hydrogenation of ketones. *Journal of Coordination Chemistry, 77*, 1118–1131. https://www.semanticscholar.org/paper/e581f5947c8da14a45ce9df35f36097efe097f8a (accessed on 13th September 2025).

38. Montini, T., Melchionna, M., Monai, M., & Fornasiero, P. (2016). Fundamentals and catalytic applications of CeO_2-based materials. *Chemical Reviews, 116*, 5987–6041. https://doi.org/10.1021/acs.chemrev.5b00603 (accessed on 13th September 2025).

39. Müller, A. V., & Polo, A. S. (2018). Mechanistic insights into the stepwise assembly of ruthenium(II) tris-heteroleptic compounds. *Inorganic Chemistry, 57*, 13829–13839. https://doi.org/10.1021/acs.inorgchem.8b02337 (accessed on 13th September 2025).

40. Newton, C. G., Wang, S., Oliveira, C. C., & Cramer, N. (2017). Catalytic enantioselective transformations involving C–H bond cleavage by transition-metal complexes. *Chemical Reviews, 117*, 8908–8976. https://doi.org/10.1021/acs.chemrev.6b00692 (accessed on 13th September 2025).

41. Nierengarten, J. F., Nierengarten, I., Holler, M., Sournia-Saquet, A., Delavaux-Nicot, B., Leoni, E., Monti, F., et al. (2019). Dinuclear copper(I) complexes combining bis(diphenylphosphanyl)acetylene with 1,10-phenanthroline ligands. *European Journal of Inorganic Chemistry, 2019*, 2665–2673. https://doi.org/10.1002/ejic.201900335 (accessed on 13th September 2025).
42. Nijdam, F. B., Hof, M. A. J., Blokzijl, H., Bakker, S. J., Hak, E., Hopfgartner, G., Klont, F., et al. (2025). Pharmacometabolomic enables real-world drug metabolism sciences. *Metabolites, 15*. https://www.semanticscholar.org/paper/8cda24bd3f45f0eeb6f73c2e832dc891a54855c0 (accessed on 13th September 2025).
43. Pantea, V., Pavlovschi, E., Sardari, V., Protopop, S., Macari, V., Ambros, A., Curlat, S., et al. (2025). Investigating the *in vivo* effects of copper coordination compounds with thiosemicarbazones on erythrocyte redox balance. *Journal of Experimental and Molecular Biology*. https://www.semanticscholar.org/paper/336add4e4baecb7342c04b80c89110de429a5a51 (accessed on 13th September 2025).
44. Pawlus, K., Stolarczyk, A., Jarosz, T., Polis, M., Szydło, K., Hawełek, Ł., Waśkiewicz, S., et al. (2024). Energetic coordination compounds: Investigation of aliphatic ligands and development of prototype detonators. *International Journal of Molecular Sciences, 25*. https://www.semanticscholar.org/paper/4456e1eb40cb1e1b228163f17e97a746d96da53b (accessed on 13th September 2025).
45. Pfeifer, L., Engle, K. M., Pidgeon, G. W., Sparkes, H. A., Thompson, A. L., Brown, J. M., & Gouverneur, V. (2016). Hydrogen-bonded homoleptic fluoride–diarylurea complexes: Structure, reactivity, and coordinating power. *Journal of the American Chemical Society, 138*, 13314–13325. https://doi.org/10.1021/jacs.6b07501 (accessed on 13th September 2025).
46. Schreier, M. R., Guo, X., Pfund, B., Okamoto, Y., Ward, T. R., Kerzig, C., & Wenger, O. S. (2022). Water-soluble tris(cyclometalated) iridium(III) complexes for aqueous electron and energy transfer photochemistry. *Accounts of Chemical Research, 55*, 1290–1300. https://doi.org/10.1021/acs.accounts.2c00075 (accessed on 13th September 2025).
47. Skvortsova, S. V., Verkhov, F., Nikolaenkova, E. B., Rakhmanova, M. I., Kokina, T. E., Sukhikh, T. S., Shekhovtsov, N., et al. (2025). Interplay of the Cu□Cu distance and coordination geometry as a factor affecting the quantum efficiency in dimeric copper(I) halide complexes with derivatives of 4-pyrazolylpyrimidine-2-thiol. *Dalton Transactions*. https://www.semanticscholar.org/paper/

d68bd1354590b6c11e75dc78f44ff1065e40426a (accessed on 13th September 2025).

48. Slater, M., Kim, D., Lee, E., & Johnson, C. S. (2012). Sodium-ion batteries. *Advanced Functional Materials, 23*, 947–958. https://doi.org/10.1002/adfm.201200691 (accessed on 13th September 2025).
49. Song, M., Tan, H., Chao, D., & Fan, H. J. (2018). Recent advances in Zn-ion batteries. *Advanced Functional Materials, 28*. https://doi.org/10.1002/adfm.201802564 (accessed on 13th September 2025).
50. Sukardi, S. R., Dagang, A. N., Salleh, H., Wan Hassan, W. H. W., Nik Ali, N. A., & Sujod, M. Z. (2025). Analysis of photodegradation and reaction kinetics of textile organic dyes on *Caesalpinia sappan*, *Lawsonia inermis* and *Areca catheu* using spectroscopic methods for solar cell application. *Journal of Advanced Research in Micro and Nano Engineering*. https://www.semanticscholar.org/paper/d725521c78ff009a54f3ec88ce117e155c84de2c (accessed on 13th September 2025).
51. Valekar, A. H., Lee, M., Yoon, J. W., Kwak, J., Hong, D. Y., Oh, K. R., Cha, G. Y., et al. (2020). Catalytic transfer hydrogenation of furfural to furfuryl alcohol under mild conditions over Zr-MOFs: Exploring the role of metal node coordination and modification. *ACS Catalysis, 10*, 3720–3732. https://doi.org/10.1021/acscatal.9b05085 (accessed on 13th September 2025).
52. Venkatraman, R. K., & Orr-Ewing, A. J. (2021). Solvent effects on ultrafast photochemical pathways. *Accounts of Chemical Research, 54*, 4383–4394. https://doi.org/10.1021/acs.accounts.1c00549 (accessed on 13th September 2025).
53. Yan, D., Li, Y., Huo, J., Chen, R., Dai, L., & Wang, S. (2017). Defect chemistry of nonprecious-metal electrocatalysts for oxygen reactions. *Advanced Materials, 29*. https://doi.org/10.1002/adma.201606459 (accessed on 13th September 2025).
54. Zhan, X., Li, M., Zhao, X., Wang, Y., Li, S., Wang, W., Lin, J., et al. (2024). Self-assembled hydrated copper coordination compounds as ionic conductors for room temperature solid-state batteries. *Nature Communications, 15*. https://www.semanticscholar.org/paper/0190294572aa2ba566595dbda9a1b6196f48a858 (accessed on 13th September 2025).
55. Zhang, Q., Wang, Y., Braunstein, P., & Lang, J. (2024). Construction of olefinic coordination polymer single crystal platforms: Precise organic synthesis, *in situ* exploration of reaction mechanisms and beyond. *Chemical Society Reviews*. https://www.semanticscholar.org/paper/765b

abd457ddae3dca009f1afd99a7afa95c1238 (accessed on 13th September 2025).

56. Zhou, H. C., & Kitagawa, S. (2014). Metal–organic frameworks (MOFs). *Chemical Society Reviews, 43*, 5415–5418. https://doi.org/10.1039/c4cs90059f (accessed on 13th September 2025).

CHAPTER 9

BIOINORGANIC CHEMISTRY

Contents

INTRODUCTION

Bioinorganic chemistry is an interdisciplinary discipline, located at the boundary of biology and inorganic chemistry, and is concerned with the role played by metals in biological processes. It involves a broad array of topics including metalloproteins, metalloenzymes, and metal ions in biochemistry, with special focus on the basic roles played by metals in living organisms. Notably, transition metals such as iron, copper, and zinc are key players as cofactors in enzymatic catalysis, facilitating processes that are crucial to biological function. In their ability to participate in more than one oxidation state, they can display varied reactivity, enabling participation in electron transfer, catalysis, and protein structural stabilization. By understanding these interactions, bioinorganic chemists aim to decipher the complicated processes of life's biochemical mechanisms and the consequences of metal ion deficiencies or toxicities. One of the intriguing uses of bioinorganic chemistry is in the design of metal-based drugs that are able to target specific biological pathways. For instance, the development of chemosensors utilizing metal ions for the detection of heavy and transition metal ions is indicative of innovative strategies adopted in the field. The application of a pyrene appended ratiometric fluorescent chemosensor, as reported in recent work, is an example of how metal ion sensing will bring about conformation change leading to measurable fluorescence modification. These technologies have enormous potential for environmental testing and clinical diagnosis. On top of that, the finding of dicarba-closo-dodecarboranes in half-sandwich complexes has opened up new avenues for drug discovery, particularly in hitting biologically important proteins and creating Boron Neutron Capture Therapy agents (Barry et al., 2012). The above note shows conjugation of inorganic chemistry rules with biological aspirations. Bioinorganic chemistry studies contribute significantly to overcoming the health and environmental sustainability challenges of the day. Exposure to heavy metals has serious impacts on human health, and this necessitates efficient detection methods, as attested by current studies on the development of chemosensors. Moreover, insights into the interaction between metal ions and biomolecules can lead to new therapeutic strategies toward the treatment of cancer and neurodegeneration. The application of sophisticated metal complexes in drug chemistry exhibits a prospective route in the discovery of drugs, enhancing the specificity and efficacy of the therapy. By the combination of intrinsic principles of inorganic chemistry with bio-related useful applications, bioinorganic chemistry not only improves our understanding of life at the molecular level but also offers insights towards new solutions for priority health issues of the world, making the field a priority area of research in scientific and medical environments.

Definition and Scope of Bioinorganic Chemistry

Bioinorganic chemistry is a significant interdisciplinary science that examines the involvement of inorganic elements in biological molecules. Intersecting biology, chemistry, and physics, bioinorganic chemistry examines the way metal ions are engaged in biological systems, particularly in metalloenzymes and metalloproteins. These molecules have vital roles in many biological processes such as respiration, photosynthesis, and metabolism. Moreover, bioinorganic chemistry investigates how these metal ions facilitate critical biochemical processes, providing information concerning enzymatic efficiency and selectivity. Bioinorganic chemistry is not only fascinating in itself but also imperative in its applications to medicine, environmental science, and materials science, offering valuable insights into health-mediating metal ion deficiencies and toxicities that can greatly influence human welfare. Bioinorganic chemistry is applied more extensively than the mere finding of the biological roles of metals. It encompasses the synthesis and use of metal-based drugs and therapeutics. For instance, specific targeting of metal ions within the biologic system has enabled the development of new therapeutic drugs such as platinum drugs employed in chemotherapy during cancer treatment.

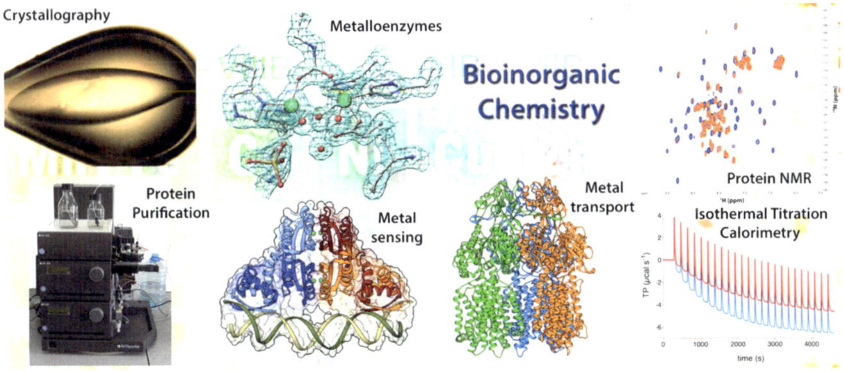

Figure 9.1. *Bioinorganic chemistry.*

Source: https://site.unibo.it/bioinorgchem/en/@@eod.tiles.textimage/90e223bd897e4bb19d4ba4a2e39beb53/@@images/ac0188b7-53db-47b9-8001-ea04b-b1a5a96.png.

Moreover, radioisotopes, as represented by the discussion in *in vivo* generators, exemplify the merging usages of bioinorganic chemistry for the purposes of medical diagnostic and therapeutic interventions. Radionuclide generators utilize the unique properties of specific isotopes for targeting and reacting to causative factors of diseases for successful cures, attesting to the practical implications of bioinorganic chemistry in current medical practices. Through the integration of a multi-disciplinary approach, the field brings the metallurgic effect of metals to disease and health into focus, which renders the field more applicable in modern-day research and development. Additionally, bioinorganic chemistry addresses tough issues and problems within other areas of science. Unraveling the complex interaction of metal ions with biological macromolecules is perhaps the most significant aspect of the field, which primarily requires advanced modeling and analytical techniques. As seen in optimal transport problem explorations present in advanced theoretical work, understanding of these interactions' behavior across different scales is vital to predicting biochemical outcome and stability in cellular environments. The synergy offered by the integration of computational tools into bioinorganic chemistry enables the synthesis of novel biomimetic materials and pharmaceutical candidates with high similarity to natural processes. This approach represents the dynamic function of bioinorganic chemistry to foster novel solutions and advance scientific discovery, thus demonstrating its essential place both in basic research and applied sciences, future advances and findings in a variety of fields (Figure 9.1).

Historical Development and Significance

The development of the history of bioinorganic chemistry has been dominated by critical milestones that work to mark the discipline as a complete field of study within the broader science of chemistry. Metals in biological systems were not of much concern early on, with attention instead focused on organic compounds. However, significant discoveries in the late 19th and early 20th centuries highlighted the inreplaceable position metals play in biological functions. Studies on metalloproteins, enzymes, and the interactions of these with metal ions started to occur, which showed complex interrelations responsible for biological activity. This change of emphasis paved the way for the eventual recognition of bioinorganic chemistry as an independent discipline, showcasing the importance of the appreciation of metal behavior in the biological environment. These developments were an indicator of growing awareness of the complexity of life at a molecular level and paved the way for advancement in both inorganic chemistry and biochemistry. As development in the field picked up pace in the latter half of the 20th century, the combination of spectroscopic procedures and computational methods completely revolutionized the study of bioinorganic systems. Significant progress was made in the elucidation of structures and functions of metal-containing biomolecules, which promoted increased understanding of enzyme mechanisms and metabolism. The investigation of transition metal ions, such as iron, copper, and zinc, illustrated their importance as key components in electron transfer, transport of oxygen, and catalysis in living organisms. Apart from that, the identification of techniques such as X-ray crystallography allowed scientists to visualize these intricate biomolecules in more detail than ever. This crossing of experimental practices and computational methodologies not only boosted the scientific narrative of bioinorganic chemistry but also encouraged collaboration across disciplines, leading to the promotion of metals in biochemistry. The applicability of bioinorganic chemistry extends well beyond academic interest; it carries profound significance in the development of new therapeutic methods and biotechnological methods. Information regarding metal ion homeostasis and metalloprotein functioning in disease has led to new therapies for diseases such as anemia, cancer, and neurodegenerative diseases. As cautioned by recent reviewers, e.g., Bertini et al. (1994), the emerging science still provides useful information about the targeting or manipulation of metalloproteins for the achievement of intended therapeutic effects. Finally, the history of bioinorganic chemistry behind the discipline emphasizes the multidisciplinary character of the task, which is thus subject to input from biochemists, chemists, and biomedical scientists equally. The history of bioinorganic chemistry is not just a description of the coming together of scientific inquiry but of hope for sustainable solutions in the face of global healthcare problems.

Interdisciplinary Nature of the Field

The interdisciplinary nature of bioinorganic chemistry basically bridges the gap between the conventional chemistry and intricate mechanisms of life processes. This is a field that employs the principles of inorganic chemistry to probe the roles played by metal ions in biological processes and hence necessarily involves the convergence of a broad array of scientific disciplines. Metal ions such as iron, copper, and zinc play pivotal roles in enzyme catalysis, electron transfer, and the stability of proteins' structure. By the interfolding of data from analytical chemistry, biochemistry, and molecular biology, researchers can dissect metalloproteins and enzymes and thus acquire new insights in disease and health. The interfolding is not only helpful in expanding our understanding of cellular processes but also plays a key role in propelling therapeutic agents as well as diagnostic reagents that are founded upon the intrinsic property of metal ions in biological systems (Booth et al., 2014). Furthermore, the approach used in bioinorganic chemistry is a reflection of the value of interdisciplinary research since approaches from various fields of science converge. Computational chemistry and biophysical tools, for instance, allow scientists to model metal–ligand interactions accurately and monitor how metal binding influences protein function and folding. Results obtained can result in new drug design and materials science, reflecting the applicability of research in bioinorganic chemistry. Apart from the institutional education systems, as captured in materials such as those of the Claremont Colleges, stress multisciplinary learning communities encompassing a number of disciplines that encourage understanding of various applications and usefulness of bioinorganic processes (N/A, 1989). In doing so, the diversified approach not only enhances scholarship but also enhances the capacity of prospective researchers. Additionally, the applied significance of bioinorganic chemistry extends beyond a theoretical consciousness, to influence genuine environmental and medical applications. With the recognition of heavy metal toxicity and biological effect gaining momentum, bioinorganic chemistry has a significant part to play in investigating how living organisms respond to metal exposure. By examining processes like bioremediation and metal detoxification, researchers work towards developing countermeasures against pollution and promoting the health of ecosystems. Concurrent with this, an awareness of metallodrugs in the clinic, such as platinum and gold drugs, points towards the promise of metal-based medicines as therapeutic drugs to cure diseases like cancer. Their synthesis and derivatization are characteristic of the interface between inorganic and organic chemistry, where the multidisciplinary approach of bioinorganic chemistry not only enriches the science but also has a direct payback into society through innovations that address real-world challenges (Figure 9.2).

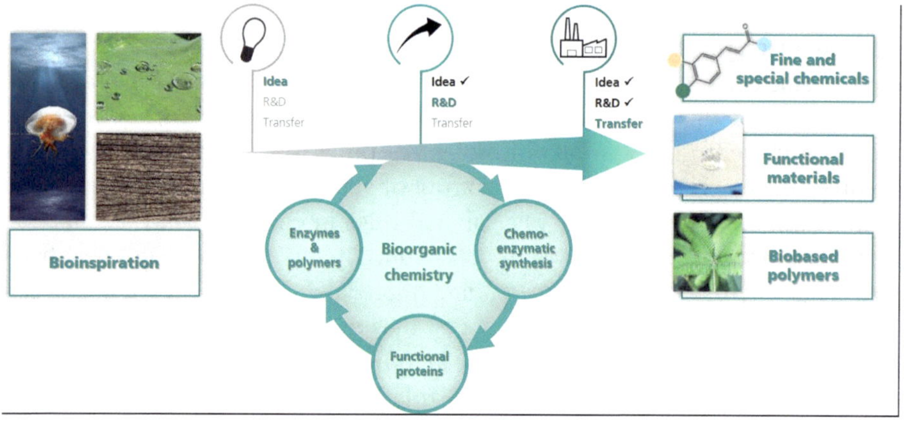

Figure 9.2. *The interdisciplinary nature of bioinorganic chemistry.*

Source: https://www.igb.fraunhofer.de/en/research/bioinspired-chemistry/bioorganic-chemistry/jcr:content/contentPar/sectioncomponent/sectionParsys/wideimage/imageComponent/image.img.jpg/1643908062056/Grafik-EN-Bioorganic-Chemistry.jpg.

9.1. ROLE OF METALS IN BIOLOGICAL SYSTEMS

Metals are integral parts of biological processes, actively participating in numerous biochemical processes essential to life. Through the ability to show special properties, they prove efficient cofactors in enzymatic processes, imparting stability and increasing molecular interaction. Transition metals such as iron, zinc, and copper are particularly important due to their ability to form coordination complexes with organic molecules. These interactions are most often linked with increases in enzyme activity through stabilization of reaction intermediates or direct participation in catalysis. Besides this, metals are required for structural integrity of many proteins, anchoring them in a particular conformation enabling biological function. The diverse functional roles of metals underscore the subtle balance engaged in with living organisms such that metal homeostasis must be preserved to prevent deficiencies or toxicities interfering with cellular processes. The ribosome, the central macromolecular complex that is involved in protein synthesis, most accurately reflects the role of metals in biological functions. Research has established that metals like Mg^{2+}, Fe^{2+}, and Mn^{2+} profoundly influence ribosomal structure and function. The divalent cations not only stabilize the rRNA but also catalyze the translation of mRNA to functional proteins. Experimental investigation with selective 2-hydroxyl acylation analyzed by primer extension (SHAPE) highlights that Mg^{2+}, Fe^{2+}, and Mn^{2+} impose astonishingly similar effects on rRNA folding and stability. This implies that the evolutionary past of the ribosome was governed by the availability of these metals in early environments, and their utilization in crucial biological processes has been sealed. Metals, thus, dictate structural and functional competence in critical cellular machinery, demonstrating their indispensable place in bioinorganic chemistry. Moreover, the interface of metals with biological processes paints the picture of a co-evolutionary history since the early days of life. The hypothesis that Fe^{2+} and Mn^{2+} might have facilitated the emergence of the ribosomal system of translation suggests early biological process was intricately linked to the geochemical conditions of Earth. In an era where oxygen levels were low and divalent metal ions were abundant, these metals would have been the cause of primordial metabolic pathway evolutionary fitness. In addition, the ability of Fe^{2+} and Mn^{2+} to replace Mg^{2+} during translation indicates that the presence of metal ions was largely responsible for the evolution of ribosomal function. This evolutionary perspective illuminates us to the broader implications of metals in the living organism, with their central importance not just in contemporary biochemical reactions but also in ancestral biochemistry that evolved life as we know it (Bertini et al., 1994; Bowman et al., 2018).

9.1.1. Essential Trace Elements and Their Functions

Trace elements of key importance in bioinorganic chemistry are highlighted through their key functions within various biochemical processes. Trace elements such as iron, zinc, copper, and manganese are essential cofactors in enzyme reactions, facilitating biological conversions vital to life. Iron, for instance, is a keystone of hemoglobin and its involvement with oxygen transport in biological systems, and zinc is vital in the structure and function of enzymes. These trace elements are typically found in infinitesimal quantities, but their absence can result in serious physiological illness, emphasizing the importance of these trace elements in metabolism and nutrition. The unique functions of these trace elements illustrate the intricate biochemical mechanisms underlying life, showing the profound interconnection between inorganic chemistry and biology. Research on the trace element activity surpasses enzymatic activity by showing they have a role in affecting ecological environments as well. Trace elements, for example, are critical in microbial environments that enable greenhouse gas emission from various environments. Some microbes, such as methanogenic archaea, depend on trace metals like nickel and cobalt for their enzyme activity in methanogenesis, a distinctive characteristic of methane emissions that bring about climate change. Conversely, deficiency in these metals can potentially significantly limit microbial activity and thus influence greenhouse gas cycling and atmospheric concentrations. This interaction between trace elements, microbial life, and greenhouse gas processes is an indication of the complex relationship in ecosystems and the necessity for multidisciplinary research on bioinorganic chemistry and environmental science (Glass et al., 2012). Trace elements' evolutionary characteristics are also an aspect worth exploring within bioinorganic chemistry. The theory proposes that the biochemical reactions which utilize essential trace elements developed as a bootstrapping process wherein less-than-metal, low periodic table non-metals laid down the groundwork for systems dependent on higher-energy elements. This model of evolution propounds that trace elements became increasingly necessary for sustaining life's more advanced forms as more energetic metabolic processes arose. As mentioned, the original functions of elements such as carbon, nitrogen, and oxygen trace their origins back to these ancient biogenic processes, a testament to their enduring relevance in biological evolution. Therefore, understanding the significance of trace elements not only deepens our knowledge of biochemical activity but also offers insights into the mechanisms of evolution that are at work, demonstrating the abiding connection between inorganic chemistry and biological advancement (Kletzin et al., 2007).

9.1.2. Mechanisms of Metal Ion Transport in Organisms

In bioinorganic chemistry, it is important to understand the mechanisms of metal ion transport in organisms as a way of explaining their functions in physiology and in cellular processes. Metal ions such as iron, zinc, and copper are essential in various biochemical processes through their roles as cofactors for enzymic reactions or structural functions as components of proteins. To facilitate their transport, organisms have sophisticated systems regulating ion uptake, distribution, and excretion. Such regulatory systems not only maintain essential metal ion homeostasis but also act as a defense mechanism against noxious metal ion accumulation. The sophistication with interaction among transport proteins, i.e., metallochaperones and membrane transporters, indicates the importance of metal ion specificity and regulation to enable organisms to respond to fluctuating environmental conditions and fluctuating metal availability. At a molecular level, metal ion transport is controlled by several channels and transporters that each possess specific selectivity and function. The physiological relevance of zinc transport is exemplified by the Zrt/Irt-like protein (ZIP) family, for instance, through facilitating zinc uptake according to cellular requirements.

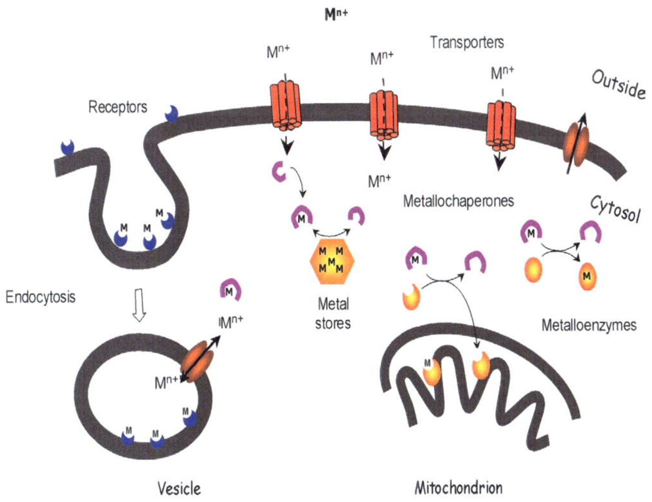

Figure 9.3. *The mechanisms of metal ion transport within organisms.*

Conversely, efflux transporters, such as the P-type ATPases, also have a major responsibility in eliminating excess metal ions, and this portrays the bidirectional nature of metal ion transport. Significantly, the development of metallomics, which examines the habit and placement of metals in biological systems, has provided insights on the dynamic and complex nature of metal homeostasis. Using new methods, including multiple dimensional liquid chromatography-inductively coupled plasma mass spectrometry (LC-ICPMS), researchers uncover the mechanisms and regulatory control on metal ion transport and its impact on human disease and health (GB et al., 2013). Moreover, the mechanism of metal ion transport within microbial systems has been the subject of much attention due to its relevance to human health and environmental sustainability. Microorganisms employ diverse mechanisms of gaining and controlling metal ions, generally towards balance between essential uptake and harmful accumulation. Research in metallomics has unveiled the intricate metabolism of metal trafficking in microorganisms, e.g., the competitive iron acquisition through siderophore production as an example of a strategy under nutrient stress. Such mechanisms not only highlight the crucial roles played by metal ions in microbial physiology but also expose potential bioremediation opportunities. Through the application of the natural potential of microorganisms for metal ion transport and sequestration, scientists aim at developing new technologies for minimizing metal toxicity and promoting ecological sustainability in various environments (Bertini et al., 1994) (Figure 9.3).

9.1.3. Impact of Metal Ion Deficiency and Toxicity

It cannot be overstated that metal ions are of central importance in biological systems, where they play key roles in many of the cell's processes. Essential metal ion deficiencies, such as those of iron, zinc, and copper, can lead to severe physiological malfunctioning. Iron deficiency is, for instance, long recognized to be the cause of anemia, which adversely affects oxygen transport and overall cellular respiration. Similarly, zinc is at the heart of immune function, and its deficiency can compromise the body's defense mechanisms, leading to increased vulnerability to infections. Moreover, the biological importance of these metal ions highlights the delicate balance required for ultimate well-being. With decreasing concentrations of metals below those necessary, the resulting deficiency is likely to initiate a chain of health issues, and this underscores the necessity of having knowledge on bioinorganic chemistry of necessary metals and their functions in living organisms (GB et al., 2013). Conversely, toxicity due to excessive metal ion concentrations should also be considered. Heavy metals, including lead, cadmium, and mercury, are known to cause damage to human health via neurological impairment, organ damage, and even carcinogenicity. The toxicity of metals typically involves interference with cellular functions, e.g., oxidative stress, which can be translated into inflammation and cell death. For instance, mercury ions are capable of preventing the action of essential enzymes and proteins by binding to sulfhydryl groups, thereby inhibiting crucial biochemical reactions. Moreover, the detoxification strategies used by organisms to eliminate metals have basic significance in avert their toxic effects. Understanding the mechanisms by which cells deal with deficits as well as surfeits of metal ions is important in designing therapeutic strategies under the aegis of bioinorganic chemistry and public health (Bertini et al., 1994). Metallomics offers a new avenue to probing the influence of metal ion deficiency and toxicity at an unparalleled level of sophistication. It involves studying the distribution, location, and speciation of metal ions in biological systems, providing clues about their physiological functions. Advanced analytical equipment such as multi-dimensional liquid chromatography-inductively coupled plasma mass spectrometry (LC-ICPMS) and synchrotron X-ray fluorescent spectroscopy enable scientists to elaborate on the role of metal ions in physiological functions under various conditions. This new field not only sheds light on the fine-tuned balance of metal ions but also has significant bearings in terms of health in the environment. By learning about how microorganisms manage to deal with metal uptake and toxicity, we are given information that could be translatable to human health and disease management, forming a significant link in the area of bioinorganic chemistry (GB et al., 2013).

9.2. METALLOPROTEINS AND ENZYMES

The study of metalloproteins and enzymes occupies a prime position in the general framework of bioinorganic chemistry, emphasizing the intricate interrelationship between metal ions and bioactivity. Metalloproteins, which incorporate metal ions into their structure, are responsible for a variety of physiological functions including catalysis, electron transfer, and oxygen transport. Notably, these proteins can catalyze the transformation of low-molecular-weight substrates to high-molecular-weight biomolecules through direct metal center intervention, making available new reaction mechanisms that are otherwise unavailable through energetics. The fine-grained coordination of metal ligands in metalloproteins underscores their significance in maintaining health and the potential for disease-related pathology. This doubleness indicates the necessity of serious study of metalloproteins and their functional functions within cellular processes, emphasizing the convergence of inorganic and biochemical perspectives in understanding lifes processes. Enzymes, as a group of metalloproteins, illustrate additionally the essentiality of metal ions in catalysis. Many enzymes employ metal cofactors, such as iron, copper, or zinc, to attain optimal function. These metal ions often play significant functions in transition state stabilization in enzyme catalysis, consequently lowering activation energy and increasing the rate of reactions. For instance, zinc is a cofactor in carbonic anhydrase-like metalloproteins to catalyze hydration of CO_2, which is essential in physiological bicarbonate homeostasis. The understanding of enzyme mechanisms at the molecular level not only shows the nature of metabolic processes but also shapes the design of drugs that are meant to modulate enzyme action. According to the scope of bioinorganic chemistry, the double role of metal ions as structural support and as catalytic performance in enzymes is well established, showing the functions of such biomolecules in sustaining life processes. Moreover, the emerging field of metallomics sheds more light on metalloproteins and enzymes through the investigation of metal distributions and speciation within biological systems. Techniques such as multiple dimensional liquid chromatography-inductively coupled plasma mass spectrometry (LC-ICPMS) and synchrotron X-ray fluorescent spectroscopy (XFS) have been used to elucidate the functions and interactions between various metals in cellular contexts in real time. This integrative approach not only helps establish important metal cofactors of specific metalloproteins but also sheds light on dysregulation of metal homeostasis causing pathologies. Ongoing new advances in analytical methods further extend the potential for metallomics to drive areas of environmental health and therapeutic intervention, underscoring the far-reaching interconnexions between the topic of bioinorganic chemistry and overall health outcomes (Bertini et al., 1994; GB et al., 2013) (Figure 9.4).

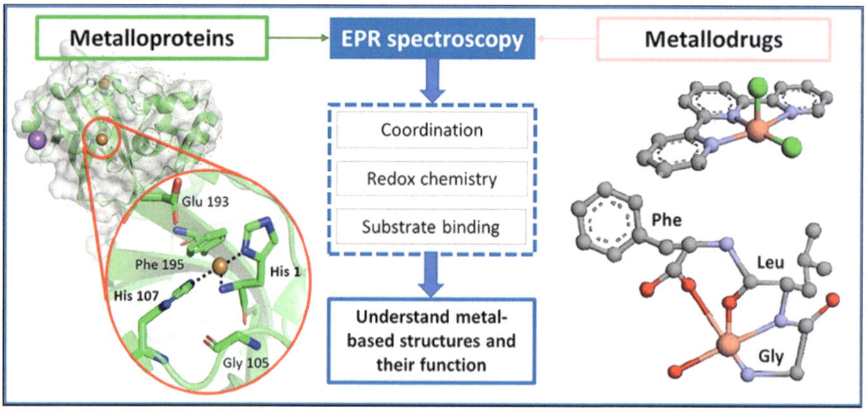

Figure 9.4. *The study of metalloproteins and enzymes.*

Source: https://ars.els-cdn.com/content/image/1-s2.0-S1367593120301575-fx1_lrg.jpg.

9.2.1. Structure and Function of Metalloproteins

Metalloproteins are a crucial class of biomolecules classified based on the presence of metal ions, which perform critical roles in biological processes. The structural frameworks of metalloproteins vary, with the designs to accommodate certain metal ions contributing to their stability as well as functionality. Transition metals like zinc, copper, and iron are the centerpieces of this show since they have unique chemical attributes, the ability to adopt varying oxidation states as well as coordination geometries. These attributes enable metalloproteins to catalyze vital biochemical reactions, like electron transfer within cytochromes or oxygen binding in hemoglobin, emphasizing the absolute necessity of metalloproteins in cellular metabolism and bioenergetics. In addition, the structural complexity of metalloproteins can yield insights into their function, fueling advances in fields like drug design and environmental biochemistry. Metalloproteins' functional diversity arises due to a synergy between their metal co-factors and the protein matrix environment. For instance, incorporation of the metal ion into the protein fold stabilizes the global fold as well as enhances the efficiency of enzymatic catalysis. Metalloproteins utilize varied mechanisms to take advantage of the unique redox properties of metal ions and facilitate electron transfer and substrate activation. Redox metalloproteins, for example, ferredoxins, utilize the high rate of electron transfer catalyzed by their iron-sulfur clusters, showing the exquisite harmony between structure and function. Furthermore, advances in metallomics in recent times have revealed the complex processes of metal ion transport and distribution within biological systems, allowing for a better understanding of how metalloproteins act within cellular regimes. These examinations highlight the significance of metalloproteins across basic and applied sciences, especially with regard to their utility in human health and disease. Crystallography and spectroscopy become critical tools for the resolution of the structure-function relationship of metalloproteins so that scientists can visualize metal interaction in a three-dimensional framework. Structural analysis of metalloproteins often encounters challenges, such as the appropriate inclusion of metal ions and the appropriate crystallization environmental conditions. However, these drawbacks are outbalanced by the useful properties of metals in crystallography to allow techniques like anomalous dispersion to determine metal site positions with high precision. Secondly, the rich electronic signatures due to transition metals allow for deeper spectroscopic examination and favor studies of enzymatic activity as well as metal coordination. With ongoing research in bioinorganic chemistry, synergistic application of these approaches is likely to enhance metalloprotein dynamics knowledge and unlock possibilities to realize their potential towards therapeutic and biotechnological applications, thereby proving their relevance in the modern scientific community (Bowman et al., 2015; GB et al., 2013).

9.2.2. Examples of Metalloenzymes and Their Catalytic Roles

Metalloenzymes are the essential enzymes which catalyze a number of biochemical processes, and nearly half of all enzymatic reactions utilize metal ions. Biological catalysts utilize the unique electronic properties of metals to facilitate reactions ranging from oxidation-reduction to hydrolysis. Carbonic anhydrase is one such classic example, which utilizes zinc as a cofactor to facilitate the reaction of carbon dioxide and water to bicarbonate and protons. This enzyme is a vital component in biological pH homeostasis and carbon dioxide homeostasis. Carbonic anhydrase catalytic activity is largely attributed to the coordination of water molecules around the zinc ion, lowering the activation energy of the reaction. The incorporation of metal ions typically results in enhanced reaction rates, which demonstrates how metalloenzymes utilize such ions to improve vital biological processes and therefore demonstrating their importance in bioinorganic chemistry. A second fascinating example includes phosphoesterases, such as glycerophosphodiesterase from *Enterobacter aerogenes* (GpdQ) and the organophosphate-degrading agent from *Agrobacterium radiobacter* (OpdA). Such metalloenzymes favor the use of metal ions, such as manganese or zinc, to catalyze the hydrolytic cleavage of phosphoester bonds and play a key role in signal transduction and nucleic acid metabolism. Metal ions stabilize transition states and enable nucleophilic attack on substrates. Increased knowledge of the structural features and binding of metal ions in phosphoesterases has significant biotechnological significance, especially in the construction of agents that will remediate organophosphate pollutants. Combining the research on these metalloenzymes with model complexes improves our understanding of their structure and function, and shows how there are complex relationships between metal coordination and catalytic activity that is a basic tenet of bioinorganic chemistry (Daumann et al., 2014). When looking at the evolutionary histories of metalloenzymes, one realizes that the conservation of metal-binding sites highlights how crucial they are in enzymatic mechanisms. Research emphasizes that the role of non-redox metals is not significantly altered when structural change occurs at the catalytic metal site, whereas redox-active metals possess functional diversification that will cause their catalytic characteristics to alter. The observation demonstrates the intricate relationship among structure of the enzyme, metal ion coordination, and function and emphasizes the driving force of catalytic efficiency. These findings are critical to the understanding of enzyme adaptation as well as the *in vivo* evolutionary significance of metalloenzymes. Metal ion catalytic activities are central to metabolic processes and illuminate the versatility and robustness of biochemical pathways, necessitating further growth of the field of bioinorganic chemistry (Andreini et al., 2017).

9.2.3. Techniques for Studying Metalloproteins

Metalloproteins—a complex group of proteins characterized by the presence of metal ions—are investigated because of advanced methodologies that allow their structure and function to be deciphered. A number of techniques, which collectively contribute to metallomics as a discipline, have emerged to address these needs. These include multiple dimensional liquid chromatography-inductively coupled plasma mass spectrometry (LC-ICPMS), gel electrophoresis-laser ablation-inductively coupled plasma mass spectrometry (GE-LA-ICPMS), and synchrotron X-ray fluorescent spectroscopy (XFS). Both techniques are used to reveal the distribution, speciation, and interactions of metal ions in biological systems.

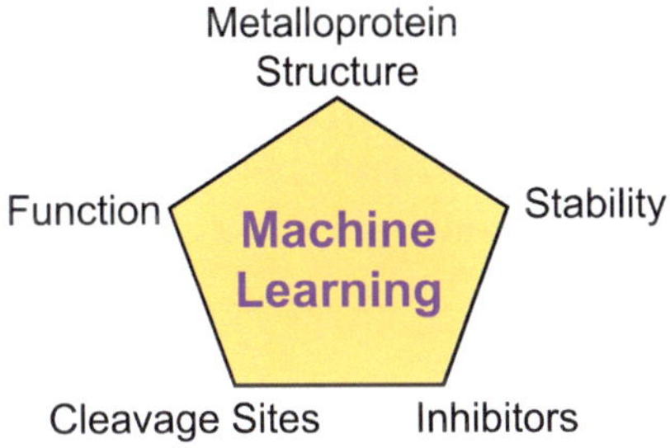

Figure 9.5. *Techniques for studying metalloproteins.*

Source: https://pub.mdpi-res.com/molecules/molecules-27-01277/article_deploy/html/images/molecules-27-01277-ag.png?1645261168.

The ability to analyze trace metals in complex biological matrices makes it possible to better understand their role in enzymatic activities and the potential effects for human health, specifically the growing importance of metallomics for biomedical research and environmental studies, as summarized in the most recent developments in methodological progress (GB et al., 2013). Structural examination of metalloproteins is central to their biochemical significance. X-ray crystallography, among others, is an important technique that allows visualization of the metal ions and metallocofactors in protein structures. Aside from common crystallization problems, such as correct inclusion of the metal ions and presentation of a satisfactory protein environment, metal ions' unique attributes enhance the crystallographic process. The ion metal resonance with regard to some wavelengths of X-rays enables useful techniques like anomalous dispersion. The techniques provide information about the three-dimensional structure of metalloproteins, enabling the solving of their catalytic mechanism and function. Progress in synchrotron technology, including improved beamline brilliance and advanced detection systems, has further accelerated metalloprotein structure determination, hence boosting the science of bioinorganic chemistry (Bowman et al., 2015). Aside from X-ray crystallography, spectroscopic techniques are also in the forefront of studying metalloproteins and their catalytic activities. Ultraviolet-visible (UV-Vis) spectroscopy, for instance, is essential in studying electronic transitions of metal ions in metalloproteins and their impact on their catalytic activities. K-edge X-ray absorption spectroscopy also gives information on the coordination environment and oxidation states of metal centers in proteins. These spectroscopic methods yield complementary information such that researchers can examine the interactions between metal ions and their protein partners in a context of functionality. The more advanced understanding of metalloproteins creates a need for combining these diverse methods, which demonstrates the diversity and versatility of metalloproteins within biological systems. All these developments show a continued fascination with unraveling the sophisticated roles played by metal ions in health and disease, again underscoring the significance of bioinorganic chemistry (Bowman et al., 2015) (Figure 9.5).

9.3. METAL-BASED DRUGS AND THERAPEUTICS

The uncovering of metal-based medicines in the field of bioinorganic chemistry has led to enormous developments in the treatment approach, particularly against cancer (Bowman et al., 2015). Conventional treatments have relied primarily on platinum-based medicines, such as cisplatin, which, although useful, come with an array of side effects. Scientists are therefore moving with great zeal to find alternative metal-based medicines that will offer diverse modes of action and avoid these side effects. New avenues, such as prodrug design and targeting of delivery systems, have emerged to enhance the therapeutic value of these drugs. This research not only places into perspective the limitations of existing remedies but also evokes a broader realization regarding the use of metals of different kinds in anticancer drug development, a sign of a shift in direction for medicinal bioinorganic chemistry (Aird et al., 2009). One of the foremost developments in metal drugs is the coordination of dicarba-closo-dodecarboranes with half-sandwich transition metal complexes like ruthenium, osmium, rhodium, and iridium. This hybridization holds the potential to produce new classes of compounds that can be applied in drug-related applications, more towards enhancing the efficacy of established treatments. The new properties of these complexes can also provide new areas for applications in addition to the conventional ones, potentially incorporating them into new treatment protocols such as Boron Neutron Capture Therapy. Synthesis and design of such compounds entail intricate approaches to facilitate varied metal geometries as well as functionalities to allow targeted drug development targeting biologically relevant proteins directly (Barry et al., 2012). In summary, the search for emerging metal therapeutics is a testament to the influential role bioinorganic chemistry has taken and continues to take in modern medicine. As researchers strive to circumvent the limitations of traditional therapy, the study of various metal complexes promises much for future effectiveness in therapy. The ongoing development of synthetic techniques, coupled with a greater insight into metal interactions within the body, is the basis on which future medicines are being constructed to be both stronger and less harmful. This dynamic science holds the potential for significantly improving therapeutic success and expanding the armory of therapy against diseases, emphasizing the central role of metal-based drugs in contemporary drug research (Aird et al., 2009; Barry et al., 2012).

9.3.1. Overview of Metal-Based Pharmaceuticals

Metal-based pharmaceuticals is a significant branch of bioinorganic chemistry, characterized by the convergence of metals and biological systems toward medical purposes. Such complexes frequently result from efforts at improving existing treatments, for instance, in cancer therapy where standard chemotherapy drugs like cisplatin produce unwanted side effects. Researchers have redirected their focus to other metal complexes that exert new mechanisms of action, thus possibly lowering poisonous levels of known drugs. Furthermore, prodrug and targeted drug delivery systems seem to be promising paths towards transferring the therapeutic potential of these metal complexes. With the advancement in science, how these metal complexes interact at a molecular level becomes crucial concerning developing safer and more effective treatments (Aird et al., 2009). Aside from their applications in oncology, metal-containing drugs more and more often include metals like palladium and gold, which possess catalytic properties. Not only do these metals demonstrate their significance when used therapeutically, but also the medical use of these metals provides insights into key issues in the challenges of pharmacology and drug design. For example, palladium complexes have recently also come under the limelight due to their involvement in green practices, especially for their recyclability and environmental footprint. New methods, such as the biological synthesis of palladium nanoparticles, provide potential avenues in optimizing drug performance and minimizing environmental concerns. Through the exploitation of the unique properties of these metals, researchers are able to find novel paths towards effective drug design aligned with the principles of sustainability (Boon et al., 2011). Metal-containing drugs increasingly gain significance, particularly because research reveals their intricate functions within organisms. Their identification is emphasized by growing understanding of the interactions between metal ions and biological macromolecules and the concomitant potential for therapeutic application and mechanistic insight. Advance in bioinorganic chemistry not only promises novel therapeutic agents but also aspires to address significant clinical issues with drug delivery and selectivity. With continued investigation of new metal complexes, the outlook for metals as drugs is excellent. This dynamic setting fosters collaborative research among disciplines by chemists, biologists, and clinicians, ultimately driving the search for new and effective treatments that take advantage of both the unique properties of metals and our understanding of biological systems (Aird et al., 2009; Boon et al., 2011) (Figure 9.6).

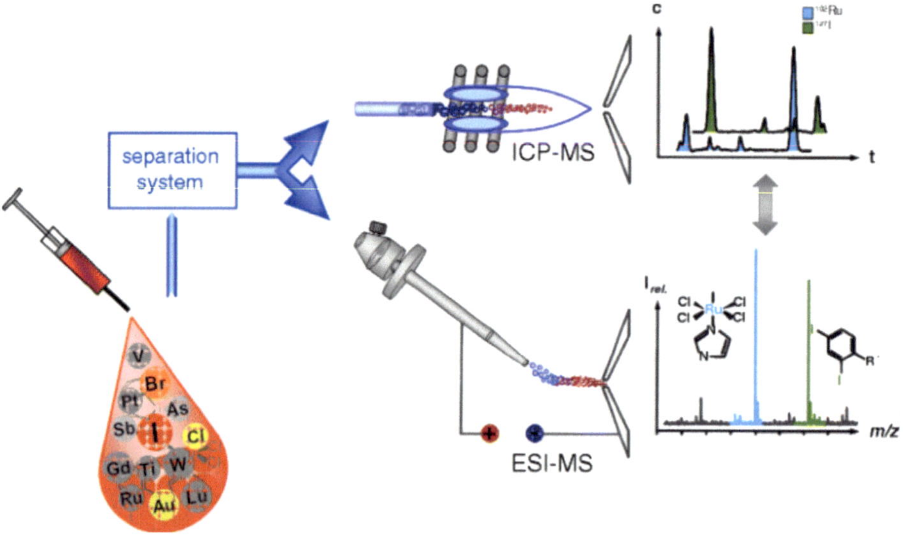

Figure 9.6. *The study of metal-based pharmaceuticals.*

Source: https://media.springernature.com/lw685/springer-static/image/art%3A10.1007%2Fs00216-012-5915-9/MediaObjects/216_2012_5915_Figa_HTML.gif.

9.3.2. Mechanisms of Action of Metal Drugs

Metal drug research has significantly expanded the field of bioinorganic chemistry, particularly in the therapeutic area. Traditional metal-based medications, such as platinum complexes, have been widely used in anticancer medicine.

The drugs mentioned above prefer to show strong side effects and resistance. Research has been directed to prepare novel metal complexes with different mechanisms of action, which can counter the drawbacks associated with conventional therapy. For instance, the design of new pharmacologically active metal species is aimed at optimizing their bioavailability and therapeutic efficacy while minimizing toxicity. It is a novel approach that highlights the importance of unravelling the intricacy of biochemical cascades involved in metal drug action, allowing more accurate targeting of cancerous conditions and better patient outcomes (Aird et al., 2009). In addition to the targeting of pathways traditionally mapped onto anticancer drugs, new metal complexes have been designed to induce novel cell stress responses. For example, the copper(I) complex HydroCuP has emerged as a promising replacement for platinum drugs. It acts through the induction of endoplasmic reticulum (ER) stress, leading to cell death in a mechanism known as paraptosis. This noncanonical action is particularly worthwhile in that it possesses specificity for cancer cells with minimal damage to normal tissues. Antitumor activity of HydroCuP was illustrated by profound inhibition of tumor growth in both oxaliplatin-sensitive and resistant models, illustrating the worth of metal drugs in avoiding mechanisms of resistance that typically limit chemotherapy efficacy. Hence, exploration of other metal drug mechanisms can provide important insights into better and safer cancer treatment (Ceresa et al., 2017). The therapeutic potential for metal drugs is also increased by their high biodistribution and low toxicity. HydroCuP depicts this through minimal toxicity in murine models but pronounced inhibition of tumor growth. The absence of neuro- and nephrotoxicity on administration of HydroCuP points towards the significance of designing metal complexes not just with therapeutic activity but with maximum tolerance of the organism as well. Such studies present valuable information that could be used to design drugs based on metals in the future, such as drug targeting and prevention of side effects. The ongoing quest for mechanisms of metal drug action is a central feature of bioinorganic chemistry and holds the potential to revolutionize cancer treatment paradigms in the next several years (Ceresa et al., 2017).

9.3.3. Case Studies: Cisplatin and Other Metal-Based Therapies

Metal-based treatments in relation to cancer have been researched with growing pace over the past decade or so, with much focus on agents such as cisplatin, a platinum-based chemotherapeutic agent. Cisplatin works through the mechanism of DNA cross-linking, halting replication and transcription, leading to apoptosis in cancer cells.

The intricate process underscores the importance of understanding bioinorganic chemistry because metal complex interactions with biological molecules are central to therapeutic success. Given the growing incidence of cancer globally, as named by its position as the second most prevalent cause of death (Dudek-Różycki et al., 2013), metal-based treatments, like cisplatin and its derivatives, are essential in driving oncological therapy further. Aside from cisplatin, other approaches being pursued using metal-based therapy include the utilization of copper complexes, which have been found to exhibit cytotoxic activity against many cancer cell types. Their therapeutic potential lies in the extremely unique ability to induce oxidative stress, which leads to cell damage and apoptosis. The mechanisms underlying these phenomena are crucial for the discovery of new drugs because they determine the efficacy and selectivity of the drugs. The elucidation of the structure-activity relationships of copper(I) complexes is shown in current researches as a point of relevance to ensure the optimization of the antitumor activity.

An attempt to design such complexes is also supported by evidence showing the versatility of transition metals to show various oxidation states and coordination geometries, which enhance their biological interactions (Dudek-Różycki et al., 2013). Apart from cisplatin and copper complexes, α-N-Heterocyclic thiosemicarbazones (N-TSCs) have also proven to be among the hopeful candidates among antitumor metal-based agents.

Such compounds are potent inhibitors of the enzyme ribonucleotide reductase, which plays a crucial role in DNA repair and synthesis, and are a promising anticancer target. By structural characterization and mechanism-of-action analysis, researchers have begun to elucidate the parameters governing their therapeutic properties. This. Further research is important not only because it illuminates the governing principles of bioinorganic chemistry but also because it guides the purposeful design of novel therapeutic agents which take advantage of the favorable properties of metal complexes. As this emerging field advances, it once more places the metals at the focal point of advancing the therapeutic approach to cancer with the promise of new hope for clinicians and patients alike (Figure 9.7).

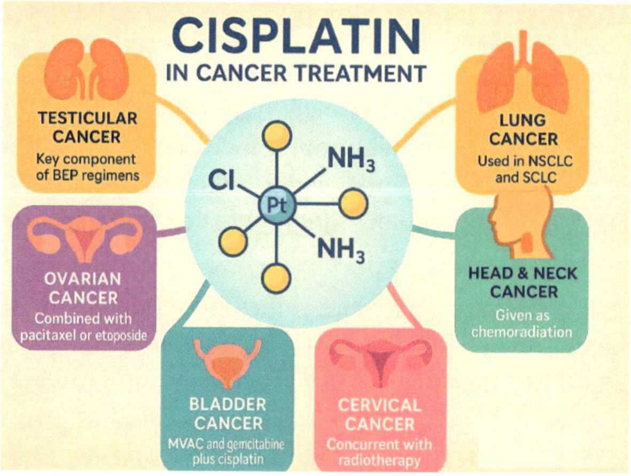

Figure 9.7. *Cisplatin in cancer treatment: Mechanism, resistance, toxicity, and future directions.*

Source: https://oncodaily.com/pub/uploads/2025/07/An_infographic_created_by_OncoDaily_visually_illus.jpg.

9.4. BIOINORGANIC CHEMISTRY IN ENVIRONMENTAL SYSTEMS

The interface of bioinorganic chemistry with environmental systems provides an intriguing model for the study of how inorganic components influence biological function in varying ecosystems. At the center of this relationship is the role of metals in biochemistry, especially transition metals, in enabling the function of most enzymes and metalloproteins. These biomolecules facilitate important biological processes, like photosynthesis and respiration, processes that essentially underlie energy transfer within ecosystems. The energy input variations to these systems, as indicated in historical observations, are proportional to wholesale transformations wrought by human activities, most significantly since the Industrial Revolution. Through the examination of how such technological advances have altered biological organization and natural material cycling, scientists can better understand not only the beauty of ecological processes but also the urgency of sustainability for both our energy consumption and material uses (Lenton et al., 2016). New research within the discipline of microbial bioinorganic chemistry has been a focus in recent years with its potential application in environmental remediation. Certain microbes possess the remarkable capacity of mobilizing and altering inorganic elements that plays a pivotal function in biogeochemical cycles. The application of bio-mineralization processes, for instance, has been viewed as an advanced technique for renovating and reinforcing construction material such as concrete. This technique utilizes the inherent nature of bacteria to precipitate calcium carbonate, not only mending cracks in concrete but also lessening the utilization of environmentally destructive chemical treatments (Berenjian et al., 2016). Additionally, understanding these microbiota interactions with metals and their subsequent impact on material properties is essential, as this validates the suggestion that bioinorganic structures are central to the creation of ecological well-being in addition to material sustainability. The concept of uniting bioinorganic chemistry with current environmental uses is not only timely but necessary in the aftermath of issues plaguing modern society, such as pollution and resource depletion. As the need intensifies for greener materials, progress in bioinorganic applications is increasingly sought after for their promise to create sustainable technologies. For example, microbial biomineralization could be a sustainable pathway for manufacturing sustainable building materials with strength and low environmental impact. The studies on these biological systems continue to emphasize the need for interdisciplinary collaboration to enhance understanding and application of bioinorganic chemistry. The future of the earth's environmental systems, in fact, hangs in the balance of our capacity to tap these natural processes and create a paradigm that is in consonance with human activities as well as the ecological requirements of the earth (Berenjian et al., 2016; Lenton et al., 2016).

9.4.1. Role of Metals in Biogeochemical Cycles

Metals play a critical role in the biogeochemical cycles of the earth, primarily acting as essential micronutrients to various biological processes. They take part actively in enzyme-catalyzed reactions, catalyzing metabolic transformations essential for life. For instance, trace metals such as iron, copper, and zinc are structural components of catalytic proteins that catalyze a range of biochemical transformations. Cycles and availability of these metals are controlled by geological processes, atmospheric processes, and living organisms, creating a complicated set of interactions within ecosystems. When such organisms incorporate these metals, they are biotransformed and the distribution and availability of metals in the environment change. The balance between metal input and output is important because any disruption may lead to nutritional deficiencies or toxic accumulations and demonstrates the delicate balance in biogeochemical cycles. Understanding such dynamics assists in the appreciation of larger ecological implications and the health of marine and terrestrial ecosystems.

The redox reactions have a strong control over the biogeochemical cycles, with metals playing a vital role as electron donors and acceptors. Cycling of metals such as chromium and molybdenum is of interest because it yields valuable insights into ancient oceanic conditions. It has been studied that metal enrichment records provide an opportunity to study changes in seafloor redox environments at a geological timescale. For instance, the changing oxidation states of trace metals are indicative of conditions within the deep oceans at different times, as based on sediment core research. Such indications point to the fact that, during the Proterozoic Eon, the sea was predominantly anoxic and had impacts on microbial life and biogeochemical processes on a grand scale. Moreover, metal reservoir sensitivity to ocean redox states defines the ability of metals in deciphering past and current environmental changes, including those likely responsible for future ecological transformations due to global warming and pollution (Bekker et al., 2013). New advancements have brought forward the concept of a metal-limited biosphere as a crucial element with the ability to influence marine ecosystems productivity and health. As an example, the reaction of molybdenum with nitrogen from seawater has been identified as being of prime significance to primary production, particularly for those organisms that rely on such nutrients. When metal availability is reduced, nitrogen and other major nutrient cycles can be restricted, leading to reduced biological productivity. This association possesses paramount importance in understanding the marine ecosystem response to changes in environmental conditions, such as anthropogenic nutrient loading. Metal cycling's, particularly under changing redox conditions, thus signal intricate associations with nutrient cycling, thus impacting the ecological balance. Through the exploration of these critical interactions, scientists can better predict future trends in biodiversity and ecosystem function in the context of ongoing climate change and human manipulation of the environment (Bowden et al., 2014).

9.4.2. Impact of Heavy Metals on Ecosystems

Heavy metal presence in ecosystems presents severe challenges to environmental health and biogeochemical processes. Mercury, cadmium, and lead are metals that are toxic to various organisms, disrupting metabolic functions and arresting biological processes. This is not confined to a sole species but extends to having cascading effects within the ecosystem. For instance, metals required for microbial processes, iron and copper, are crucial for microbial life, which in turn affects greenhouse gases. Research has indicated that low bioavailability of these metals limits processes like methanogenesis to avoid conversion of organic matter into methane, a potent greenhouse gas. These interactions are significant to comprehend because metallic pollutants can alter microbial community structures, and this changes ecosystems' capacity to work correctly and maintain carbon cycling (Glass et al., 2012). In addition, the action of heavy metals with biological targets has the potential to impose heavy influences on microbially-mediated processes in the environment, such as denitrification. Heavy metals have the potential to influence the microbial populations involved in nitrogen cycling, shifting the pathways in a direction that is supportive or prohibitive of the formation of nitrogen gas. This is particularly so under anaerobic conditions, in which metal requirements define the efficiency of key enzymes involved in nitrogen conversions. For instance, the formation of nitrous oxide—a very powerful greenhouse gas—is influenced by metal limitation, and its accumulation in settings where copper is in short supply. From this perspective, a keen interest is given to elucidating the intricate interactions between heavy metals and microbial metabolism in natural ecosystems because it can demonstrate the operational mechanisms that govern ecosystem service modification and greenhouse gas flux dynamics (Glass et al., 2012). Research on the bioremediative capacity of organisms that have been cultivated under heavy metal exposure points to their adaptation processes towards environmental pollutants. For example, research on the saprotrophic fungus *Penicillium griseofulvum* reveals it to be tolerant to contaminated environments, where it develops biotransformation processes that reduce the toxic effect of heavy metals. This organism's ability to catabolize hexachlorocyclohexane along with metals illustrates the complex interactions between heavy metals and organic pollutants in a contaminated environment. Such findings point towards the importance of studying microbial functions in xenobiotic-laden environments. By employing such natural metabolic pathways, it becomes possible to come up with strategies for remediating contaminated ecosystems in an effective manner, ultimately enhancing ecosystem resilience against heavy metal toxicity and ensuring sustainability by virtue of bioremediation practices (Figure 9.8).

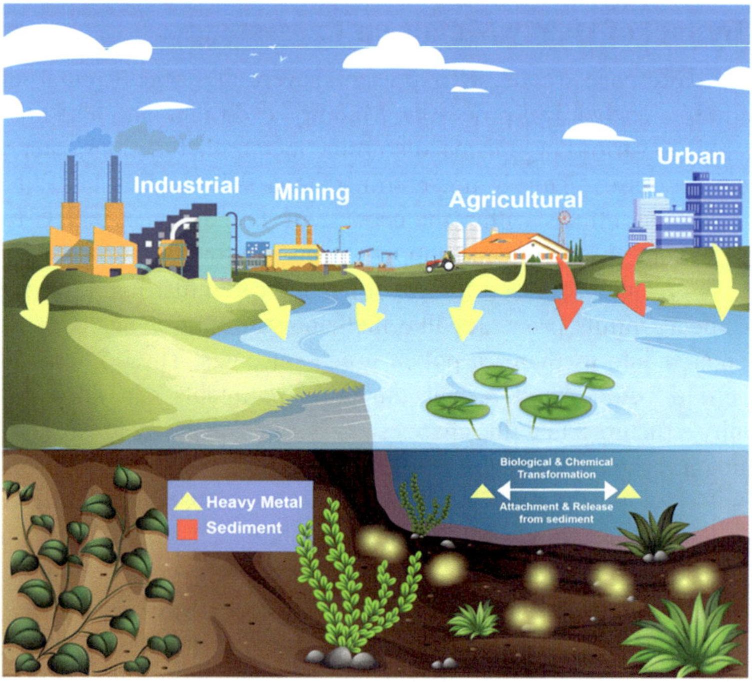

Figure 9.8. *Impact of heavy metals on ecosystems.*

Source: https://ars.els-cdn.com/content/image/1-s2.0-S1018364722000465-gr1.jpg.

9.4.3. Bioremediation Strategies Using Bioinorganic Principles

The intricate interactions between bioinorganic chemistry and bioremediation strategies depict a complex means of addressing environmental pollution. Behind these strategies lie metal ions and metalloenzymes that play a key role in mediating the biochemical interactions necessary in detoxifying hazardous pollutants. For instance, microbes utilize metal ions as a co-factor in enzymatic reaction processes facilitating breakdown of complex organic pollutants.

Scientists can improve the efficiency of bioremediation by leveraging on the inherent potential of such microorganisms. Bioinorganic strategy evolution assists not only in the explanation of metal ion interactions but also in the design of the best conditions for microbial growth, hence minimizing pollutant degradation. Emerging techniques using carbon-based nanomaterials (CBNs) are the latest innovation in bioremediation. The nanostructures can either be used separately or in combination with bacteria, providing a new nanobiotechnological solution to water purification. As established in recent studies, CBNs possess superior adsorption capacity, effectively removing a broad range of industrial contaminants from wastewater, from heavy metals to organic dyes. Bacterial combination enhances their performance, resulting in better pollutant degradation through synergistic interactions. As they grow, their adherence to sustainability principles becomes increasingly crucial to propel circular economy activities without decreasing the bioremediation efficiency required to address worldwide environmental challenges (Octaviano et al., 2024). Use of bioinorganic approaches in bioremediation not only addresses current ecological requirements but also stimulates deeper understanding of environmental interaction at the molecular level. Labs like the Environmental Molecular Sciences Laboratory (EMSL) are essential to providing scientists with access to leading-edge facilities in this field of science. These labs enable research that uncovers the sophisticated interactions between pollutants and metal ions and microbial systems and thereby constructs the pillars of successful bioremediation strategies. These experimental results, highlighted in EMSL's quarterly reports, document the growing significance of bioinorganic chemistry in environmental management. Effectively managed bioremediation processes not only restore contaminated sites but also assist in enhancing the overall understanding of ecological recovery and resilience, proving the utmost significance of ongoing research and development in this sector (Beckman et al., 2010).

9.5.ANALYTICALTECHNIQUESINBIOINORGANIC CHEMISTRY

The field of bioinorganic chemistry encompasses a variety of analytical strategies necessary to understand the complex interaction between metals and biological systems. Spectroscopic methods such as infrared (IR) spectroscopy are of vital importance in identifying metal-binding sites within proteins. IR spectroscopy can identify structural alteration occurring on the interaction of metal species with biomolecules. For instance, IR spectroscopy can be used to trace the formation of bioinorganic assemblies from proteins like lysozyme and aluminum species. These aggregates demonstrate the presence of some charge-stabilized complexes among the species and elucidate the pivotal role of acidic surface residues like aspartate and glutamate in facilitating interaction. Such findings establish the key role played by analytical techniques to unveil the nuances of metal-protein interactions and the attendant implications for biological function and stability, thus improving our appreciation of bioinorganic chemistry as a discipline. Other than spectroscopy, chromatography also serves a complementary role in the analysis of bioinorganic compounds. Separation and identification of various metal complexes or biomolecules are crucial to determine their biological activity and modes of action. Column chromatography, for example, can be used to separate fractions with varying bioactive compounds, as some studies of *Schinus molle* leaf extracts have shown. Employment of chromatography to isolate separating fractions allows researchers to study the cytotoxicity of a single compound against human cell lines and hence establish a connection between bioactive compounds and their potential therapeutic applications. It gives greater information on the interaction of metal ions and coordination complexes with biological molecules, which is important for identifying their roles in medicinal chemistry and the entire field of bioinorganic chemistry. Finally, mass spectrometry (MS) is a powerful analytical method in bioinorganic chemistry, particularly for metal-bearing biomolecule identification of mass and structural information. The development of MS techniques, such as tandem mass spectrometry, provides sophisticated analysis capable of identifying the occurrence and structure of metal complexes in complex biological matrices. Such an approach is particularly relevant in the study of metalloproteins and their multifaceted role in biology. By the deployment of mass spectrometry sensitivity and specificity, scientists can determine the differences in metal-binding affinity and stability among proteins. Overall, the synergy of a range of analytical instruments including infrared spectroscopy, chromatography, and mass spectrometry is essential to expanding our knowledge of bioinorganic chemistry and its consequences in the fields of molecular biology all the way to pharmacology, and hence metal ion interaction complexity in biological systems.

9.5.1. Spectroscopic Methods for Metal Detection

Spectroscopic detection of metals is a significant aspect of bioinorganic chemistry. Such methodologies enable scientists to follow metal interactions within the biological system, as is required to comprehend their role in enzymic processes as well as physiological functions. A common technique employed is that of fluorescence spectroscopy, as it is highly sensitive to metal ions. For instance, the preparation of fluorescent chemosensors like L1 has been highly effective in the detection of metals like Cu^{2+}. Upon introduction with the ion, L1 exhibited an increase in the intensity of the fluorescence, pointing towards the responsiveness and specificity of the sensors towards copper. These advancements demonstrate the strength of fluorescence in enhancing our comprehension of trace metal dynamics and their impact on biochemical processes, thus making it a necessary method of research in bioinorganic chemistry (Georghiou et al., 2016). Aside from applying fluorescence methods, other spectroscopic techniques such as nuclear magnetic resonance (NMR) and ultraviolet-visible (UV-Vis) spectroscopy are imperative to the study of metal coordination environments. NMR reports information about metal complex electronic and structural properties and hence supports analysis of binding phenomena. For instance, NMR titration experiments can elucidate interaction forces between some metal ions and organic ligands.

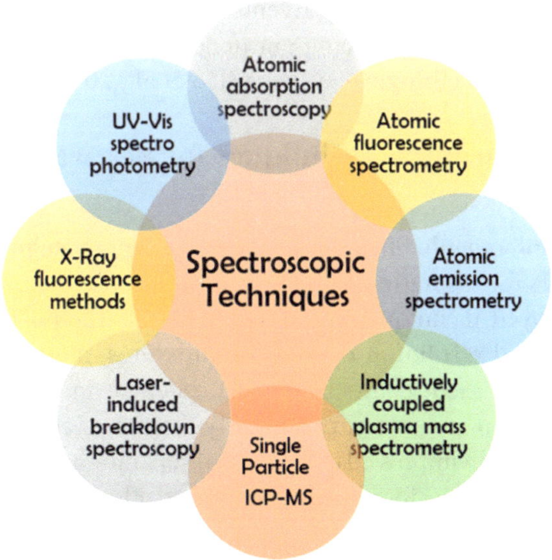

Figure 9.9. *Spectroscopic methods for metal detection.*

Source: https://media.springernature.com/lw685/springer-static/image/chp%3A10.1007%2F978-3-030-80334-6_1/MediaObjects/512416_1_En_1_Fig2_HTML.png.

Moreover, UV-Vis spectroscopy is critical in the determination of metal-induced electronic state changes and electronic transitions, revealing key information on metal-ligand coordination. Together, these methods provide an multi-faceted means of studying metal ions in bioorganisms and the development of targeted sensors and chelators for use in the environment and in the clinic. This careful investigation illuminates the fundamental functions which metals play in biomolecular processes (Do et al., 2011). Furthermore, ongoing investigations of structural properties of metal-binding sites continue to put the preeminence of spectroscopic techniques in the spotlight of the science of bioinorganic chemistry. Studies have been focused on the coordination environments and geometries of metal centers, particularly in models of biological importance like diiron complexes. By using diverse spectroscopic methods, researchers can access the transient intermediates and measure the reactivity of the centers with molecules like dioxygen. The utilization of spectroscopic techniques to study these complexes has revealed electronic properties that dictate the intermolecular interactions of metals with substrates in biology. The advancements in the techniques of stabilizing wanted molecular species and preventing unwanted metal clustering are fundamental to the development of new models imitating biological function and reactivity. Eventually, these will drive the field of bioinorganic chemistry by advancing our understanding of metal behavior and their incorporation into biological systems (Do et al., 2011) (Figure 9.9).

9.5.2. Chromatographic Techniques for Separation and Analysis

In the field of bioinorganic chemistry, chromatographic techniques have been fundamental tools for separation and analysis of biomolecules. Among the most frequently employed techniques, High-Performance Liquid Chromatography (HPLC) has enabled resolution of complex mixtures, allowing one to detect and analyze metal-based compounds that interact in biological systems. Development in chromatography, particularly the use of core-shell particles in HPLC, has significantly improved efficiency of separation. This new technology replicates UHPLC-like performance under the deployment of regular HPLC instrumentation. Such methodologies make HPLC a sensible option in various analytical scenarios, including pharmaceuticals, environmental analysis, and toxicology analysis, where the work is the separation of trace levels of analytes efficiently and precisely (Preti et al., 2016). Employment of chromatographic techniques is not only limited to mere separation but, instead, they contribute fundamentally towards understanding neonicotinoid insecticide migration and retention under varied conditions. For instance, the migration of neonicotinoid insecticides such as clothianidin and thiamethoxam can be more effectively explored using chromatographic techniques in an attempt to examine their elution characteristics in various soils. The findings point towards the fact that the soil composition has a significant influence on the absorption and desorption of such active ingredients, which are crucial in measuring their environmental impacts. On the basis of analytical precision provided by chromatography, scientists have demonstrated that soil organic matter and water solubility control the environmental fate of such compounds. The chromatographic techniques, thus, not only serve as a separation technique but also serve as a critical tool in ecological risk assessment and regulatory measures (András Székács et al., 2016). Further, the addition of chromatographic techniques to bioinorganic chemistry offers better insight into metal ion-biomolecule interactions. By using methods like size-exclusion chromatography and ion-exchange chromatography, researchers are able to break down the complicacies of metal ion functions in biological systems. These approaches give insight into metal agent speciation and binding capacities toward several biologic ligands, giving us a better understanding of metal-mediated biochemistry. Through the deconstruction of these complex interactions, researchers can guide metallotherapeutic development that utilizes targeted metal coordination and binding principles. In brief, chromatographic techniques form a fundamental cornerstone of bioinorganic chemistry that enables crucial discoveries both in analytical as well as practical contexts.

9.5.3. Advances in Imaging Techniques for Biological Samples

The relentless development of imaging methods has greatly improved our knowledge of biological samples, especially that of bioinorganic chemistry. Techniques like synchrotron X-ray fluorescence spectroscopy (XFS) offer unparalleled levels of information on the localization and distribution of metal ions in cells and tissues. It works at the nanoscale to enable scientists to see where the metals are accumulating and how they relate to other biological processes. Having the capability to chart the spatial distribution of metals is important to studying their place in metabolic pathways and their significance in disease. New imaging capabilities provide a thorough study of metallomics, which deals with the relationship between metals and biological systems. This merging of imaging technology is useful to clarify metal ion-induced biochemical pathways and hence contribute to an expansion of our understanding of their inherent biological functions as well as their possible applications in therapeutic terms (GB et al., 2013). Increased improvement in imaging methods has also been observed with the innovations in multiaxis dimensional liquid chromatography combined with inductively coupled plasma mass spectrometry (LC-ICPMS). This approach increases resolution and sensitivity, enabling scientists to distinguish and quantify metal species in intricate biological matrices with excellent precision. The incorporation of such sophisticated imaging tools facilitates the study of metal interactions at the cellular level, with a specific emphasis on their dynamic behavior and trafficking within organisms. Since metal ions may either function as requisite cofactors for enzymatic processes or as harmful substances, their accurate characterization plays a central role in pharmacology and toxicology investigations. Such thorough examination opens the door to the discovery of new therapeutic interventions for counteracting metal toxicity while taking advantage of the positive attributes of essential trace metals (GB et al., 2013). In medical research, imaging technology progress also comes in the form of new radionuclide generator applications, e.g., the 212Pb/212Bi generator. The technology is a prime example of where imaging and therapy can merge in bioinorganic systems. The generator takes advantage of the half-life of 212Pb to enable targeted delivery of the therapeutic isotope 212Bi, which has major cancer-treatment implications. Application of such alpha emitters with high therapeutic promise highlights the significance of accurate imaging for tracking their distribution and effects in biological tissues. Additionally, knowledge of the chemical processes involved in radionuclide decay is needed to assess the generator's clinical effectiveness. This overlap of imaging and therapeutics demonstrates the game-changing potential of advanced methods in promoting bioinorganic chemistry and improving biomedical applications (Figure 9.10).

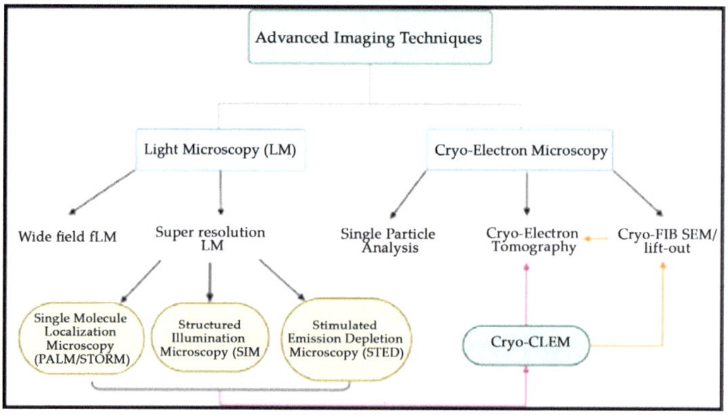

Figure 9.10. *Advances in imaging techniques for biological samples.*

Source: https://ars.els-cdn.com/content/image/1-s2.0-S0065216423000011-f01-01-9780443192708.jpg.

9.6. CURRENT RESEARCH TRENDS IN BIOINORGANIC CHEMISTRY

There has been a growing importance of bioinorganic chemistry in recent years as an interdisciplinary field crossing the boundaries of significant environmental concerns, particularly through its advances in knowledge regarding biogeochemical cycles. Trends in research have been centered more on the function of metal ions in life, for example, their participation in enzymatic reactions that help to cycle nitrogen. Denitrification, for instance, oxidizes biologically available nitrogen to nitrogen gas, a reaction which is necessary for ensuring ecological equilibrium. Research points to extensive loss of nitrogen in oceanic environments, where other processes using manganese oxides are being considered. Such other processes complicate the conventional understanding of the nitrogen cycle and highlight the sophistication of microbial interaction in aquatic ecosystems. Therefore, as scientists discover these new mechanisms, there is increased appreciation of how bioinorganic chemistry can offer information on sustainability in the environment and nutrient cycle management (Aller et al., 2005). In addition, the inclusion of nanotechnology in bioinorganic chemistry is revolutionizing existing research patterns by providing new answers to environmental issues. The potential for nanoscale manipulation of materials offers particularly valuable opportunities for building more efficient bioremediation schemes, specifically in the area of pollution management. Research finds that nanoparticles have the potential to increase detection and removal of contaminants from water supplies, which is ever more vital given current increases in global pollution. Scientists are exploring how the characteristics of nanoparticles contribute to their reactivity and efficiency in breaking down pollutants. This not only illustrates the applied aspects of bioinorganic chemistry but also illustrates the interdisciplinary aspect of the field as it draws on principles from nanotechnology, environmental science, and materials engineering (Chandrakar et al., 2023). Lastly, the study of metal-dependent enzymes and mechanisms has taken substantial momentum in bioinorganic chemistry studies. These enzymes, usually metal cofactors, are responsible for important oxidative and reductive processes, essential for energy transfer and metabolism in living organisms. Future research is investigating the catalytic role of these metal ions, looking at how they affect enzymatic activity and stability with different environmental conditions. It not only facilitates an understanding of biological processes but also opens avenues for biotechnological applications, including the design of biocatalysts for industrial processes. Ongoing research seeks to emphasize the subtle interplay between metal ions and biological systems and reiterate the relevance of bioinorganic chemistry as a unifying force for both basic scientific inquiries and applied problems in diverse areas (Aller et al., 2005).

9.6.1. Nanotechnology and Its Applications in Bioinorganic Research

The developing branch of nanotechnology holds immense promise for bioinorganic research, especially in creating new therapeutic agents. Nanoscale materials allow for improved targeting and delivery of bioactive molecules with the potential for optimizing therapeutic effects while reducing off-targeting. For example, the creation of nanoparticles capable of drug encapsulation facilitates the use of controlled release mechanisms to improve the bioavailability of therapeutic agents. Additionally, progress in bioinorganic nanomaterials, like gold and silver nanoparticles, has been promising in enhancing imaging methodologies and enhancing the sensitivity of diagnostic tests. The materials are not only capable of acting as carriers for therapy agents but also as biological imaging probes, hence spanning bioinorganic chemistry into state-of-the-art biomedical developments, inherently solving issues in conventional treatment regimens (Mitchell et al., 2018). In addition, the interaction between nanomaterials and biological systems offers rich soil for creative research in bioinorganic chemistry. The embedding of metal ions into nanostructured materials has been found to have a profound effect on enzymatic behavior. Metal nanoparticles, for instance, can be mimics of natural enzymes, enabling biochemical transformations under mild conditions and providing alternatives to traditional catalytic processes. This is directly applicable in artificial metalloenzyme design, which may construct new routes toward complex organic conversions. These advances point toward the hope offered by combining nanotechnology with bioinorganic concepts to yield new biomimetic catalysts that will change synthetic chemistry (Brown et al., 2015). As such, it is critical to understand these interactions to leverage the complete therapeutic value of nanomaterials. In the treatment of cancer, nanotechnology's application has provided new avenues for treatment modalities through pioneering the mode of administration of therapies. Chimeric Antigen Receptors (CARs) is an innovative technology whereby nanoparticles are able to encapsulate genetic material that guides immune cells to destroy and target cancer cells specifically. The technology was designed to reduce the unwanted damage to normal tissues usually observed in conventional cancer therapies, leading to improved patient outcomes. In addition, CAR-expressing exosomes developed from nanoparticles provide a more biocompatible vector compared to conventional vectors, with the hope of decreasing side effects while enhancing the therapeutic agents' efficacy (Mitchell et al., 2018). By connecting nanotechnology capability to bioinorganic science, scientists are able to move forward towards more targeted and potent cancer therapies that conform to the principles of personalized medicine.

9.6.2. Development of New Metal-Based Biomaterials

The creation of novel metal-based biomaterials is a great leap in the practice of bioinorganic chemistry, especially because they have prospects for their use in regenerative medicine and drug delivery systems. They usually consist of metals like titanium, aluminum, and silver, which are each important for favorable biological response to implants as well as enhancing the therapeutic activity of drug formulations. As an example, titanium is known to be highly compatible with living tissues and strong mechanically, and therefore is found often in orthopedic and dental implantation applications. Improved surface modifications, such as alloying and coating, seek to advance the osseointegration of such materials further, overcoming the limitations that are presented by the body's reaction to foreign agents.

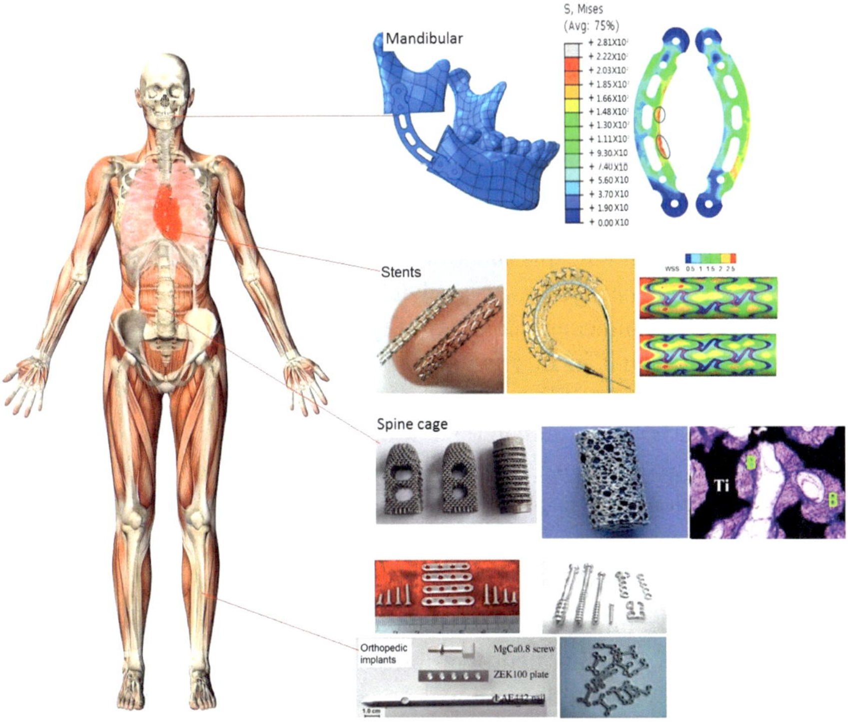

Figure 9.11. *Development of new metal-based biomaterials.*

Source: https://www.mdpi.com/ijms/ijms-19-00024/article_deploy/html/images/ijms-19-00024-g001.png.

Such novel strategy not only enables the integration of implants but also opens avenues for the creation of advanced biomaterials that take advantage of the physical and chemical properties of metals to enhance patient outcomes. Recent studies have explored the creation of bioinorganic assemblies that marry protein interactions with metal species, providing insight into the dynamics of metal-based biomaterials. Experiments have shown that the protein-metal species interactions, such as between lysozyme and BSA with aluminum species, produce stable complexes with characteristic structural properties controlled by the charge and composition of the proteins surface. For instance, Deschaume et al. (2008) indicates that there are particular interactions between aluminum species and lysozyme that lead to distinct structural assemblies that are quite distinct from gel-like aggregates with BSA. Knowledge of these interactions provides meaningful information that can inform the design of better biomaterials, ensuring they tap the synergies between inorganic metals and biological molecules to create greater biocompatibility and performance within medical applications. Another potentially fruitful frontier in metal-based biomaterials research is the investigation of the material-microorganism interface, particularly for uses in biocatalysis and environmental remediation. Advances in hybrid technologies that couple inorganic semiconductor light harvesters to biological systems show the potential in metal-based materials to power biochemical reactions and hence improve basic processes like photosynthesis. As Cestellos-Blanco et al. (2018) has pointed out, this junction creates important questions regarding energy transfer and charge processes between the two different entities, the optimization of bio- and chemocompatibility being thus exceedingly important. The continued research in this field not only works towards optimizing the productivity and sustainability of chemical synthesis but also broadens the function of metal-based biomaterials into new areas beyond conventional medical fields, making them key players in the future of bioinorganic chemistry and green technology (Figure 9.11).

9.6.3. Future Directions and Challenges in the Field

As bioinorganic chemistry continues to grow, emerging directions will be directed more toward the incorporation of advanced materials and novel methodologies for sustainable chemical processes. An area with great promise is the synthesis of semiconductor nanowires, which have novel properties that further increase their effectiveness in photoelectrochemical applications. By replicating the process of photosynthesis in nature, these nanostructures provide a means of obtaining chemicals from sunlight, air, and water in pursuit of global sustainability objectives. Additionally, research is presently seeking to answer fundamental questions regarding interfacing of the nanowires with other materials to optimize photoelectrochemical responses. Strategies to develop integrated nanosystems to mimic closed cycles of artificial photosynthesis will certainly push the frontier, eventually leading to the development of solar-driven chemical plants to produce renewable energy and minimize environmental footprint (Deng et al., 2019). No less challenging are the issues from the accurate modeling and description of open-shell metal complexes in quantum chemistry frameworks. Although density-functional theory (DFT) is still a dominant approach in this field, its constraints require ongoing refinement to improve predictive power. As the intricacies of transition metal clusters and complexes become increasingly clear, researchers are challenged to revisit the principles of DFT, especially with regard to applying it to open-shell systems. Studies on the density of spin and the development of exchange-correlation energy functionals will set the stage for more precise quantum mechanical models to accurately predict reactions and interactions in bioinorganic systems. These developments will be crucial to the understanding of vital biochemical processes and in the construction of novel biomimetic catalysts, which will validate the application of bioinorganic chemistry in green technologies (Aquino et al., 2012). Future research efforts should also take into account inter-disciplinary collaborations between bioinorganic chemists, material scientists, and computational chemists to address such priority challenges. Through their development of a multi-disciplinary approach, researchers are able to combine insights from diverse disciplines, leading to a deeper understanding of intricate biochemical interactions as well as the creation of innovative technologies. Synergistic efforts may result in new bioinspired catalysts that efficiently mediate critical biochemical reactions, vital to medicine and environmental sustainability. While underscoring this interdisciplinarity, the bioinorganic chemistry community is able to address the challenges of creating sustainable solutions while overcoming the scientific issues involved in modeling and experimentation. As the discipline evolves, a concerted effort will promote breakthroughs that assure its dominance in developing environmentally friendly solutions and elucidating the underlying biochemical processes that control life (Aquino et al., 2012).

SUMMARY

In Summary, the discipline of bioinorganic chemistry has a central role in expanding our knowledge of the complex interactions between metal ions and biological systems. The study of metal-containing biomolecules not only brought to light their inherent biological activity but also helped in designing new therapeutic strategies. Through the analysis of metal ion interactions with proteins, nucleic acids, and other biomolecules, scientists are able to uncover life processes' underlying mechanisms. This information not only moves the scientific community's understanding of biochemistry forward, but it is also at the forefront of medical development, notably the development of targeted treatments for disease states involving metal ion dysregulation. Additionally, recent research highlights the significance of incorporating molecular dynamics simulations into bioinorganic studies in order to improve our ligand binding affinities and selectivities predictions. For example, the study of zinc-dependent matrix metalloproteinases (MMPs) emphasizes how mobility within an enzymes active site can greatly dictate the effectiveness of putative inhibitors. The bioinorganic methodology employed in the discovery of new inhibitors, like AM-6, emphasizes the importance of taking protein dynamics into account to make precise predictions. The results indicate that considering only static structures can provide an incomplete representation of molecular interactions and highlight the importance of looking at the bigger picture in drug design. Last but not least, such integrative studies bear not only fundamental implications but also pave the way to innovative treatments of metal ion-related pathologies. As mentioned, MMPs are also vital in many diseases, such as cancer and cardiovascular diseases, where their function is stringently controlled by zinc. The selectivity of the observed AM-6 towards MMP-3 over MMP-2 shows how specific inhibitors can be used to improve therapy. By continuing to investigate the dynamics of metalloprotein interaction within biological systems, the bioinorganic chemistry community can foster the development of more specific and effective treatments, illustrating the fields potential to impact health and disease profoundly. In this light, the intersection of inorganic chemistry and biological sciences not only enriches our scientific knowledge but also serves humanity by addressing critical medical challenges.

In the bioinorganic chemistry field, a close scrutiny of the inorganic elements' interactions with biological systems unearths a terrain filled with novel chemical mechanisms and properties. The conversations tend to revolve around the metal ions' roles in biological catalysis, electron transfer mechanisms, and structural stabilization in biomolecules. Main themes cover the role of metalloproteins and metalloenzymes, which emphasize the role of metal ions like iron, zinc, and copper in biological processes through catalyzing redox reactions or as structural elements in enzymes. Complexity of such interactions provides insight into nature and has implications for applications from biocatalysis to medical treatments. These conversations highlight the need for a solid grasp of both inorganic chemistry and biochemistry, moving beyond classical discipline divides to guide current research (Bertini et al., 1994). Another area of emphasis in bioinorganic chemistry is the novel exploration of enzyme systems, most notably halogenases, which have become increasingly prominent over the past few years. These enzymes exhibit the wonderful mechanisms by which nature utilizes halogens in organic molecules, a critical process for the biosynthesis of natural products. New research reveals the varied mechanisms employed by halogenases through computational work that supplements experimental research, hence enhancing understanding of enzymatic activity. Challenges still exist, particularly in outlining the specific mechanistic pathways and in optimizing computational methods for the study of such reactions. The major progress in enzymology enhances appreciation of how basic principles of bioinorganic chemistry are able to solve intricate biological processes, opening up possibilities for potential biotechnological applications (Senn et al., 2014). The metal complex synthesis and functional incorporation into biological systems is another critical area of bioinorganic chemistry that has been given attention. Research on synthetic metallodrugs and their behavior in biological systems emphasizes not only the therapeutic potential but also the difficulties these substances encounter, such as bioavailability and selectivity. Commonly, debates over such research stress the need for interdisciplinary interaction, with chemists, biologists, and materials scientists cooperating to tap the properties of bioinorganic compounds for therapeutic applications. With the progression of the field, cooperative efforts are necessary to close the gap between theoretical knowledge and applications, so that the findings in bioinorganic chemistry lead to real benefits in terms of health and technology. Such a multidisciplinary effort represents the promising future of bioinorganic research and its capacity to revolutionize different scientific fields (Bertini et al., 1994).

REFERENCES

1. Aird, A., Ang, A., Barnard, A., Barnes, A., Bednarski, A., Bergamo, A., Bergamo, A., et al. (2009). Current applications and future potential for bioinorganic chemistry in the development of anticancer drugs. https://core.ac.uk/download/46355.pdf (accessed on 13th September 2025).
2. Aller, R., Curwood, R., Canfield, D. E., Dalsgaard, T., Engström, P., et al. (2005). Nitrogen removal in marine environments: Recent findings and future research challenges. https://core.ac.uk/download/12043452.pdf (accessed on 13th September 2025).
3. Andreini, C., Furnham, N., Rosato, A., Thornton, J., et al. (2017). To what extent do structural changes in catalytic metal sites affect enzyme function? https://core.ac.uk/download/141439366.pdf (accessed on 13th September 2025).
4. Aquino, A., Autschbach, J., Ayers, P., Baerends, E., Bagus, P., Baron, R., Baym, G., et al. (2012). Spin in density-functional theory. http://arxiv.org/abs/1206.2234 (accessed on 13th September 2025).
5. Barry, N. P. E., & Sadler, P. J. (2012). Dicarba-closo-dodecarborane-containing half-sandwich complexes of ruthenium, osmium, rhodium and iridium: Biological relevance and synthetic strategies. https://core.ac.uk/download/9053752.pdf (accessed on 13th September 2025).
6. Beckman, M. T., Carper, R. R., Manke, K. L., Showalter, J., et al. (2010). *EMSL Quarterly Highlights Report Second Quarter, Fiscal Year 2010 (January 1, 2010 Through March 31, 2010).* https://digital.library.unt.edu/ark:/67531/metadc847203/m2/1/high_res_d/1025097.pdf (accessed on 13th September 2025).
7. Bekker, A., Gill, B. C., Konhauser, K. O., Lalonde, S., et al. (2013). Proterozoic ocean redox and biogeochemical stasis. https://core.ac.uk/download/12817719.pdf (accessed on 13th September 2025).
8. Berenjian, A., Samani, A. K., & Seifan, M. (2016). Bioconcrete: Next generation of self-healing concrete. https://core.ac.uk/download/85165317.pdf (accessed on 13th September 2025).
9. Bernardini, A., Ceci, A., Ceci, A., Ceci, A., Ceci, A., Chanda, A., Mishra, A., et al. (2018). Metabolic synergies in the biotransformation of organic and metallic toxic compounds by a saprotrophic soil fungus. https://core.ac.uk/download/132208310.pdf (accessed on 13th September 2025).
10. Bertini, I., Gray, H. B., Lippard, S. J., & Valentine, J. S. (1994). *Bioinorganic Chemistry*. https://core.ac.uk/download/4890319.pdf (accessed on 13th September 2025).

11. Boon, N., De Corte, S., De Gusseme, B., Hennebel, T., et al. (2011). Bio-palladium: From metal recovery to catalytic applications. https://core.ac.uk/download/55890952.pdf (accessed on 13th September 2025).

12. Booth, C., Brecher, D., Lowe, M. S., Stone, J., et al. (2014). *Chemistry Curriculum Map 2013–2014*. https://core.ac.uk/download/70981645.pdf (accessed on 13th September 2025).

13. Bowden, S., Parnell, J., Spinks, S., Still, J., et al. (2014). Cadmium sulfide in a Mesoproterozoic terrestrial environment. https://core.ac.uk/download/29151116.pdf (accessed on 13th September 2025).

14. Bowman, J. C., Bray, M. S., Glass, J. B., Haynes, K., et al. (2018). Multiple prebiotic metals mediate translation. https://core.ac.uk/download/323063003.pdf (accessed on 13th September 2025).

15. Bowman, S. E. J., Drennan, C. L., & Rabb, J. (2015). Metalloprotein crystallography: More than a structure. https://core.ac.uk/download/85123347.pdf (accessed on 13th September 2025).

16. Brown, D., Estrada, D., Fielding, A. L., Heishman, J., et al. (2015). Production and characterization of graphene and other 2-dimensional nanomaterials: An AP high school inquiry lab (Curriculum Exchange). https://core.ac.uk/download/61750667.pdf (accessed on 13th September 2025).

17. Ceresa, C., Esposito, G., Gandin, V., Indraccolo, S., et al. (2017). Therapeutic potential of the phosphino Cu(I) complex (HydroCuP) in the treatment of solid tumors. https://core.ac.uk/download/141699493.pdf (accessed on 13th September 2025).

18. Cestellos-Blanco, S., Kornienko, N., Lim, J., Liu, Y., et al. (2018). Physical biology of the materials-microorganism interface. https://core.ac.uk/download/287623640.pdf (accessed on 13th September 2025).

19. Chandrakar, S., Sahu, J., Sahu, S., Sharma, S., Pandey, K., & Pandey, M. (2023). *Nanotechnology in Environmental Applications: Nanoparticles Synthesis for Water Pollution and Wastewater Treatment*. https://core.ac.uk/download/660319431.pdf (accessed on 13th September 2025).

20. Cotar, C., Friesecke, G., & Pass, B. (2013). Infinite-body optimal transport with Coulomb cost. https://core.ac.uk/download/pdf/81873673.pdf (accessed on 13th September 2025).

21. Daumann, L. J., Gahan, L. R., Ollis, D. L., Schenk, G., et al. (2014). Spectroscopic and mechanistic studies of dinuclear metallohydrolases and their biomimetic complexes. https://core.ac.uk/download/43351314.pdf (accessed on 13th September 2025).

22. de Oliveira, C. A. F., Durrant, J. D., & McCammon, J. A. (2011). Pyrone-based inhibitors of metalloproteinase types 2 and 3 may work as conformation-selective inhibitors. https://core.ac.uk/download/323066581.pdf (accessed on 13th September 2025).
23. Deng, J., Liu, B., Liu, C., Liu, Y., et al. (2019). Nanowire photo electrochemistry. https://core.ac.uk/download/287623654.pdf (accessed on 13th September 2025).
24. Deschaume, O., Fournier, A. C., Perry, C. C., Shafran, K., et al. (2008). Interaction of aluminum hydrolytic species with biomolecules. https://core.ac.uk/download/30628580.pdf (accessed on 13th September 2025).
25. Do, L. H., & Lippard, S. J. (2011). Evolution of strategies to prepare synthetic mimics of carboxylate-bridged diiron protein active sites. https://core.ac.uk/download/83196474.pdf (accessed on 13th September 2025).
26. Dudek-Różycki, K., Kyzioł, A., & Płotek, M. (2013). Selected copper(I) complexes as potential anticancer agent. https://core.ac.uk/download/53112848.pdf (accessed on 13th September 2025).
27. Edem, P. E., Fonslet, J., Herth, M., Kjær, A., et al. (2016). *In vivo* radionuclide generators for diagnostics and therapy. https://core.ac.uk/download/84003420.pdf (accessed on 13th September 2025).
28. Garzoli, S., Masci, V. L., Ovidi, E., Tiezzi, A., et al. (2019). Chemical investigation of a biologically active *Schinus molle* L. leaf extract. https://core.ac.uk/download/231748429.pdf (accessed on 13th September 2025).
29. Georghiou, P. E., Jiang, X., Jin, C. C., Kowser, M., et al. (2016). Fluorescent turn-on sensors based on pyrene-containing Schiff base derivatives for Cu^{2+} recognition: Spectroscopic and DFT computational studies. https://core.ac.uk/download/151161639.pdf (accessed on 13th September 2025).
30. Glass, J. B., & Orphan, V. J. (2012). Trace metal requirements for microbial enzymes involved in the production and consumption of methane and nitrous oxide. https://core.ac.uk/download/4894185.pdf (accessed on 13th September 2025).
31. Hu, L., He, B., G. B., J., Sun, Y., et al. (2013). Metallomics in environmental and health related research: Current status and perspectives. https://core.ac.uk/download/38022173.pdf (accessed on 13th September 2025).
32. Jin, C. C., Mou, L., Ni, X. L., Redshaw, C., et al. (2015). Synthesis of a ditopic homooxacalix[3]arene for fluorescence enhanced detection of heavy and transition metal ions. https://core.ac.uk/download/151156815.pdf (accessed on 13th September 2025).

33. Kletzin, A., Anbar, A. D., Alberts, B., Pullman, B., de Duve, C., de Duve, C., Myers, C. R., et al. (2007). Bootstrapping the energy flow in the beginning of life. https://core.ac.uk/download/pdf/15453002.pdf (accessed on 13th September 2025).
34. Lenton, T. M., Pichler, P. P., & Weisz, H. (2016). Revolutions in energy input and material cycling in Earth history and human history. https://core.ac.uk/download/43098286.pdf (accessed on 13th September 2025).
35. Matesanz, A. I., & Souza, P. (2009). α-N-heterocyclic thiosemicarbazone derivatives as potential antitumor agents: A structure–activity relationships approach. https://repositorio.uam.es/bitstream/handle/10486/665374/heterocyclic_matesanz_mrmc_2009_ps.pdf?sequence=1 (accessed on 13th September 2025).
36. Mitchell, P., Pierotti, M., & Piro, M. (2018). Production of living nanoparticles for blood cancer therapy. https://core.ac.uk/download/215442652.pdf (accessed on 13th September 2025).
37. N/A. (1989). *Focal Spot, Summer 1989.* https://core.ac.uk/download/129069275.pdf (accessed on 13th September 2025).
38. Octaviano, A. L. M., Aguiar, E. M., Lima, E. N. C., Lima, E. N. C., Piqueira, J. R. C., Justo, J. F., & Souza, L. H. M. (2024). Can bacteria and carbon-based nanomaterials revolutionize nanoremediation strategies for industrial effluents? https://core.ac.uk/download/615443433.pdf (accessed on 13th September 2025).
39. Preti, R. (2016). *Core-Shell Columns in High-Performance Liquid Chromatography: Food Analysis Applications.* https://core.ac.uk/download/54530354.pdf (accessed on 13th September 2025).
40. Senn, H. M. (2014). Insights into enzymatic halogenation from computational studies. https://core.ac.uk/download/42356666.pdf (accessed on 13th September 2025).
41. Székács, A., Darvas, B., Győri, J., Mörtl, M., Kereki, O., Klátyik, S., & Vehovszky, Á. (2016). Study on soil mobility of two neonicotinoid insecticides. https://core.ac.uk/download/42945110.pdf (accessed on 13th September 2025).

CHAPTER 10

APPLICATIONS OF COORDINATION COMPOUNDS

Contents

INTRODUCTION

Coordination chemistry is an area of importance to the satisfaction of contemporary scientific demands. Coordination complexes are formed through the reaction of metal ions with organic ligands and exhibit varied structures and properties that support varied applications. They are noted for their ability to confer unique features to the resulting complexes, which are of direct concern in such varied fields as industrial catalysis to biomedicine. The vast scope of the applications of coordination compounds is in how they can be customized, making it more convenient for researchers to create these compounds for specific use in, e.g., drug delivery, diagnostics, and environmental remediation. This introduction will delve into the applications of coordination compounds, highlighting how their influence has transformed sectors while discussing their elementary chemical characteristics. The versatility of coordination compounds' use is also complemented by ongoing research and innovation. For instance, it has been proven through research that coordination of metal ions with organic ligands significantly enhances the antioxidant potential, and hence such molecules become of immense value in medicinal and pharmaceutical chemistry. As a fact proven by recent scientific breakthroughs, the ability of coordination compounds to catalyze electron transfer reactions positions them as key players in redox processes. In addition, their role of trapping free radicals justifies their importance in maintaining human health. The sheer diversity in uses of these compounds, from their use in food storage and dietary supplements, illustrates the advantages accrue-ing in multilateral applications following fundamental research and development in this area. With the progress in coordination compound research, knowledge of mechanisms governing their function assumes maximum priority. Studies have extensively documented the relationship between the structure of the ligand and resulting properties of the coordination compounds, which extensively expands the scope of uses. An understanding of these interactions not only facilitates the synthesis of novel metal complexes and ligands but also optimizes the efficacy of existing compounds for use in practical applications. This overview of coordination compounds will tap into learnings gleaned from over two decades of research, namely the discoveries of researchers dedicated to expanding our understanding of these compounds. As we work our way through their extensive array of applications—from organic ligands to biological systems—we begin to realize the potential that these versatile molecules hold for science and industry development and sustainability (Beltran et al., 2025; El-Lateef et al., 2023).

Definition of Coordination Compounds

Understanding the concept of coordination compounds entails immersion in their intricate structures and bonding mechanisms. Coordination compounds or complex compounds are formed by a metal atom or ion coordinating with the surrounding molecules or ions, referred to as ligands. Ligands may be neutral molecules, e.g., ammonia or water, or charged ions, e.g., hydroxide or chloride. The metal ion in the center is typically from the d-block of the periodic table, i.e., the transition metals with variable oxidation states and exhibiting complexing behavior. This bonding nature, where the metal atom accepts ligand electron pairs, results in the formation of a coordination sphere that is crucial in describing the properties and behaviors of these compounds in applications like catalysis, chemicals, biological entities, and pigments. Other than their definition, coordination compounds contain diverse geometries that influence their chemical properties and behavior. The geometrical figures of the ligands around the central metal atom can be octahedral, tetrahedral, or square planar, among others. The geometries hugely define the reactivity and activity of the compounds. Octahedral complexes, for example, have distinctive optical and electronic properties that make them an ideal choice to apply in catalysis or as colorants in dyes and pigments. Understanding these structural variations is significant for chemists, as it allows design and synthesis of coordination compounds specifically targeted for particular applications, such as drug delivery systems or sensors. As elucidated by recent breakthroughs, the exploration of the coordination sphere is at the center of advancing the boundaries of materials science and inorganic chemistry. Moreover, the interdisciplinary use of coordination compounds renders them important in other fields of study, ranging from medicine to environmental science to materials engineering. Coordination mechanisms can elucidate the coordination interaction in biological systems, for instance, in metalloenzymes in which the metal ions play a crucial role in catalysis. From an environmental chemistry perspective, coordination compounds can even be the focus of remediation efforts, whereby they facilitate the removal of poisonous ions from wastewater or soil. The recent discovery of covalent organic frameworks (COFs), in which coordination compounds become part of their structure, is evidence for a quantum leap in the ability to design porous materials with a target of desired properties for filtration or storage applications (Diercks et al., 2017). As per the modern studies glossary, terminological distinctiveness in the situation of coordination compounds assists in giving rise to interdisciplinarity collaboration and understanding (Минкин, 2016).

Importance in Chemistry

The significance of coordination compounds to the field of chemistry cannot be overstated. Coordination compounds are simple building blocks in vast fields of chemical usage based on their unique structural and electronic properties. Coordination compounds contain metal ions as a central atom bonded to ligands surrounding it, with the ability to vastly modify their reactivity and stability. This reactivity is a critical element in catalysis, a cornerstone of inorganic and organic chemistry. Tailoring the properties of coordination compounds, particularly by choosing ligands, allows chemists to engineer materials with desired functionalities. Such flexibility offers grounds for new synthetic pathways, ultimately resulting in enhanced efficiency for chemical reactions. The future application of such compounds extends beyond basic science, impacting industrial processes, environmental chemistry, and biochemistry, thus guaranteeing their contribution to the advancement of both theoretical and applied fields in chemistry. Coordination compounds have been very useful in drug discovery and design in pharmacy. One of the finest examples of this is the application of heterocyclic compounds, which form a part of about 80% of medicines, obviously proving the importance of coordination chemistry in medicine. Of these, benzimidazole is a drug scaffold with wide versatility along with a range of therapeutic activities like antiviral and anticancer. This places in perspective the analogy between coordination compounds and biologically active compounds, within a general trend where substituted coordination complexes are more stable and bioavailable within the biological environment. In response to the demand for more effective and targeted drugs, research into the coordination compounds becomes ever more significant. Scientists are compelled to investigate synthesis and design pathways to optimize the activity of these compounds so they become precious tools for modern medicine (Mahurkar et al., 2023). Yet another significant feature of coordination compounds is their potential for green applications. Schiff-base ligand synthesis, for instance, has been researched for their multimodal character. These ligands, produced through the condensation reaction of carbonyl and amino molecules, are not only stable metal complexes but also exhibit varied bioactive properties such as antifungal and antibacterial activity. The variety thereof places Schiff-base complexes in a strong position to address environmental issues, from bioremediation and water treatment to wastewater treatment. Their selective binding and removal of poisonous metal ions from the environment are the typical example of where chemistry intersects with sustainability. With the advancement of research, these ligands' functions in coordination chemistry will expand further, opening up new solutions to pressing issues of environmental concerns. The intricate nature of coordination compounds with the environment highlights their significance in the overall purview of chemistry (Soroceanu et al., 2022).

Overview of Applications

Applications of coordination compounds are varied and multifaceted, extending across such disciplines as catalysis, materials science, and biomedicine. In catalysis, these molecules significantly increase the rate of reactions and selectivity and, therefore, induce effective chemical processes. The development of single-atom catalysts (SACs) based on individual metal atoms trapped within a coordination compound is one of the foremost examples of leading-edge innovations in this field. By ultrahigh atom density via controlled synthesis methods, researchers have unlocked enhanced catalytic performance (Li et al., 2022). These single-atom structures enable better understanding of the catalytic processes, ultimately leading to maximized industrial processes and reducing resource usage, which is crucial in sustainable growth. Coordination compounds have revolutionized the synthesis of new materials with specific functions in materials science. The combination of coordination chemistry and two-dimensional materials has stimulated tremendous interest, particularly in the development and fabrication of heterostructure devices. These devices exploit the unique properties of a single two-dimensional crystal, creating new functionalities unachievable with traditional materials. As recent studies affirm, the characteristics of these new materials, including electronic and optical content, play a fundamental role in the development of next-generation electronics, like tunneling transistors and light-emitting diodes (Novoselov et al., 2016). These advancements show coordination compounds' ability to address technological issues through innovative material design. Biomedicine is also a fundamental field of application for coordination compounds, particularly in drug and delivery system development. The multifunctionality of coordination compounds allows the inclusion of therapeutic entities, targeting ligands, or imaging agents, thereby enhancing the efficacy of medical treatments. For example, coordination chemistry is utilized by metal-based drugs such as cisplatin to bind to DNA and induce cytotoxicity in cancerous cells. Furthermore, the incorporation of coordination compounds in nanocarriers provides effective controlled release and improved bioavailability of the drugs. Such multifaceted use of coordination compounds in biomedicine illustrates that they can drive innovations to achieve improved therapeutic results, and hence their importance to contemporary medical research and practice.

10.1. COORDINATION COMPOUNDS IN MEDICINE

Therapeutic and diagnostic use of coordination compounds varies from a whole array of therapeutic drugs to diagnostic reagents, underpinning the pivotal role of coordination compounds in contemporary medicine. Characterized by their unique structural properties, the compounds deliver metal ions in a limited way that optimizes the efficiency of some biological processes. For instance, transition metals like copper and iron are involved in numerous biochemical reactions as crucial cofactors for enzymic activity. Moreover, the synthesis of these compounds allows for the design of targeted drug agents, along with increased specificity and reduced side effects in treatment. In particular, target-based synthesis and design of metal drug agents are proving to be very beneficial for their diagnostic use, as they can selectively interact with biological targets and provide imaging contrast in medical interventions, thereby increasing diagnostic precision and patient outcomes (Kostova, 2023). Bismuth coordination complexes are a good example of great progress in the treatment of numerous diseases, such as gastroenterology and oncology. They are applied for the treatment of infections, including *Helicobacter pylori* eradication therapy, with good tolerance and efficacy in patients. Recent studies have further highlighted the potential of bismuth complexes as anticancer drugs, where their efficacy in inhibiting cancer growth in various types of cancers, including breast and colon cancer, has been proven (Kowalik et al., 2017). Such dual application as therapeutic and diagnostic medicine further establishes the performance of bismuth coordination compounds since they can also be used in radioimmunotherapy applications, further enhancing the curative effects of radiative therapy. The ongoing research of these compounds in pharmaceutical research indicates fruitful avenues for improving treatment regimens and patient outcome in complex disease models. Beyond their uses in cancer and infectious disease, coordination compounds are crucial to more general biomedical uses, including vaccine delivery systems and biosensors. Their inclusion in nanomaterials has ushered in new avenues for the improvement of vaccine efficacy with improved stability and targeted delivery systems. Similarly, coordination compound-based biosensors have revolutionized the analysis of biomarkers for various diseases in such a way that diagnostic results are now easily acquired and sensitive. The technical advancements are due to the intrinsic nature of coordination compounds, whose properties can be accurately tuned by modifying their ligands and metal centers. As multidisciplinary research progresses, the potential of coordination compounds in medicine will widen with hopes to introduce new solutions for emerging healthcare problems and to advance therapeutic methods (Kostova, 2023).

10.1.1. Role in Chemotherapy

Coordination compounds possess a significant use in chemotherapy, mainly owing to the fact that they possess unique structural characteristics to allow for selective bio-molecule binding. One of the most optimistic applications of such compounds lies in the development of cancer-targeted treatments. These are drug treatments designed to specifically attack cancerous cells with less harm to regular tissues, with the benefit of fewer side effects when compared to traditional chemotherapeutic agents. For instance, the identification of small-molecule targeted drugs has revolutionized treatments for cancer with higher patient survival and better outcomes. The strength of these drugs is their ability to interfere with specific pathways of cellular relevance to cancer cell proliferation and survival, highlighting the tactical value of coordination compounds as agents against cancer (Coghi et al., 2021). Organoboron compounds have also emerged as key agents in modern oncology, illustrating the range of coordination compounds as a category of agents beyond traditional chemotherapeutics. These drugs have been active in a wide range of clinical uses, from antibacterial to anti-fungal to anti-cancer therapies. Most intriguingly, boron-based drugs such as bortezomib have been used to treat multiple myeloma, showing they can change the treatment approach for a disease.

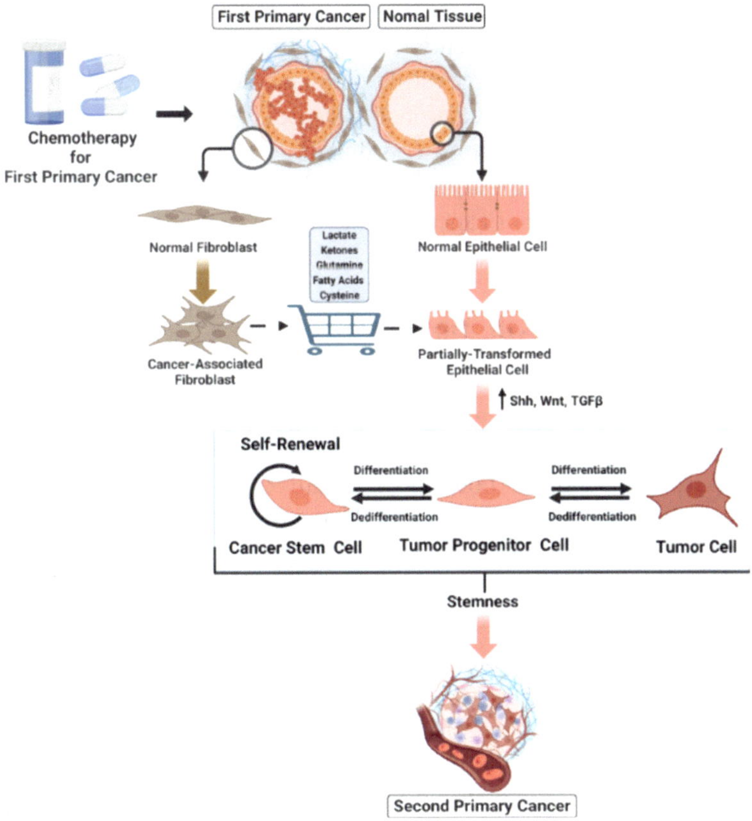

Figure 10.1. *Role of chemotherapy induced cancer associated fibroblast (CAF) in transforming partially.*

Source: https://www.researchgate.net/publication/388361337/figure/fig2/AS:1143 1281305442641@1737775363472/Role-of-chemotherapy-induced-cancer-associated-fibroblast-CAF-in-transforming-partially.png.

The mechanism of these drugs is to interfere with molecular pathways central to cancer cell survival and growth and eventually lead to apoptosis in neoplastic cells. Moreover, boron neutron capture therapy (BNCT) is a new approach in utilizing the coordination properties of boron compounds, particularly in the management of glioblastoma and head and neck cancers, thereby expanding the therapeutic outcome against aggressive cancers (Coghi et al., 2021). Over the past few years, chemotherapy as a discipline has been updated day by day, particularly with the progress observed with targeted therapy, which in most instances has proven improved efficacy and safety profiles compared to conventional chemotherapeutic drugs. 89 small-molecule targeted antitumor drugs registered by the regulatory agencies are clear evidence of the future of coordination compounds in anticancer treatment. These developments are phenomenal against the challenges still facing oncology, such as drug resistance and lack of response with conventional therapy. As researchers go deeper into the complexities of cancer biology, coordination compounds will likely remain in the forefront of the design of next-generation drugs, thereby enhancing the efficacy and selectivity of anticancer therapy in the years to come (Zhong et al., 2021) (Figure 10.1).

10.1.2. Use in Diagnostic Imaging

The contribution made by coordination compounds to diagnostic imaging is so crucial, particularly given their potential to enhance the specificity and clarity of imaging techniques. These compounds are typically contrast agents, allowing certain areas of the body to be visualized during imaging procedures such as magnetic resonance imaging (MRI) and computed tomography (CT). With metal ion complexation, these agents can be designed to alter the magnetic properties or the X-ray attenuation of tissues in such a manner that there will be enhanced discrimination between disease and normal state. They are applied extensively across a variety of areas, such as oncology and cardiology, where imaging is significant in diagnosis and treatment planning. Furthermore, the specificity of the coordination compounds can be engineered to target specific tissues or cancer types, further extending the boundaries of personalized medicine and diagnostic technology in clinical settings. Advances in diagnostic imaging technology have indicated that coordination compounds play a revolutionary part in nuclear medicine with the application of radiolabeled agents. These agents are typically made up of a central metal ion coordinated to a set of ligands, which can be designed to recognize disease-specific receptors on target cells. Such procedures allow for the study of cellular function and detection of diseased tissues at an early stage. Improvements in the manufacture of radiometals, such as technetium-99m and gallium-68, have revolutionized methods of imaging by providing ultra-sensitive and specific means of visualizing physiological function. Therefore, coordination compounds have been crucial in tumor localization and characterization, particularly in oncology, where coordination compounds enable clinicians to make treatment decisions based on real-time imaging data. In order to maximize the usefulness of coordination compounds for diagnostic imaging, research is now focused on devising these agents for targeted applications, overcoming limitations that are usually posed by traditional imaging modalities. For example, issues like poor bioavailability and the potential for serious side effects associated with some imaging agents are addressed using novel design strategies. Drug encapsulation in coordination complexes or conjugation with targeting ligands are methods employed to achieve maximal specificity while preventing undesirable interactions in the body. Recent innovations have also demonstrated the potential to utilize biodegradable coordinating agents which would hopefully minimize the risk of toxicity without forgoing sharp image contrasts. Such innovations aim to optimize the diagnostic performance of coordination compounds so that, eventually, patient outcomes in a clinical context are improved as has been explored in recent reviews and research on the subject (Dorothée Liebschner et al., 2019; Senapati et al., 2018) (Figure 10.2).

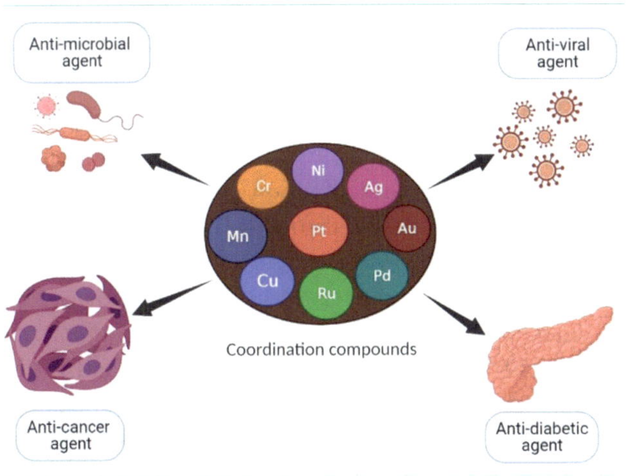

Figure 10.2. *The significance of coordination compounds in diagnostic imaging.*

Source: https://ars.els-cdn.com/content/image/1-s2.0-S0277538723002073-ga1_lrg.jpg.

10.1.3. Development of Pharmaceuticals

Pharmaceutical development has evolved to utilize coordination compounds, which are central to therapeutic effectiveness and enhancing the design of drugs. They are typically metal ions that are coordinated with organic ligands and offer new chemical characteristics which can be accountable for new pharmacological applications. One of the most important applications is in antimicrobial compounds, to which silver coordination complexes have emerged as strong contenders. Silvers' healing properties have been exploited in medicine since time immemorial, as in the case of research on the clinical uses of silver and silver complexes. A particular field of potentiality is the overcoming of the rising resistance to conventional antibiotics, with silver compounds possessing a wide range of antimicrobial activity and thus creating new therapeutic lines in pharmaceutical sciences (Dominik Żyro et al., 2023). Coordination chemistry is equally important in enhancing antioxidant activity in pharmaceutical compounds. Organic ligand complexes of various metal ions have shown an impressive capability to augment antioxidant activity, which is of utmost significance in the development of drugs against oxidative stress-related disorders. Chelation allows one to increase the stability and activity of such compounds, underlining the useful benefits gained by coordination compounds over traditional synthetic antioxidants. As reported in a comprehensive review of coordination compound applications, not only do these metal chelates exhibit free radical scavenging activity, but they also protect living organisms from oxidative damage. The extensive activity confirms the importance of coordination compounds in treating various health conditions and pharmaceutical formulation development (El-Lateef et al., 2023). While the pharmaceutical potential of coordination compounds is enormous, further research will be required to surmount difficulties associated with their synthesis and therapeutic use. Research efforts are currently directed toward modifying the physicochemical properties of such compounds to enhance bioavailability and therapeutic effects. Studies have also determined that there are limitations to the translatability of laboratory results to the clinic, and further study into safety and efficacy of these agents will be necessary. As an example, studies of silver(I) nitrate complexes, which have recently been evaluated for possible use in clinical applications, show the iterative nature of this activity. With scientists continuing to examine the intricacies of coordination chemistry, the ultimate goal is the easy integration of these compounds into effective pharmaceutical remedies that fulfill the needs of contemporary medicine (Dominik Żyro et al., 2023; El-Lateef et al., 2023).

10.2. INDUSTRIAL APPLICATIONS

Industrial applications of coordination compounds are vast and varied, attesting to their status as a ubiquitous participant in most industries. One of the first is in the development of new materials, with metal-organic frameworks (MOFs) as a prime example. Coordination compounds possess wonderful properties such as high surface area and tunable porosity, ideal for processes such as gas separation and catalysis. The method of coating MOFs onto soft substrates is a fine demonstration of this application, whereby coordination replication method is employed in the synthesis of stable fibrous MOF/textile composites. The method not only offers the mechanical stability of the material but also brings impressive practical applications, for example, the extraction of organosulfur compounds like dibenzothiophene from fuel, exemplifying the central role of coordination compounds in imparting industrial efficiency and sustainability (Ma et al., 2019). Furthermore, coordination compound development has impacted the electronics industry to a very significant degree, particularly with the emergence of next-generation communication technology. As mentioned, the transition towards 5G technology necessitates drastic material property advancements, including new coordination compounds capable of facilitating high-density device and antenna integration. Coordination compounds facilitate the synthesis of advanced semiconductors and conductive polymers that are essential to achieving 5G networks' required high-frequency performance and energy efficiency. Application of these compounds surmounts key issues confronting the telecommunication sector, pushing the boundaries of materials further with evolving technology demands while ensuring a smooth user experience across different platforms (Andrews et al., 2014). In addition, catalysis has also seen spectacular development with the use of coordination compounds, which serve as catalysts to numerous industrial operations. Because of their unique electronic properties, they can selectively activate substrates and thus attain improved reaction rate and yield. Particularly, coordination compounds find their application in the manufacture of fine chemicals, drugs, and agrochemicals, where their ability to facilitate complex reaction mechanisms is of utmost concern. These catalysts' scalability is also significant, as testified by continuous flow synthesis processes that leverage coordination complex stability and reusability. Thus, the employment of coordination compounds as catalysts not only enhances the efficiency of production but also supports the demand for emerging green practices in the production of chemicals.

10.2.1. Catalysts in Chemical Reactions

In the world of chemicals, catalysts are an essential component in speeding up the rates of reactions without being permanently changed themselves. The ability to reduce the activation energy required for a reaction makes them vital in operations in industry and processes in the environment. In coordination compounds, catalyst design and application have grown significantly, particularly in providing more sustainable processes. For instance, improving catalytic efficiency is crucial in the production of ammonia due to the fact that traditional Haber-Bosch methods result in excessive energy consumption and greenhouse gas emissions. New electrocatalysts, particularly those derived from coordination compounds, have been discovered to be promising in order to achieve high yield rates of ammonia with enhanced Faraday efficiency, and hence resolving the concern of both efficiency and the environment on industrial scales (Liu et al., 2023).

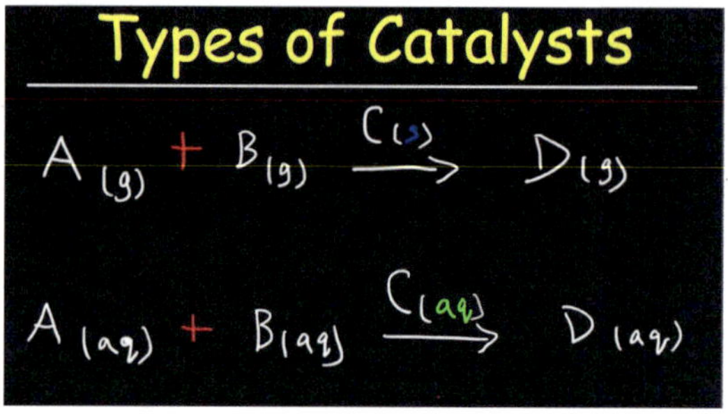

Figure 10.3. *Catalysts in chemical reactions.*

Source: https://i.ytimg.com/vi/Tb0nzaj7bSA/maxresdefault.jpg.

The coordination environment and metal-ligand interaction of coordination compounds also significantly influence their catalytic activity. Most surprisingly, novel single-atom catalysts have been made that exhibit enhanced activity compared to their conventional homogeneous counterparts. The catalysts are distinguished by the atypical coordination of metal atoms within a support material, which optimizes their electronic properties for some reactions. For instance, a set of Cu-based single-atom catalysts have been demonstrated to have improved performance in the process of azide-alkyne cycloaddition, a reaction critical in drug synthesis. Through a rigorous design, such catalysts were demonstrated to have improved selectivity and activity, initiated by an expert modulation of the metals coordination sphere (Vilé et al., 2022). This emphasizes the imperative of diligent structural research in the design of effective catalytic systems based on coordination chemistry. In addition, research into novel strategies in catalyst design can propel innovations in chemical efficiency and sustainability. Surface modification and defects engineering are some of the techniques employed to enhance the inherent activity of coordination compounds. Focusing on the coordination environment at metal sites, scientists are opening new fronts towards optimizing catalytic performance. The argument on the electrochemical nitrate reduction reaction (e-NO_3 RR) captures this trend, demonstrating how carefully designed catalysts can lead to greener routes of ammonia production. With advancing science, the interface between coordination chemistry and the use of catalysts is an immense potential for altering industrial operations with reduced harm to the environment, a significant step towards sustainable chemical processes (Liu et al., 2023) (Figure 10.3).

10.2.2. Role in Dye Production

The role of coordination compounds in the production of dyes has also assumed significance in the textile field, wherein colorants are the major contributors towards imparting beauty to fibers. Coordination complexes that contain metal ions coordinated to organic ligands possess bright colors, so they become ideal candidates to be utilized as dyes. The presence of varying metal ions in these complexes can cause a bathochromic shift that generates deep and richer tones. This aspect is useful for achieving the required color depth, as textile manufacturers can manipulate the metal-ligand interactions to generate dyes with specific colors and properties. Consequently, coordination compounds not only bring aesthetic quality to textile materials but also provide a design and production center for innovative exploration of new colors and product differentiation in a competitive market. Moreover, the employment of metal-containing dyes has significant environmental implications, particularly with respect to wastewater management. The production process of textiles generates significant amounts of effluent with heavy metal ions like Cr (VI), Pb (II), and Cd (II) because dyeing is a water-intensive process. These metal-containing effluents are severe toxicological issues in that they have the potential to contaminate natural water bodies and lead to sickness among residents. The use of coordination compounds as pigments must hence be investigated along with wastewater treatment processes. Here, the development of novel materials for the adsorption of heavy metal ions from textile wastewater has emerged as a matter of utmost significance so that the radiance of scintillating dyes does not take place at the cost of environmental integrity. Novel solutions such as graphene oxide absorbers have exhibited excellent efficiency towards fulfilling these environmental challenges while maintaining the merits of coordination compounds in dye synthesis. Considering these aspects of the environment, ecologically sustainable development of coordination compounds in the production of dyestuffs is crucial. Advances in recent times in the field of chemical engineering and materials science have opened the way for the production of cleaner dyes, like those derived from natural resources as substitutes for synthetic ones. Through the application of more environmentally degradable coordination compounds, manufacturers of textiles are able to prevent the adverse effects accompanying heavy metal contamination of dyes. Research into biodegradable ligands and lower toxicity metals within coordination complexes not only addresses regulatory demands but also addresses consumer demand for environmentally friendly production. Lastly, the dual role of coordination complexes as dyes and products of environmental concern necessitates a balanced approach that includes innovation with sustainability while reasserting their central role in future textile industry dye manufacturing (Figure 10.4).

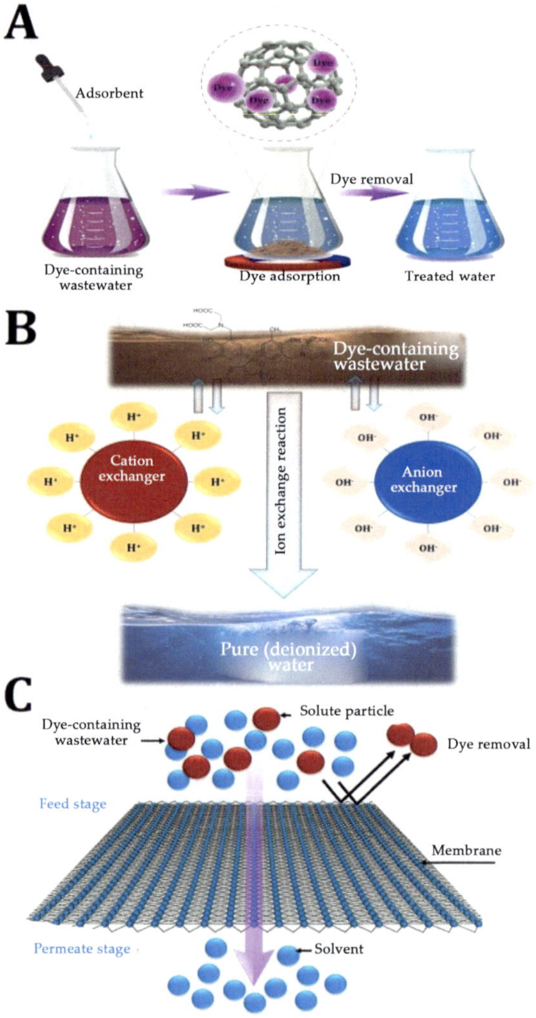

Figure 10.4. *The utilization of coordination compounds in dye production.*

Source: https://ars.els-cdn.com/content/image/1-s2.0-S0147651321012720-gr6.jpg.

10.2.3. Use in Electroplating Processes

The use of coordination compounds in electroplating operations is a good overlap between coordination chemistry and practical application. Electroplating is the formation of a metal layer on a substrate via the use of electrolytic methods, and the use of coordination complexes has the potential to enhance both the quality and deposition efficiency. The existence of coordination compounds can stabilize metal ions in solution, hence their reduction and deposition on the target surface. The stabilization can be a determinant of the morphology and adhesion of the deposited metal, which are essential parameters in applications from the automotive sector to electronics.

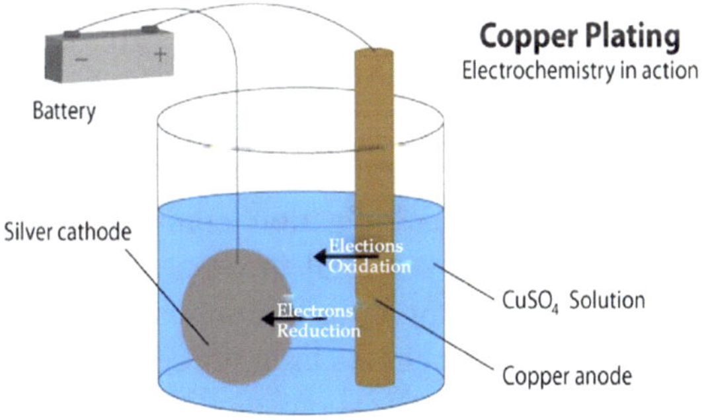

Figure 10.5. *The utilization of coordination compounds in electroplating processes.*

Source: https://sensorex.com/wp-content/uploads/2021/05/page_1-e1621309989205.jpeg.

Furthermore, the complexity of these compounds facilitates their adjustment of electroplating parameters, thus optimizing properties like hardness, corrosion resistance, and esthetic finish of the deposit layer. Apart from raising deposition efficiency, application of coordination compounds is also critical in order to achieve desired properties for the electroplated layer. For instance, nickel coordination complexes are generally employed to improve the uniformity and integrity of nickel deposits. The application of such compounds can minimize the formation of defects such as porosity and roughness, which are detrimental to the performance of electroplated products. Regulation of these settled properties through coordination chemistry underscores the importance of the method in electroplating technology. Advances in this field can be able to give harder and more visually appealing surfaces, which is essential in markets demanding high-quality finishes. Research into coordination compounds' interactions with target metals remains a subject of interest to continue innovation regarding the application of electroplating. Moreover, the environmental impacts of coordination compounds throughout electroplating processes cannot be overlooked. Traditional electroplating practices have been followed by a large amount of waste formation and the use of toxic chemicals. However, the attainment of sustainable electrolyte solutions via green coordination compounds is unlocking avenues to greener electroplating methods. Not only do they enhance the electrochemical characteristics but also align with general world directions towards sustainability in production. Applying next-generation electrolysis technologies, as seen in the new research in alkaline water electrolysis systems, is a paradigm change. Scientists are researching the integration of electroplating operations with green sources of energy, showing promising avenues for reducing energy usage and wastage of resources in electroplating (Lee et al., 2024) (Figure 10.5).

10.3. ENVIRONMENTAL APPLICATIONS

Environmental applications of coordination compounds have been in the limelight a lot in terms of their potential to suppress heavy metal pollution. The textile and mining sectors produce enormous quantities of wastewater containing harmful metal ions such as chromium, lead, and cadmium. These harmful compounds have drastic effects on aquatic life and human health. The contribution of coordination compounds toward the formulation of efficient remediation protocols offers a new path to combating these environmental challenges. For instance, recent studies indicate the application of graphene-based materials as adsorbents for heavy metal ions; their high specific surface area and surface function that can be tuned make efficient removal easy. Applying coordination compounds in wastewater treatment helps us minimize the adverse effects of industrial effluents on natural water bodies (Velusamy et al., 2021). In addition to the treatment of wastewater, coordination compounds also play a significant role in the sustainable remediation of heavy metal-polluted soils such as that of lead. Lead contamination, generally an industrial process by-product as well as farm runoff, not only poses a threat to human health through food webs but also destroys soil biota. Research indicates that lead bioavailability is the most significant parameter affecting its toxicity in plants and animals. Therefore, various remediation methods with the application of coordination compounds are being explored to remediate lead contamination in soils. Specifically, microbe-assisted phytoremediation with the incorporation of plant growth and microbial assistance for metal uptake exhibits promising approaches to decreasing the concentration of lead at contaminated areas. Such emerging remediation methods underscore the application of coordination compounds in promoting environmental recovery and food security (Kumar et al., 2020). Moreover, the applications of coordination compounds extend beyond environmental remediation to involve the manufacture of sensors for the detection of heavy metal ions in the environment. Early detection of pollutants can be a significant advancement in environmental monitoring since it offers the chance to intervene in timely manner. Coordination compounds, with their ability to selectively bind specific metal ions, can be utilized for the design of effective and sensitive sensor systems. These sensors can provide real-time data regarding heavy metal composition in water and soil to enable active pollution control. This confluence of coordination chemistry with sensor science is a giant leap in environmental protection, thus contributing significantly to overall strategies for ecosystem maintenance and public health. In total, the novel usage of coordination compounds in various environmental contexts highlights their significant role in assisting in resolving the issues brought about by heavy metal contamination.

10.3.1. Heavy Metal Ion Removal

Decontamination of heavy metal ions from water systems is presently a vital research focus due to the established harmful effect of metal contamination on human health and the environment. Coordination compounds have solved this issue to a great extent through the presence of stable metal ion complexes. Among most coordination ligands, Schiff bases have gained tremendous attention owing to their ability to coordinate with heavy metals very strongly. Their effectiveness in scavenging poisonous ions like lead, cadmium, and mercury lends credence to the purification of polluted water bodies. The synthesis and characterization of such compounds demonstrate their utility in environmental applications, specifically in the removal of heavy metals from industrial effluents, and thereby exemplifying the crucial role of coordination chemistry in the promotion of sustainable practices (Godwin et al., 2024). Further, dithiocarbamate complexes show excellent efficacy in removing heavy metal ions, providing promising ways for environmental remediation. These dithiocarbamate complexes are synthesized through substitution reactions between metal salts and carbon disulfide, and these resulting ligands can give stable metal complexes with metal ions such as FeII, CoII, and ZnII. Studies show that if these dithiocarbamate complexes are introduced into contaminated water systems, they can attain a rate of scavenging of metal ions as high as 98% and thereby mitigate metal ion poisoning in aquatic organisms (Al-fahdawi et al., 2024). The structural analysis of such complexes shows robust coordination chemistry, tetrahedral and octahedral geometry favoring coordination towards many heavy metal ions. These aspects reinforce the inordinate utility of these coordination compounds toward averting environmental hazards. In conclusion, the applications of coordination compounds toward the removal of heavy metal ions attest to their significance in water pollution resolution. The combination of Schiff bases and dithiocarbamate ligands in remediation is an example of the importance of chemical creativity for tackling important environmental issues. There is still a need to simplify the synthesis and scale-up of such materials for real-life applications, so that they can be utilized effectively to combat heavy metal contamination. With growing environmental awareness, the coordination compound synthesis will be critical to the delivery of clean and safe water resources, thus safeguarding public health and ensuring ecological balance. The uninterrupted synergy between chemistry and environmental science will ensure the effective delivery of solutions for generations (Figure 10.6).

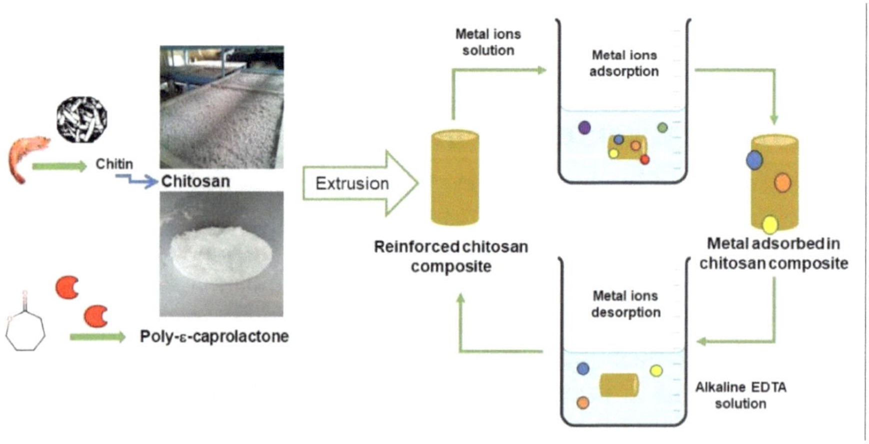

Figure 10.6. *The removal of heavy metal ions from water systems.*

Source: https://pub.mdpi-res.com/polymers/polymers-14-05196/article_deploy/html/images/polymers-14-05196-ag.png?1669718676.

10.3.2. Role in Water Treatment

The use of coordination compounds to treat water grows more relevant as global water scarcity and pollution worsen. The compounds are effective agents in operations such as coagulation and flocculation, where they function to ensure suspended matter is eliminated. Coordination compounds induce the clumping of fine particles by complexing with contaminants, allowing them to settle effectively and be drained from the water system. For instance, ferric and aluminum salts, used widely as coagulants at water treatment centers, act by coordination to achieve improved water clarity as well as reduce turbidity. Structure and composition tailoring of the compounds enables optimization for a specific pollutant, thus making them even more efficient in different water treatment situations. Moreover, coordination compounds find use in the removal of heavy metals from wastewater. Metallic ions such as lead and mercury are serious health and environmental hazards, requiring efficient remediation. Coordination compounds have the potential to selectively capture these toxic metals and react with them to form stable complexes that may either be precipitated out of solution or extracted through various means. The selectivity is important since it enables one to treat contaminated sites at will while minimizing the chances of secondary pollution. Other new approaches involving functionalized coordination compounds have also been explored to further enhance the uptake and removal efficiencies. By changing the ligands of such complexes, researchers can improve their capacity for binding with some contaminants to enhance the overall efficiency of water treatment processes, particularly in industrial applications where multiple forms of pollutants are found. Apart from their use in conventional treatment processes, coordination compounds are being utilized in new technologies such as advanced oxidation processes (AOPs) and membrane filtration systems. These systems predominantly employ the unique properties of coordination compounds to catalyze reactions that degrade organic pollutants, maximizing the entire water treatment process. For example, metal-organic frameworks (MOFs), a class of coordination compounds, have been reportedly well appreciated for their ability to adsorb a wide variety of pollutants, thereby serving dual functions as adsorption and catalytic degradation. Furthermore, scalability of such technologies, as well as coordination compounds' tunability, presents a promising path to developing environmentally friendly water treatment technologies. With the world still struggling with water pollution challenges, such integration of cutting-edge approaches will be pivotal towards ensuring access to clean water in the future.

10.3.3. Applications in Soil Remediation

Soil remediation is a significant area in the field of environmental studies, particularly in contaminant management such as heavy metals and organic pollutants. Coordination compounds are especially useful in such a case as they can be stabilized into complexes with other poisonous metals and thus their recovery from contaminated soils becomes viable. The distinctive properties of the compounds enhance the effectiveness of the remediation processes with the potential for specialized treatment based on the nature of the contaminant. For instance, certain coordination compounds that are especially designed to be capable of interaction with lead or cadmium ions have great affinity for such metals and lead to greater removal efficiency in laboratory as well as field tests. Scientists can make remediation processes compatible with environmental safety as well as sustainability objectives using such compounds. Graphene-based materials, particularly graphene oxide, have revolutionized the scenario of soil remediation. These compounds are of high surface area and easily functionalized for the adsorption of the heavy metal ions. Graphene oxide can be employed as a support matrix for the adsorption of these compounds in the case of coordination compounds, thus creating hybrid remediation agents that optimize metal ion extraction from soils. It has been confirmed through research that the functionalization of graphene oxide with specific coordination complexes enhances the adsorption significantly, resolving major issues detected in conventional remediation methods. The use of such advanced materials eliminates issues with the long-term stability of adsorbents, among others, and provides cost-effective and sustainable solutions for the rehabilitation of polluted land (Velusamy et al., 2021). Further, coordination compound potential is more than mere adsorption, which influences organic pollutant degradation of the most common occurrence in contaminated soils. Certain compounds can facilitate catalytic reactions to degrade toxic organic chemicals into less harmful products. Coordination chemistry-based methods are particularly promising where organic contaminants fail to degrade. The complexity of coordination compounds permits the development of remediation strategies not just efficient but also sustainable. Future research needs to concern itself not just with adsorption or degradation efficiency but with the design of multifunctional systems to remove various types of pollutants simultaneously. This integrated strategy would do much to advance the science of soil remediation, enabling cleaner conditions in agricultural settings as well as metropolitan areas (Qasem et al., 2021).

10.4. COORDINATION COMPOUNDS IN AGRICULTURE

One of the most significant fields where coordination compounds exhibit immense utility is in agriculture, i.e., in enhancing plant nutrition and soil renewal. Coordination compounds, with their central metal ion bonded to multiple ligands, play a crucial role in the delivery of bioavailability of nutrients. Iron chelates, a coordination compound, is employed to cure iron chlorosis in plants due to deficiency of iron in the soil. Chelates facilitate the absorption of iron by creating stable complexes and thereby supporting healthy plant growth. Coordination compounds are also employed in fertilizers, whereby they improve nutrient efficiency and reduce environmental impacts as compared to traditional fertilizers. This coordination compound approach not only optimizes crop yield but is also ecologically friendly, thus addressing the growing market demand for environmentally friendly products in agriculture. Other than nutrient management, coordination compounds are crucial in pest management in agriculture as efficient agrochemicals. Many coordination complexes are antimicrobial and can be employed for combating several agricultural diseases and pests. For example, certain organometallic compounds have been found effective in the control of fungal diseases while being least toxic to non-target species. This dual effectiveness promotes a sustainable process by potentially reducing dependency on conventional pesticides, which often kill pest-destroying insects and cause loss of diversity.

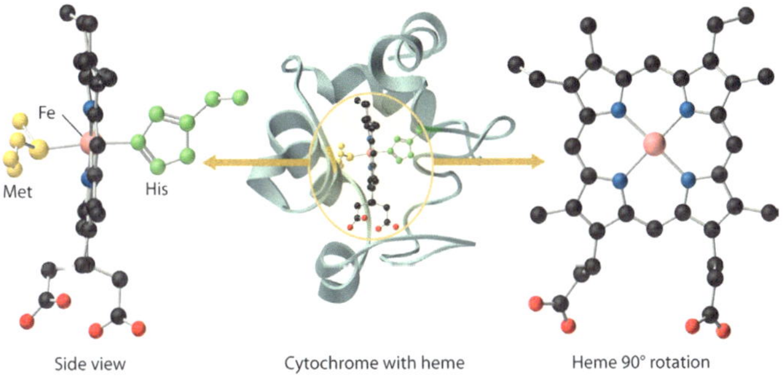

Figure 10.7. *Coordination compounds in agriculture.*

Source: https://files.mtstatic.com/site_4334/50605/0/webview?Expires=175679-8476&Signature=N6hsLbgnDheUsBfyCoVH6SgzQnzkcLWoiCzg2lpmU6mOTDAmq1s-XZslTgvdICP5R3P4aECIZQk6EcBBzOO8wPAUJDXKDY4yFcn7gm6b6Q-iDIz9LEMefXvP51XKZxNlDHZ1ONgxhv5wc2bTrMAY184dFgXFePqbaKeSn-lc7-gJc_&Key-Pair-Id=APKAJ5Y6AV4GI7A555NA.

The application of coordination compounds in pest control practices not only guarantees the safety of crop yields but also facilitates sustainable development goals—particularly those that provide food security without sacrificing well-being to the environment. Indeed, innovative application of such compounds in crop protection validates their indispensable position within modern agricultural practices. The entry of digitalization even further enhances the effectiveness of coordination compounds for application in agriculture. With the use of smart technologies and data analysis, farmers can monitor the content of soil nutrients more accurately and utilize coordination-based fertilizers better. This precision agriculture maximizes the potential of coordination compounds to facilitate better decision-making, leading to higher crop yields and better resource utilization. Further, as seen from ongoing discussions on sustainable practices, digital platforms' convergence presents opportunities to deal with the onslaught of climate change and biodiversity loss (Calvin et al., 2023). The holistic approach advocated by digitalization, compound coordination compounds, has prospects to bolster the food-water-energy nexus (Mondéjar et al., 2021). Such multidisciplinary combination not only enhances the farm produce but also complements the efforts of food system sustainability, hence supporting global initiatives to produce a healthier world (Figure 10.7).

10.4.1. Use as Micronutrient Fertilizers

Application of coordination compounds as micronutrient fertilizers is of great significance in enhancing plant development and nutrition content. These compounds are capable of increasing the bioavailability of trace essential elements such as iron, zinc, and manganese that are significant in many physiological processes in plants. Soils are devoid of these micronutrients, despite their importance, which limits crop yields and results in lower nutritional quality of food crops. Research has proven that coordination complexes effectively increase nutrient absorption in plants, particularly under stress, e.g., in poor soil or undesirable climate conditions. Coordination complexes' unique properties enable enhanced nutrient retention and slow release, making them an important resource in modern agriculture, particularly as global warming changes the nutrition profile of farm produce (José Soares et al., 2019). In addition to enhancing nutrient uptake, climate change can also be alleviated by coordination compound-based micronutrient fertilizers in plant nutrition impacts. Based on studies, elevated CO_2 levels and other climate change factors have been shown to harm the mineral content and protein concentration of crops. Such stressors can be overcome by farmers using coordination compounds as micronutrient fertilizers. These chemicals not only improve the nutritional availability but also have been related to an increased tolerance for environmental stress. The use of effective strategies through the application of such fertilizers can guarantee that crops maintain their nutritional value under changing climatic conditions and hence guarantee food security on a global scale (José Soares et al., 2019). In addition, the use of biochar produced from nutrient-rich substances into the soil as a coordination compound is an applied process of enhancing soil fertility. Chicken dung and coffee husk biochars enrich the soil with micronutrients like calcium and potassium, enabling enhanced retention of nutrients and ensuring maximum availability of nutrients to plants. The unique physicochemical properties of biochars render them appropriate for application as a liming agent in acidic soils and as an organic fertilizer supplement to conventional agriculture. These synergistic approaches not only increase agronomic yields, but contribute to sustainable agriculture by waste organic recycling and reducing greenhouse gas emissions associated with fertilizer production. For this purpose, coordination compounds are not merely micronutrient fertilizers but also depict sophisticated practices that are consistent with sustainable agricultural practices.

10.4.2. Role in Pest Control

Utilization of coordination compounds in pest control is increasingly becoming important in addressing agriculture challenges, particularly those associated with invasive pests and plant diseases. These complexes typically function as biological control agents that are able to significantly reduce reliance on traditional synthetic pesticides, which are health and environment risky. For instance, a few coordination complexes have proved active against infamous field pathogens such as *Phytophthora infestans*, the cause of late blight of tomatoes and potatoes. The synergistic action of the complexes enhances plant resistance by triggering suppression of disease by biochemical means, leading to healthier crops with minimal chemical deposits. Use of these biocontrol techniques conforms to sustainable agricultural practice, hence contributing to integrated pest management (IPM) strategies that prioritize ecologically based strategies to crop protection (Wang et al., 2023). Incorporation of coordination compounds in pesticides promises to achieve higher efficacy under varying environmental conditions. The new molecular architectures of the compounds enable them to control biological systems in novel modes of action, which ensures action against a specific pest with minimal non-targeting effects. For example, biocontrol agents derived from coordination compounds can be employed to disrupt the life cycles of pests or diseases and, therefore, lower their populations before they achieve threshold levels. In addition, digitalization innovations also hold promise for precision agriculture, and coordination compound application can be optimized with smart technologies. Such preventive measure of pest control can favor improved utilization of resources while ensuring ecosystem and human health (Mondéjar et al., 2021). Furthermore, ongoing research into the modes of action of the coordination compounds is also promising for future pest management. Molecular level interactions have been tested in experiments that reveal complex biochemical processes leading to crop resistance and pathogen resistance. Designing new formulations and application methods can increase the consistency of these biocontrol agents to be more mainstreamed in farms. Through coordination among scientists, agriculturalists, and technologists, coordination compounds can be best utilized in IPM programs for field-based practical applications. The ultimate goal is to create environmentally sustainable pest management methods that not only enhance agricultural productivity but also promote environmental health and biodiversity conservation (Wang et al., 2023).

10.4.3. Enhancing Plant Growth and Yield

Use of coordination compounds in agriculture is a promising future direction for enhancing plant growth and yield. The complexes can assist in allowing plants to absorb necessary nutrients such as nitrogen, phosphorus, and potassium effectively, all of which play significant roles in plant growth. By forming stable complexes with the nutrients, coordination compounds not only make them more soluble but also accessible to plants. This is particularly significant in soils low in nutrients or when plants experience high competition for nutrients. In conclusion, increased bioavailability of nutrients leads to increased biomass production and significantly high crop yields, indicating the useful application of coordination chemistry in agriculture.

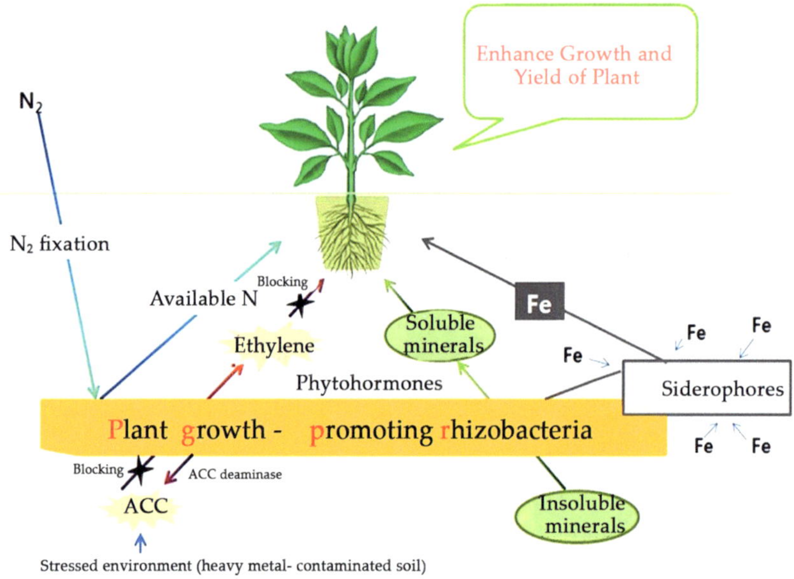

Figure 10.8. *Enhancing plant growth and yield.*

Source: https://www.researchgate.net/profile/Rc-Dubey/publication/230636447/figure/fig1/AS:646081201324032@1531049170512/Functions-of-plant-growth-promoting-rhizobacteria-Sources-http-wwwenvitopcokr.png.

In addition to nutrient solubilization, coordination compounds can be implicated in plant resistance to stress. Abiotic stresses such as drought, salinity, and high/low temperatures can adversely affect the growth of plants and yield. Coordination compounds can also control the physiological responses of plants towards such stresses and assist them in acclimatizing and surviving. For instance, such compounds stabilize protein and membrane structures such that the plants are able to maintain cellular integrity while under unfavorable conditions. Further, some coordination complexes are able to boost antioxidant enzyme activity, thereby minimizing oxidative stress in plant cells. The ability of these compounds to promote resilience to environmental stress is most crucial in maximizing agricultural productivity in light of climate change and resource constriction. Moreover, the valorization of lignocellulosic biomass through coordination complexes is critical for environmentally friendly agricultural processes, particularly for enhancing soil fertility and health. The major constituent of plant material, lignocellulose, can be transformed into valuable chemical feedstocks that recycle nutrients back into the environment. As explained in recent literature, exploitation of the structural functions of lignin by coordination chemistry can lead to effective nutrient formulations and soil amendments. Not only does this exercise contribute to improved plant nutrition but also guarantees a circular economy where waste is avoided and agricultural inputs are derived from sustainably sourced materials. At scale, integration of coordination compounds into agricultural systems highlights their triple function to promote not only plant growth and yield but also environmental stewardship, which is paramount to sustainable agriculture today (Isikgor et al., 2015; Rinaldi et al., 2016) (Figure 10.8).

10.5. COORDINATION CHEMISTRY IN MATERIAL SCIENCE

The union of coordination chemistry and material science has brought tremendous development in numerous applications, particularly in the creation of functional materials. Coordination complexes, characterized by the interaction between metal ions and ligands, possess unique properties which can be tailored for some specific purposes. For instance, the use of metal-organic frameworks (MOFs), in which metal ions as nodes and organic ligands as linkers, has transformed gas storage, separation, and catalysis. The large surface areas and tunable porosities render them excellent candidates for applications ranging from carbon capture to drug delivery. By the selective selection of metal ions and organic ligands, researchers are able to tailor materials with target functionalities and hence enhance their performance in many areas, ranging from environmental science to biomedical engineering. The versatility of coordination compounds can also be exploited in the design of new materials for organic electronics and solar cells. The introduction of heterocycles in coordination chemistry has been at the forefront of novel organic semiconductor synthesis, which are primary materials utilized in organic light-emitting diodes (OLEDs) and organic solar cells. The materials leverage metal-ligand complex electronic character for efficient charge carrier transport. Recent advances in this area have demonstrated how personal ligand-metal interactions can fine-tune the electronic properties of these materials to enhance the performance and longevity of devices. Further, studies of novel coordination chemistry routes to the synthesis of energetic materials promise to lead to the development of next-generation batteries and supercapacitors and thereby make coordination compounds significant contributors to the search for sustainable energy technologies. The role of metal ions is rather different in metal-phenolic material science, emphasizing the broad application of coordination chemistry in material science. Syntheses of these materials involve metal ion complexation by phenol compounds to form new hybrid materials with fascinating properties. Understanding the complicated interactions—such as cation-π, coordination, and redox interactions—is able to allow scientists to regulate the material properties effectively. As well-documented in recent literature, metal-phenolic materials possess vast biological promise, ranging from drug delivery systems and tissue engineering to environmental remediation technology. The insight and knowledge gained from the study of these coordinative interactions not only advance the fundamental understanding of the chemistry involved but also provide directions for next-generation material synthesis with customized functions, highlighting the significance of coordination chemistry in contemporary material science (Charushin et al., 2024; Geng et al., 2022).

10.5.1. Development of New Materials

New material synthesis based on coordination compounds is now a central focus in contemporary chemistry. The characteristic behavior exhibited by these compounds, particularly the reactivity and ligand versatility in bonding, forms the foundation of their extensive application. Schiff-base ligands have drawn significant attention in particular as they can both form stable metal complexes with wide structural diversity.

These complexes can exhibit a variety of properties ranging from antibacterial to antifungal and potentially could be effectively used in biomedical applications. The generality of Schiff-base ligands provides the foundation for the preparation of new materials, which has given rise to further study into their prospective bioactivity, as seen in recent advances in coordination chemistry. In addition to furthering our understanding of these molecules, such study opens new avenues through which to pursue new uses towards the creation of advanced materials with specific functionalities, enriching the discipline of material science and coordination chemistry in turn (Soroceanu et al., 2022). Moreover, the synthesis and use of coordination compounds are central to the emergence of molecular magnetism, a field characterized by its interdisciplinary breadth. Novel ligand development using low-coordinate elements has been central to the reengineering of the electronic properties of transition metal and f-element compounds. With scientists exploring the impact of nonconventional ligands—particularly those that involve heavier p-block elements—new materials with novel magnetic characteristics await. These compounds, including single-molecule magnets (SMMs), possess magnetic field stability and therefore the potential to be utilized in data storage and quantum computing devices. The convergence of ligand design with that of the resultant magnetic properties underlines the potential of coordination compounds towards the preparation of advanced materials with previously unheard-of functionality and the exploration of new dimensions in material development (Fu-Guo et al., 2019). With ongoing search to develop new materials, coordination compounds are sure front-runners in innovation in most scientific disciplines. They can serve as catalysts in chemical reactions and participate in complexation processes, thereby enabling various applications in environmental, industrial, and pharmaceutical sectors. The future of coordination chemistry promises to continue developing tunability in these materials, allowing their properties to be tailored to meet specific needs. The incorporation of organometallic chemistry into coordination systems is also a promising pathway for the development of novel materials, beyond traditional uses. Their incorporation into real technologies not only proves their universality but also testifies to the need for continuous research towards the finding and optimization of new materials for sustainable development (Fu-Guo et al., 2019; Soroceanu et al., 2022).

10.5.2. Use in Nanotechnology

The merging of coordination compounds and nanotechnology has ushered in revolutionary advances in a number of fields, particularly catalysis and energy conversion. Nanostructured materials, normally made up of coordination complexes, possess unique properties that are favorable to the promotion of catalytic processes. For instance, materials can maximize the hydrogen evolution reaction (HER), an important process in the splitting of water to generate hydrogen. Scientists have made significant progress in the optimization of HER systems' efficiency and economy using coordination compounds as catalysts. The ability to tailor the morphology and electronic structure of such compounds enables highly efficient catalysts that can operate under a variety of conditions, thus challenging the profound impact of nanotechnology on coordination compounds' real-world applications in energy systems (Zhu et al., 2019). Furthermore, coordination compound-based applications in nanotechnology have reached as far as biomedical applications, challenging their versatility. Coordination complex-derived nanoparticles are ready to be engineered as targeted drug delivery vehicles with low side effects. Their ability to encapsulate therapeutic agents allows for controlled release profiles that optimize the therapeutic response of treatment. The biocompatibility of nanoparticles is also a considerable aspect, as it improves their use in clinical applications.

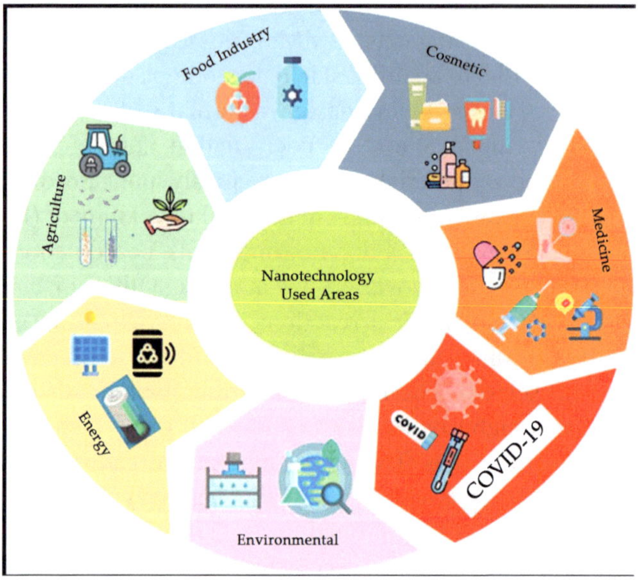

Figure 10.9. *Use in nanotechnology.*

Source: https://www.frontiersin.org/files/Articles/1059184/fmats-09-1059184-HTML/image_m/fmats-09-1059184-g002.jpg.

With the advancement of the discipline, understanding the interplay of coordination compounds with biological environments is essential, which will give rise to new nanomedicine alternatives. Such interplay illustrates how coordination compounds not only form the backbone of fundamental research in nanotechnology but also find application while designing viable solutions for real-world problems (Zheng et al., 2017). Lastly, the rapid evolution of nanotechnology has revolutionized coordination compound synthesis and application in various fields of science. The insertion of nanotechnology in coordination compound synthesis stimulates the production of materials of unparalleled precision and performance. Techniques such as self-assembly and molecular recognition are leaders, enabling the creation of complex nanostructures with the desired specific properties. This tailored functionality enhances the physical and chemical properties of coordination compounds to make them highly versatile in uses from sensors to electronics. The synergistic relationship between coordination compounds and nanotechnology gives rise to a new era of innovation in which better material properties lead to superior performance in a variety of applications, emphasizing the imperative for further work in this vibrant area (Figure 10.9).

10.5.3. Applications in Electronics

The application of coordination compounds in electronics is one of the significant achievements in materials science and engineering. The compounds possess distinctive electrochemical properties that are advantageous for a variety of electronic devices, including sensors, semiconductors, and memory storage devices. The ease with which scientists can tailor coordination compounds to meet specific needs implies that they can enhance the performance of devices without being energy-consuming. For example, certain metal complexes have the ability to be used as conductors, practically bridging the gap between inorganic and organic electronics. Therefore, the continued study of coordination chemistry promises to open up new avenues toward developing more efficient and environmentally friendly electronic technology, particularly with the demands for miniaturization and greater functionality on the rise. One of the notable applications of coordination compounds in electronics is their utilization in the production of advanced multimetallic oxide films used in semiconductor devices. Traditional approaches are found to be dependent on high-energy processes that are perhaps not environmentally friendly. Recent advances have illustrated that stimuli-responsive mixed-metal oxide thin films can be prepared from solution-processed molecular precursors under less harsh conditions. This new strategy utilizes coordination compounds, such as a CoII carboxylate complex that can be synthesized in one-pot reaction with various ligands. Heterometallic compounds formed exhibit superior thermal stability while facilitating regulated electron transport properties, which are required for efficient microprocessors and memory chips. Such accomplishments illustrate the prospect for coordination chemistry to address the demand for green electronic devices while advancing device effectiveness and stability(Seiss et al., 2024). Furthermore, coordination compounds have also been used in emerging technologies such as organic light-emitting diodes (OLEDs) and organic photovoltaics (OPVs). Incorporating metallo-organic compounds will enhance light absorption as well as carrier mobility of charges, hence optimizing device performance. Specifically, azo compounds, being a sub-class of coordination complexes, exhibit unique characteristics making them effective as high-efficiency dyes and electron-transporting materials. These compounds are accountable for a significant share of synthetic colors and have extensive use, ranging from photoelectronics to fiber coloring. On account of versatility, these compounds are not only required for applications related to beauty but also for material functionalization that improves the performance of electronic devices. With progressing research, the potential for coordination compounds to revolutionize the electronics sector is growing day by day, opening avenues for improvement in efficiency, performance, and sustainability(Al-Majdi et al., 2025).

10.6. COORDINATION COMPOUNDS IN ANALYTICAL CHEMISTRY

Coordination compounds have a serious application in analytical chemistry, particularly due to the fact that they can form stable complexes with a variety of metal ions. These complexes serve as indispensable reagents for the quantitative as well as qualitative analysis of numerous constituents of different samples. The significance of coordination compounds arises due to the fact that they have the ability to enhance the sensitivity and selectivity of analytical processes. For instance, transition metals may exhibit characteristic color variations in complexes, which can be utilized in spectrophotometric determination and provide precise endpoint determinations in titrations. Furthermore, advances in coordination chemistry have disclosed designs in ligands that provide increased metal ion selectivity, thus enhancing the analytical potential of various methods. With knowledge of these new applications, researchers can construct better trace metal detection and other analytes techniques in environmental, biological, and industrial applications. The diversity of coordination compounds brings their utility into intricate-forming reactions utilized in analytical chemistry. For example, the synthesis of various ligands, i.e., 4-Bromo-2-({2-[(5-bromo-2-hydroxy-benzylidene)-amino]-5-nitro-phenylimino}-methyl)-phenol (BSNP), has established the potentiality for the preparation of new metal complexes with high stability and selectivity of metal ion binding. The complexes have been studied by various spectroscopic methods and have been investigated in detail regarding their structural features and coordination behavior. Notably, the stability constant (Kf) values derived from these analyses allow chemists to predict the binding strength of metal and ligand, allowing for the design of specific reagents in complexation reactions. This aspect highlights the pivotal role of coordination compounds in the development of sound analytical procedures that can be applied in a variety of applications across biomedicine and environmental analysis, and thereby signifies the need for synthetic and analytical convergence. Furthermore, application of computational techniques with standard spectroscopic techniques has enhanced coordination compound understanding in analytical chemistry to a considerable extent. Scientists can predict their behavior in various environments using Density Functional Theory (DFT) calculations for insights into the electronic properties as well as reactivity of metal complexes prepared. Coordination compounds have also been demonstrated in research to exhibit antimicrobial and anticancer activity, not just demonstrating their use in analytical systems but also opening up chances for novel drugs. Applications involving molecular docking simulations also further account for the mode of action these compounds may display against biological macromolecules, which reinforces the anatomical and functional relevance of coordination chemistry to drug development. These interdisciplinarity approaches illustrate how coordination compounds can mediate differences between analytic techniques and biomedical research, emphasizing their interdisciplinarity uses in modern scientific research (Abou-Melha, 2025; Omar et al., 2025).

10.6.1. Techniques for Metal Ion Detection

Metal ion detection is a critical aspect in many industries like environmental monitoring, food safety, and medical diagnostics. Various techniques have been developed to achieve specific and sensitive detection of such ions, of which coordination compounds play a crucial role in order to provide selectivity as well as sensitivity. Among all such techniques, affinity-based methods have gained increasing popularity, especially through the use of the technique of affinity monolith chromatography (AMC). Through this technique, a monolithic support is functionalized with specific ligands, allowing the selective capture of target metal ions from complex mixtures. The primary strength of AMC is facilitating high-resolution separation and detection procedures, thereby leading innovation in metal sciences and analytical chemistry. The synthesis of binding agents, such as in immobilized metal-ion affinity chromatography, also demonstrates the innovative intersections of coordination chemistry and metal ion detection (Poddar et al., 2021). A very effective method for metal ion detection is fluorescence-based sensors, and they make use of coordination compounds to achieve high sensitivity. Fluorescent probes are often composed of metal-binding sites that undergo a conformational change when they bind with target ions and release a detectable fluorescent signal. This approach is particularly useful in the determination of trace levels of divalent metal ions in environmental samples because it allows one to make real-time monitoring with little sample preparation. These sensors are made to be tunable with varied coordination compounds that can be modified to be utilized for individual metal ions, thus improving detection efficiency. Besides, the flexibility and versatility of the fluorescent sensors make them best suited for handling different applications, including biological imaging and the tracking of ion distribution within living organisms, bearing testimony to the exceptional place coordination compounds hold in present detection (Dorothée Liebschner et al., 2019). Electrochemical detection also presents a robust alternative to the detection of metal ions, leveraging the unique behavior of coordination compounds in order to improve sensitivity and specificity. Techniques such as voltammetry and potentiometry allow ion concentration to be measured from the electrochemical signal caused by metal ions interacting with a specific ligand in solution. The use of coordination compounds as ligands not only increases the binding affinity towards the target ion but also suppresses interference from other constituents in the sample matrix. Therefore, these electrochemical sensors can detect metal ions in real-time efficiently, and so they are appropriate for numerous applications ranging from industrial wastewater testing to biological diagnosis. Generally speaking, the integration of coordination chemistry into the detection processes through electrochemistry emphasizes coordination compounds' diversity and utility towards advancing the art of metal ion sensing (Poddar et al., 2021) (Figure 10.10).

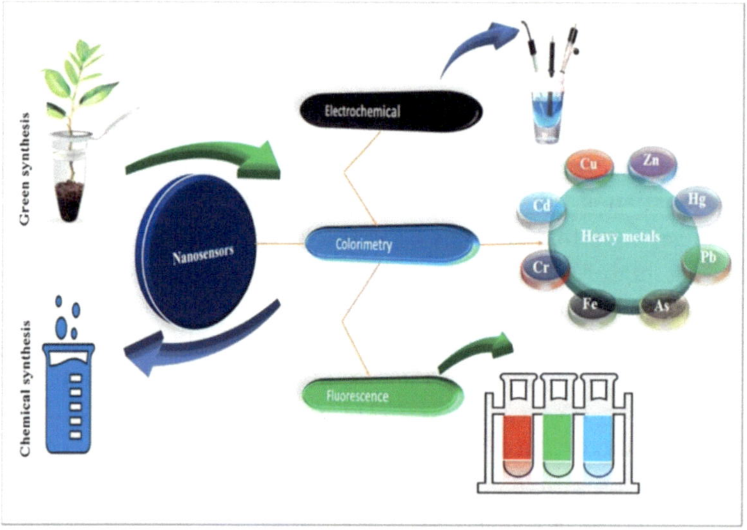

Figure 10.10. *The detection of metal ions is a crucial aspect of many fields.*

Source: https://media.springernature.com/lw685/springer-static/image/art%3A10.1007%2Fs11356-020-07865-7/MediaObjects/11356_2020_7865_Figa_HTML.png?as=webp.

10.6.2. Role in Spectroscopic Methods

Spectroscopic methods are of crucial significance to the understanding of structural and functional aspects of coordination compounds. By using different methods such as infrared (IR) spectroscopy, ultraviolet-visible (UV-vis) spectroscopy, and nuclear magnetic resonance (NMR) spectroscopy, scientists can get information about the coordination sphere around the metal centers. IR spectroscopy is best at determining functional groups in a ligand and examining metal-ligand vibrations, thereby giving binding modes and details about the nature of interactions between metals and ligands. Their responsiveness to changes in the electronic environment also makes them valuable in probing the effects of coordination on the electronic properties and worldwide stability of metal-ligand complexes, furthering our understanding of their potential applicability in catalysis and material science. In coordination compounds, UV-vis spectroscopy is a valuable technique for electronic transition in the complex to gain knowledge regarding the metal center's geometry and electronic structure. It can detect characteristic absorption bands, which change with ligand field around the metal and tend to reflect oxidation state or coordination geometry changes. Such spectral information can be interpreted to enable the researcher to make significant conclusions about the kind of bonding and electronic structure in the complex. For instance, the application of UV-vis spectroscopy in examining biguanide-metal complexes provides an insight into how the outstanding chelation properties of bidentate ligands can alter electronic transition and cast light on their catalytic and biological activity as done in the study of biguanides (Rani, 2025). Moreover, the entire spectroscopic analysis, including mass spectrometry, provides a deeper insight into coordination compounds stability and reactivity. Techniques like conductivity determination, magnetic susceptibility tests, and thermal analysis complement spectroscopic data in providing a full description of the compounds properties. These multidisciplinary approaches have led to the discovery of mixed-ligand complexes of higher biological activity, as evidenced by the characterization of FeIMCALHIS, CoIMCALHIS, and NiIMCALHIS complexes, which undergo various spectroscopic studies to analyze their structural and electronic properties (El-Lateef et al., 2024). These extensive spectroscopic investigations not only demonstrate the influence of ligand coordination on the overall stability and reactivity but also establish the therapeutic importance of coordination compounds and highlight their importance in contemporary chemistry.

10.6.3. Applications in Chromatography

Chromatography applications have been significantly advanced with the use of coordination compounds, which enable improvement in methods of separation. Chromatography relies on the interaction between the mobile and stationary phases to achieve the effective separation of substances. Coordination compounds, particularly metal-organic frameworks (MOFs), have increasingly been used as stationary phases because they are high surface area and pore size tunable. For instance, some MOFs are endowed with special properties enabling selective adsorption of specific analytes, thereby enhancing chromatography efficiency and resolution. This optimization is particularly beneficial for complex mixtures in environmental analysis, food safety analysis, and pharmaceutical applications where precise separation and characterization of compounds are necessary to ensure regulatory compliance and quality control of the product (Giovanardi et al., 2025). Such novel materials can be employed to engineer chromatography techniques that are both efficient and sustainable. Among biomedical uses, chromatography using coordination compounds is under consideration because it is being utilized in processing of vital biomolecules.

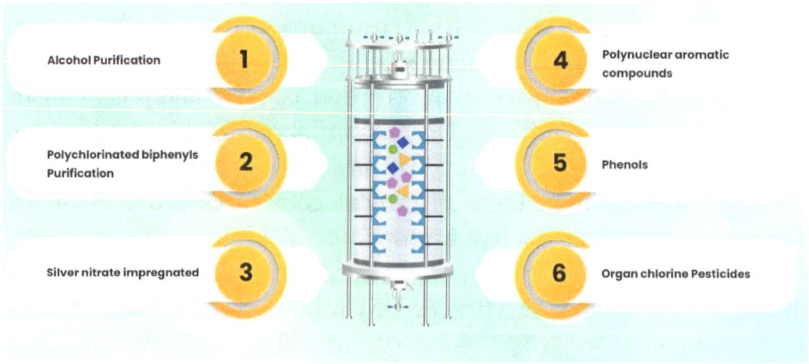

Figure 10.11. *The applications of chromatography.*

Source: https://www.column-chromatography.com/images/blog/444_application-of-column-chromatography-in-pharmacy.jpg.

Picolinic acid derivatives have also been identified as valuable ligands for coordination with metal ions for achieving improved process separation in chromatographic processes. These ligands facilitate the purification of biomolecules, including proteins and nucleic acids, through affinity chromatography, whose high specificity is ensured by the selective interaction between target molecule and ligand. Besides, the applicability of these ligands facilitates the development of complex separation procedures to accommodate the unique properties of biomolecules (Camorali et al., 2025). Therefore, all these applications in the medical field can lead to the formation of therapeutic agents and diagnostic reagents of high purity required by modern medicine, citing the revolutionary function of coordination compounds in the field. Besides, developments in chromatography have found their way into the field of environmentally friendly practices due to the applications of coordination compounds to form eco-friendly separation techniques. Separation efficiency is often solvents-dependent, and standard methods make use of harmful organic solvents. By employing coordination compounds in chromatography, researchers are coming up with innovative solutions to reduce environmental issues while enhancing or at least not depreciating separation quality. For example, some MOFs can be designed to operate efficiently using water or other less harmful solvents, resulting in greener chemistry applications. This approach contributes to the environmental interests and answers regulatory guidelines towards sustainability in analytical chemistry (Giovanardi et al., 2025). The relentless advancement in chromatography, driven by coordination compounds, speaks volumes about the emphasis on upgrading analytical techniques with a perspective towards environmental stewardship and the public good (Figure 10.11).

SUMMARY

Applications of coordination compounds go far and wide across various fields, particularly biomedicine and materials science. Over the last few years, tremendous progress has been achieved in the use of bismuth coordination complexes as potential chemotherapeutics. Studies have revealed that these complexes are not just highly antibacterial against pathogenic agents such as *Helicobacter pylori* but are also potential in curing cancer by halting the growth of tumors in types such as breast and colon cancer. This coupling of coordination chemistry with medicine demonstrates the variety of such compounds in favoring the trend of producing new therapeutic drugs. As research advances further, optimization of such sophisticated efficacy can result in new modes of treatment for current ailments (Kowalik et al., 2017). Further, the incorporation of coordination compounds in technology also emphasizes their variety of applications other than their common functions. Lithium-ion battery electrode research demonstrates how coordination chemistry is crucial to energy storage technology. Intercomparison studies of intercalation and conversion compounds, for example, lithium cobalt oxide and alloy anodes, are actively addressing major challenges such as efficiency and safety. Not only do these developments improve performance characteristics, but they also facilitate sustainable energy technology development. As the need for efficient energy solutions rises, coordination compounds are likely to be an important area of study in the production of materials capable of meeting these needs, and this depicts the vast range of their applicability in industrial as well as technological operations (Nitta et al., 2014). In conclusion, the multidirectional usability of coordination compounds depicts their relevance in scientific as well as practical use. The ongoing exploration of their uses and properties is likely to bring revolutionary innovations in medicine, technology, and more. As the broad-based dissemination of research on bismuth complexes and second-generation battery materials confirms, this field of chemistry is not only replete with potential but also full of real-world applications. In the future, it is essential that scientists continue to investigate these compounds' potential and overcome the inherent challenges that come with their application. In the end, the future of coordination compounds is promising, destined to bring about revolutionary breakthroughs that will reinterpret our methodologies across various fields (Kowalik et al., 2017; Nitta et al., 2014).

Coordination compounds have extensive and varying applications that have significantly impacted much of the work done in medicine, industry, and environmental science. Coordination compounds, particularly metal complexes, play a critical role in diagnostic imaging and therapy in medicine. Gadolinium contrast agents derived from gadolinium find frequent use in magnetic resonance imaging (MRI) as they enhance image clarity. These complexes can effectively boost the imaging of disease conditions to enable correct diagnosis. The use of transition metals in therapeutic interventions also provides channels of targeted drug delivery systems that help to attain higher therapeutic efficacy with fewer side effects, and therefore, their applications in medicine reflect their role in improving health care and patients' outcomes. In industrial processes, coordination compounds are essential reagents and catalysts and are a driving force of chemical reactions. They have the ability to catalyze a reaction and increase product yield, which makes them commercially invaluable in production and synthesis. Coordination complexes are employed on an industrial scale in petrochemical refining, where they assist in the conversion of crude oil into applicable petroleum products. In addition, the synthesis of coordination compounds in new technologies, such as Industry 4.0 processes, also signifies their relevance in current production practices. Their explicit incorporation in systems using IoT, cyber-physical systems, and cloud computing is important. This confluence of coordination chemistry and current technology enables game-changing innovations in production efficiency and sustainability, exhibiting the multifaceted characteristics of these compounds in current industrial uses (Xu et al., 2018). Environmental applications of coordination compounds are also noteworthy, as they involve pollution cleanup and resource recovery. One of the coordination compound classes, the metal-organic frameworks (MOFs), have been found to possess exceptional gas and pollutant adsorbing properties. Due to their structure being tunable, they can be synthesized with higher selectivity to adsorb carbon dioxide and other harmful compounds from the air, in order to mitigate climate change. Furthermore, coordination compounds are applied in the treatment of wastewater to eliminate heavy metals and harmful chemicals, thereby preserving water quality. The ongoing work on the application of coordination compounds in the environment shows their ability to enhance sustainability, an indicator of growing commitment to merging chemistry with environmental stewardship (Cryan et al., 2019). Through these multiple applications, coordination compounds show their key role in enhancing human welfare and environmental health.

REFERENCES

1. Abd El-Lateef, H. M., El-Dabea, T., Khalaf, M. M., & Abu-Dief, A. M. (2023). Recent overview of potent antioxidant activity of coordination compounds. *Antioxidants*, *12*, 213–213. https://doi.org/10.3390/antiox12020213 (accessed on 13th September 2025).
2. Abd El-Lateef, H. M., Khalaf, M., & Abdou, A. (2024). Preparation, characterization, *in vitro* biological evaluation, DFT calculations, and molecular docking investigations of 1H-imidazole-2-carboxylic acid and histidine-based mixed-ligand complexes. *Chemistry & Biodiversity*, *22*. https://www.semanticscholar.org/paper/d7e93f209ee3a9ec0d9bdbcc44cdc2a84e078c72 (accessed on 13th September 2025).
3. Abou-Melha, K. (2025). Sustainable synthesis of bivalent Ni^{2+}, Co^{2+}, and Cu^{2+} Schiffbasecomplexesviaballmillingasagreenapproach:Astudyofspectroscopic characterization, cyclic voltammetry, bioactivity, DFT analysis, SAR, drug likeness, and molecular docking. *Arabian Journal of Chemistry*. https://www.semanticscholar.org/paper/f01ab74f77a553157f8218577422d86be75d0a46 (accessed on 13th September 2025).
4. Al-Fahdawi, A. S., Mohamed, S. K., Potgeiter, H., & Alfutimie, A. (2024). Heavy metal ions in polluted water removable by synthesizing a new compound for dithiocarbonate derivatives. *Bulletin of the Chemical Society of Ethiopia*. https://www.semanticscholar.org/paper/8eeb8d3279e7ccdf0722c31103bfb21fc16abac2 (accessed on 13th September 2025).
5. Al-Majdi, Z. R., Al-Dahhan, W., Shihab, M., & Nazari, M. H. (2025). Azo compounds and their potential applications: Article review. *Al-Kitab Journal for Pure Sciences*. https://www.semanticscholar.org/paper/d41032dcf730799119722f4726fb59aa04cb321b (accessed on 13th September 2025).
6. Andrews, J. G., Buzzi, S., Choi, W., Hanly, S. V., Lozano, A., Soong, A. C. K., & Zhang, J. C. (2014). What will 5G be? *IEEE Journal on Selected Areas in Communications*, *32*, 1065–1082. https://doi.org/10.1109/jsac.2014.2328098 (accessed on 13th September 2025).
7. Beltran, H., Alfonso-Herrera, L. A., & Serrato-Trejo, M. F. (2025). Structure-property/activity correlations of coordination compounds, supramolecular molecular assemblies and materials, with possible industrial and biological applications. *Journal of the Mexican Chemical Society*. https://www.semanticscholar.org/paper/e2e74333676c137c1df0e38da727efb271fd31de (accessed on 13th September 2025).

8. Bonaccorso, F., Colombo, L., Yu, G., Stoller, M. D., Tozzini, V., Ferrari, A. C., Ruoff, R. S., et al. (2015). Graphene, related two-dimensional crystals, and hybrid systems for energy conversion and storage. *Science, 347*. https://doi.org/10.1126/science.1246501 (accessed on 13th September 2025).
9. Calvin, K., Dasgupta, D., Krinner, G., Mukherji, A., Thorne, P., Trisos, C. H., Romero, J., et al. (2023). *IPCC, 2023: Climate Change 2023: Synthesis Report. Contribution of Working Groups I, II and III to the Sixth Assessment Report of the Intergovernmental Panel on Climate Change* [Core Writing Team, H. Lee & J. Romero (Eds.)]. IPCC, Geneva, Switzerland. https://doi.org/10.59327/ipcc/ar6-9789291691647 (accessed on 13th September 2025).
10. Camorali, S., Tei, L., Milanesio, M., & Lopresti, M. (2025). Unraveling the crystal structures of picolinic acid derivatives: Synthesis, packing, interactions, and conformational flexibility. *ChemistryOpen*. https://www.semanticscholar.org/paper/8648de4e103bbc1833978ce9b6cec01e10de0e59 (accessed on 13th September 2025).
11. Charushin, V. N., Verbitskiy, E. V., Чупахин, О. Н., Vorobyeva, D. V., Gribanov, P. S., Osipov, S. N., Иванов, А. В., et al. (2024). The chemistry of heterocycles in the 21st century. *Russian Chemical Reviews, 93*, RCR5125–RCR5125. https://doi.org/10.59761/rcr5125 (accessed on 13th September 2025).
12. Coghi, P., Zhu, Y., Xie, H., Hosmane, N. S., & Zhang, Y. J. (2021). Organoboron compounds: Effective antibacterial and antiparasitic agents. *Molecules, 26*, 3309–3309. https://doi.org/10.3390/molecules26113309 (accessed on 13th September 2025).
13. Cryan, J. F., O'Riordan, K. J., Cowan, C. S. M., Sandhu, K. V., Bastiaanssen, T. F. S., Boehme, M., Codagnone, M. G., et al. (2019). The microbiota-gut-brain axis. *Physiological Reviews*, *99*, 1877–2013. https://doi.org/10.1152/physrev.00018.2018 (accessed on 13th September 2025).
14. Diercks, C. S., & Yaghi, O. M. (2017). The atom, the molecule, and the covalent organic framework. *Science, 355*. https://doi.org/10.1126/science.aal1585 (accessed on 13th September 2025).
15. Domingues, R. R., Trugilho, P. F., Silva, C. A., Melo, I. C. N. A., Melo, L. C. A., Magriotis, Z. M., & Sánchez-Monedero, M. Á. (2017). Properties of biochar derived from wood and high-nutrient biomasses with the aim of agronomic and environmental benefits. *PLoS ONE, 12*, e0176884–e0176884. https://doi.org/10.1371/journal.pone.0176884 (accessed on 13th September 2025).

16. El-Lateef, H. M., El-Dabea, T., Khalaf, M., & Abu-Dief, A. (2023). Recent overview of potent antioxidant activity of coordination compounds. *Antioxidants, 12.* https://www.semanticscholar.org/paper/2611beffc5fdc60ec148806d526390c458523bef (accessed on 13th September 2025).

17. Geng, H., Zhong, Q. Z., Li, J., Lin, Z., Cui, J., Caruso, F., & Hao, J. (2022). Metal ion-directed functional metal–phenolic materials. *Chemical Reviews, 122,* 11432–11473. https://doi.org/10.1021/acs.chemrev.1c01042 (accessed on 13th September 2025).

18. Giovanardi, D., Ribezzi, E., Napolitano, M., Orlandini, M., Riboni, N., Mazzeo, P. P., Bacchi, A., et al. (2025). Effect of the entanglement of microporous pillared MOFs on the uptake and release profiles of essential oil components. *Chemistry (Weinheim an der Bergstrasse, Germany), 31.* https://www.semanticscholar.org/paper/524f385d0993547ac858a600da466fb770bd14ce (accessed on 13th September 2025).

19. Godwin, J., Benson, O., Nimighaye, O., & Salvation, E. P. (2024). Schiff bases, syntheses, characterization and their application to heavy metal removal: A review. *International Journal of Research and Innovation in Applied Science.* https://www.semanticscholar.org/paper/e8e296f0a391d6cbe079fa11a9796b0ec21623d5 (accessed on 13th September 2025).

20. Guo, F.-S., Bar, A. K., & Layfield, R. A. (2019). Main group chemistry at the interface with molecular magnetism. *Chemical Reviews, 119*, 8479–8505. https://doi.org/10.1021/acs.chemrev.9b00103 (accessed on 13th September 2025).

21. Isikgor, F. H., & Becer, C. R. (2015). Lignocellulosic biomass: A sustainable platform for the production of bio-based chemicals and polymers. *Polymer Chemistry, 6*, 4497–4559. https://doi.org/10.1039/c5py00263j (accessed on 13th September 2025).

22. Kostova, I. (2023). Therapeutic and diagnostic agents based on bioactive endogenous and exogenous coordination compounds. *Current Medicinal Chemistry, 31*, 358–386. https://doi.org/10.2174/092986733066623032110018 (accessed on 13th September 2025).

23. Kowalik, M., Masternak, J., & Barszcz, B. (2017). Recent research trends on bismuth compounds in cancer chemo- and radiotherapy. *Current Medicinal Chemistry, 26*, 729–759. https://doi.org/10.2174/0929867324666171003113540 (accessed on 13th September 2025).

24. Kumar, A., Kumar, A., Cabral-Pinto, M. M. S., Chaturvedi, A. K., Shabnam, A. A., Subrahmanyam, G., Mondal, R., et al. (2020). Lead toxicity: Health hazards, influence on food chain, and sustainable remediation approaches.

International Journal of Environmental Research and Public Health, 17, 2179–2179. https://doi.org/10.3390/ijerph17072179 (accessed on 13th September 2025).

25. Lee, H. I., Kim, S., Kim, H. S., Cho, J., & Lim, D. H. (2024). Simple and scalable manufacturing process of Raney-Ni electrode via dip-coating method for alkaline water electrolysis. *ECS Meeting Abstracts*. https://www.semanticscholar.org/paper/09a64225632fb8550d019385dd1d846148445015 (accessed on 13th September 2025).
26. Li, Z., Li, B., Hu, Y., Liao, X., Yu, H., & Yu, C. (2022). Emerging ultrahigh-density single-atom catalysts for versatile heterogeneous catalysis applications: Redefinition, recent progress, and challenges. *Small Structures, 3*. https://doi.org/10.1002/sstr.202200041 (accessed on 13th September 2025).
27. Liebschner, D., Afonine, P. V., Baker, M. L., Bunkòczi, G., Chen, V. B., Croll, T. I., Hintze, B. J., et al. (2019). Macromolecular structure determination using X-rays, neutrons and electrons: Recent developments in *Phenix*. *Acta Crystallographica Section D: Structural Biology, 75*, 861–877. https://doi.org/10.1107/s2059798319011471 (accessed on 13th September 2025).
28. Liu, D., Qiao, L., Peng, S., Bai, H., Liu, C., Ip, W. F., Lo, K. H., et al. (2023). Recent advances in electrocatalysts for efficient nitrate reduction to ammonia. *Advanced Functional Materials, 33*. https://doi.org/10.1002/adfm.202303480 (accessed on 13th September 2025).
29. Ma, K., Wang, Y., Chen, Z., İslamoğlu, T., Lai, C., Wang, X., Fei, B., et al. (2019). Facile and scalable coating of metal–organic frameworks on fibrous substrates by a coordination replication method at room temperature. *ACS Applied Materials & Interfaces, 11*, 22714–22721. https://doi.org/10.1021/acsami.9b04780 (accessed on 13th September 2025).
30. Mahurkar, N. D., Gawhale, N. D., Lokhande, M. N., Uke, S. J., & Kodape, M. M. (2023). Benzimidazole: A versatile scaffold for drug discovery and beyond–A comprehensive review of synthetic approaches and recent advancements in medicinal chemistry. *Results in Chemistry, 6*, 101139–101139. https://doi.org/10.1016/j.rechem.2023.101139 (accessed on 13th September 2025).
31. Mondéjar, M. E., Avtar, R., Baños Díaz, H. L., Dubey, R. K., Esteban, J., Gómez-Morales, A., Hallam, B., et al. (2021). Digitalization to achieve sustainable development goals: Steps towards a smart green planet. *The Science of The Total Environment, 794*, 148539–148539. https://doi.org/10.1016/j.scitotenv.2021.148539 (accessed on 13th September 2025).

32. Nicolosi, V., Chhowalla, M., Kanatzidis, M. G., Strano, M. S., & Coleman, J. N. (2013). Liquid exfoliation of layered materials. *Science, 340*. https://doi.org/10.1126/science.1226419 (accessed on 13th September 2025).
33. Nitta, N., Wu, F., Lee, J. T., & Yushin, G. (2014). Li-ion battery materials: Present and future. *Materials Today, 18*, 252–264. https://doi.org/10.1016/j.mattod.2014.10.040 (accessed on 13th September 2025).
34. Novoselov, K. S., Mishchenko, A., Carvalho, A., & Castro Neto, A. H. (2016). 2D materials and van der Waals heterostructures. *Science, 353*. https://doi.org/10.1126/science.aac9439 (accessed on 13th September 2025).
35. Omar, I., Alharas, M., Feizi-Dehnayebi, M., Alharbi, S. K., Abo-Dief, H., Qasem, H. A., El-Khatib, R., et al. (2025). Design, synthesis, physio-chemical characterization, stability determination, and biomedical applications of some novel tetra-dentate imine metal chelates supported by theoretical approaches: Bridging coordination chemistry and life sciences. *Applied Organometallic Chemistry*. https://www.semanticscholar.org/paper/236e7237fefb3a995330784d1f84e84bca3c709c (accessed on 13th September 2025).
36. Poddar, S., Sharmeen, S., & Hage, D. S. (2021). Affinity monolith chromatography: A review of general principles and recent developments. *Electrophoresis, 42*, 2577–2598. https://doi.org/10.1002/elps.202100163 (accessed on 13th September 2025).
37. Qasem, N. A. A., Mohammed, R. H., & Lawal, D. U. (2021). Removal of heavy metal ions from wastewater: A comprehensive and critical review. *npj Clean Water, 4*. https://doi.org/10.1038/s41545-021-00127-0 (accessed on 13th September 2025).
38. Rani, B. (2025). Coordination chemistry of cobalt biguanide complex: Ligand interaction and metal complex stability. *International Journal for Research in Applied Science and Engineering Technology*. https://www.semanticscholar.org/paper/bcfb4b8de5ab3174c71c8f082afb4ccb19b24337 (accessed on 13th September 2025).
39. Rinaldi, R., Jastrzebski, R., Clough, M. T., Ralph, J., Kennema, M., Bruijnincx, P. C. A., & Weckhuysen, B. M. (2016). Paving the way for lignin valorization: Recent advances in bioengineering, biorefining and catalysis. *Angewandte Chemie International Edition, 55*, 8164–8215. https://doi.org/10.1002/anie.201510351 (accessed on 13th September 2025).

40. Seiss, M., Lorenz, J., Schmitz, S., Moors, M., Börner, M., & Monakhov, K. (2024). Synthesis and structures of cobalt-expanded zirconium- and cerium-oxo clusters as precursors for mixed-metal oxide thin films. *Dalton Transactions*. https://www.semanticscholar.org/paper/8c170b36dccee763d05312b8db33caa685309e77 (accessed on 13th September 2025).
41. Senapati, S., Mahanta, A. K., Kumar, S., & Maiti, P. (2018). Controlled drug delivery vehicles for cancer treatment and their performance. *Signal Transduction and Targeted Therapy, 3*. https://doi.org/10.1038/s41392-017-0004-3 (accessed on 13th September 2025).
42. Soares, J., Santos, C. S., Carvalho, S. M. P., Pintado, M., & Vasconcelos, M. W. (2019). Preserving the nutritional quality of crop plants under a changing climate: Importance and strategies. *Plant and Soil*, *443*, 1–26. https://doi.org/10.1007/s11104-019-04229-0 (accessed on 13th September 2025).
43. Soroceanu, A., & Bargan, A. (2022). Advanced and biomedical applications of Schiff-base ligands and their metal complexes: A review. *Crystals, 12*, 1436–1436. https://doi.org/10.3390/cryst12101436 (accessed on 13th September 2025).
44. Velusamy, S., Roy, A., Sundaram, S., & Mallick, T. K. (2021). A review on heavy metal ions and containing dyes removal through graphene oxide-based adsorption strategies for textile wastewater treatment. *The Chemical Record, 21*, 1570–1610. https://doi.org/10.1002/tcr.202000153 (accessed on 13th September 2025).
45. Vilé, G., Di Liberto, G., Tosoni, S., Sivo, A., Ruta, V., Nachtegaal, M., Clark, A. H., et al. (2022). Azide-alkyne click chemistry over a heterogeneous copper-based single-atom catalyst. *ACS Catalysis, 12*, 2947–2958. https://doi.org/10.1021/acscatal.1c05610 (accessed on 13th September 2025).
46. Wang, W., & Long, Y. (2023). A review of biocontrol agents in controlling late blight of potatoes and tomatoes caused by *Phytophthora infestans* and the underlying mechanisms. *Pest Management Science, 79*, 4715–4725. https://doi.org/10.1002/ps.7706 (accessed on 13th September 2025).
47. Xu, L. D., Xu, E., & Li, L. (2018). Industry 4.0: State of the art and future trends. *International Journal of Production Research*, *56*, 2941–2962. https://doi.org/10.1080/00207543.2018.1444806 (accessed on 13th September 2025).
48. Zheng, Y., Jiao, Y., Vasileff, A., & Qiao, S. Z. (2017). The hydrogen evolution reaction in alkaline solution: From theory, single crystal models, to practical electrocatalysts. *Angewandte Chemie International Edition, 57*, 7568–7579. https://doi.org/10.1002/anie.201710556 (accessed on 13th September 2025).

49. Zhong, L., Li, Y., Xiong, L., Wang, W., Wu, M., Yuan, T., Yang, W., et al. (2021). Small molecules in targeted cancer therapy: Advances, challenges, and future perspectives. *Signal Transduction and Targeted Therapy*, *6*. https://doi.org/10.1038/s41392-021-00572-w (accessed on 13th September 2025).
50. Zhu, J., Hu, L., Zhao, P., Lee, L. Y. S., & Wong, K. Y. (2019). Recent advances in electrocatalytic hydrogen evolution using nanoparticles. *Chemical Reviews*, *120*, 851–918. https://doi.org/10.1021/acs.chemrev.9b00248 (accessed on 13th September 2025).
51. Żyro, D., Sikora, J., Szynkowska, M. I., & Ochocki, J. (2023). Silver, its salts and application in medicine and pharmacy. *International Journal of Molecular Sciences, 24*, 15723–15723. https://doi.org/10.3390/ijms242115723 (accessed on 13th September 2025).

INDEX

C

D

H

I

K

L

M

N

O

P